AF333338

Conference Proceedings and Lecture Notes in Physics
Volume II

Yukawa Couplings

and

The Origins of Mass

Edited by
Pierre Ramond

International Press

International Press Incorporated, Boston
P.O. Box 2872
Cambridge, MA 02238-2872

Copyright © 1996 by International Press Incorporated

ISBN: 1-57146-025-X

Ramond, Pierre (Ed.)
  Yukawa Coupling and the Origins of Mass

Typeset using the TeX system
Printed on acid free paper, in the United States of America

Other publications from International Press

Journals:

Communications in Analysis and Geometry
The Journal of Differential Geometry
Methods and Applications of Analysis
Mathematical Research Letters
Pacific Journal in Mathematics

Books:

Mathematical Physics:

Quantum Groups: from Coalgebras to Drienfeld Algebras
Proceedings of the XI th International Conference on Mathematical Physics
75 Years of Radon Transform
Perspectives in Mathematical Physics
Essays on Mirror Manifolds
Mirror Symmetry, Volumes 1 and 2

Number Theory:

Elliptic Curves, Modular forms, and Fermat's Last Theorem

Geometry and Topology:

L^2 Moduli Spaces with 4-Manifolds with Cylindrical Ends
The L^2 Moduli Space and Vanishing Theorem
Gökova Geometry and Topology Conferences, All Editions
Algebraic Geometry and Related Topics
Lectures on Differential Geometry
Lectures on Harmonic Maps
Geometry, Topology, and Physics for Raoul Bott
Lectures on Low-Dimensional Topology
Chern, A Great Geometer
Surveys in Differential Geometry, Volumes 1 and 2

Analysis:

Tsing Hua Lectures in Geometry and Analysis
Lectures on Nonlinear Wave Phenomena
Proceedings of the Conference on Complex Analysis
Integrals of Cauchy Type on the Ball
Advances in Geometric Analysis and Continuum Mechanics

Physics:

Physics of the Electron Solid
Proceedings of the Conference on Computational Physics
Chen Ning Yang, A Great Physicist of the Twentieth Century
Yukawa Couplings and the Origins of Mass

Current Developments in Mathematics, 1995

Current Developments in Mathematics, 1996

Table of Contents

PREFACE

The Institute for Fundamental Theory at the University of Florida was created to provide and foster an environment conducive to interactions among different branches of theoretical physics and mathematics. The Institute is funded by the University of Florida.

The Institute for Fundamental Theory has sponsored, thus far, two workshops on the campus of the University of Florida on topics of current interest. The first Workshop, on February 14-16 1992, was on the subject of "Dark Matter". The second workshop, entitled "Yukawa Couplings and The Origins of Mass", took place on February 11-13, 1994. The present volume contains the Proceedings of the second workshop.

Two thirds of the parameters used to describe the Standard Model come from the Yukawa couplings: nine masses of elementary fermions, three mixing angles, and one CP-violating angle. In the hope of discovering the structure that underlies the Standard Model one may try to explain the values of these parameters, or at least uncover the systematics behind the hierarchies in the masses, and the mixing angles.

There have been many attempts to build theories that explain these numbers, but so far none have proven conclusive. There are several problems which add to the complexity of devising an organizing principle from the body of data. Most pervasive is the strong coupling nature of QCD, which makes it difficult to extract from the data knowledge about the parameters in the Lagrangian. In particular, the values of the light quark masses (u,d,s) are not known with sufficient accuracy, although their ratios are better known because the ratios are more directly related to experiment. The problem continues up to the bottom quark sector: the main source of error in determining the CKM matrix element V_{cb} is theoretical uncertainty. At the same time, not all CKM matrix elements have been measured, with the result that there are no consistency checks on the Standard Model description of family mixing.

In addition, all the parameters run with scale. In the absence of any recognizable pattern for the Yukawa parameters at experimentally accessible scales, it is natural to extrapolate to shorter distances, and hope that recognizable patterns might emerge there. However, this extrapolation requires the use of models beyond the Standard Model, which brings additional theoretical uncertainties. Since we know how to extrapolate only in the context of perturbation theory, we are practically restricted to the supersymmetric extensions of the Standard Model since these theories allow *perturbative* extrapolation over many orders of magnitudes. Many theorists hope that the mass and mixing angle patterns will become simpler when extrapolated to the Planck scale. Some theorists start from a particular theory at short distances (at GUT or Superstring scales), and then attempt to match its parameters with those of the Standard Model.

These different communities of theorists dig tunnels at opposite sides of the

Yukawa mountain; it was the purpose of this workshop to bring them together, and attempt to define a common language which may bring clarity to this important problem.

The worshop started with a summary of the experimental situation, and the extraction of the fundamental parameters from the data. Then a variety of theoretical models for fermion masses and mixing parameters were put forward; some postulated symmetries at experimental scales, some in the deep ultraviolet. At the same time, more general models, based on GUTs and on Superstrings were presented, with a special emphasis on predictions for the Yukawa sector. The rich variety of theoretical approaches showed the vitality of this field of research, which offers tantalizing glimpses of physics beyond the Standard Model.

We wish to thank the University of Florida's Division of Sponsored Research, The National Science Foundation, and the Soros Foundation for their generous support. We also owe special thanks to Mrs C. Bright and Y. Dixon for their invaluable help in the organization of this workshop.

P. Ramond
P. Sikivie

A New Bottom-Up Approach for the Construction of Fermion Mass Matrices

CARL H. ALBRIGHT
DEPARTMENT OF PHYSICS,
NORTHERN ILLINOIS UNIVERSITY[1]
DeKALB, ILLINOIS 60115

FERMI NATIONAL ACCELERATOR LABORATORY
P.O. BOX 500, BATAVIA, ILLINOIS 60510
ALBRIGHT@FNALV

SATYANARAYAN NANDI
DEPARTMENT OF PHYSICS, OKLAHOMA STATE UNIVERSITY
STILLWATER, OKLAHOMA 74078
PHYSSNA@OSUCC

Abstract: We describe a new technique which enables one to construct an SO(10)-symmetric fermion mass matrix model at the supersymmetric grand unification scale directly from the fermion mass and mixing data at the low energy scale. Applications to two different neutrino mass and mixing scenarios are given.

1 Description of the Technique

A new bottom-to-up approach proposed by us is summarized briefly in this talk with more details and references presented elsewhere.[1] We shall restrict our attention here to the construction of complex symmetric mass matrices arising with Higgs in the 10 and 126 representations of SO(10). The procedure allows us to construct mass matrices which exhibit as simple an SO(10) structure as possible with the maximum number of texture zeros allowed for the neutrino mass and mixing scenario in question.

1) Start from a set of quark and lepton masses and mixing matrices completely specified at the low scales.

2) Evolve the masses and mixing matrices to the GUT scale by making use of the one-loop renormalization group equations (RGEs) for the minimal supersymmetric standard model.

For this purpose we set $\Lambda_{SUSY} = 170$ GeV and $\Lambda_{GUT} = 1.2 \times 10^{16}$ GeV. Following Naculich,[2] we use the approximation that only the top and bottom quarks as well as the tau lepton contribute to the non-linear Yukawa terms in the RGEs. With a physical top mass expected to be near 160 GeV, we take the running mass to be $m_t(m_t) = 150$ GeV and adjust $m_b(m_b)$ and $\tan\beta$, so consistency is achieved at Λ_{GUT} which requires complete Yukawa unification with $\tan\beta \simeq 48.9$. We are working under the assumption that only one SO(10)

---

[1] Permanent address

10 of Higgs contributes to the 33 elements of the up, down and charged lepton mass matrices.

3) Construct a numerical set of complex-symmetric mass matrices M^U, M^D, M^E and $M^{N_{eff}} = M^{N_{Dirac}} M_R^{-1} M^{N_{Dirac}T}$ for the up and down quarks, charged leptons and light neutrinos by making use of a procedure due to Kusenko,[3] now applied to both quarks and leptons.

Since the quark CKM mixing matrix is unitary and represents an element of the unitary group U(3), one can express it in terms of one Hermitian generator of the corresponding U(3) Lie algebra times a phase parameter α by writing

$$V_{CKM} = U'_L U_L^\dagger = \exp(i\alpha H) \tag{1a}$$

where

$$i\alpha H = \sum_{k=1}^{3} (\log v_k) \frac{\prod_{i\neq k}(V_{CKM} - v_i I)}{\prod_{j\neq k}(v_k - v_j)} \tag{1b}$$

in terms of the eigenvalues v_j of V_{CKM}, by making use of Sylvester's theorem.[3] The transformation matrices from the weak to the mass bases are given in terms of the same generator but modified phase parameters such that

$$U'_L = \exp(i\alpha H x_q), \qquad U_L = \exp[i\alpha H(x_q - 1)] \tag{1c}$$

and relation (1a) is preserved. The complex symmetric quark mass matrices in the weak basis are then related to those in the diagonal mass basis by

$$M^U = U'^\dagger_L D^U U'^{\dagger T}_L, \qquad M^D = U_L^\dagger D^D U_L^{\dagger T} \tag{1d}$$

It suffices to expand V_{CKM}, U'_L and U_L to third order in α in order to obtain accurate numerical results for the mass matrices M^U and M^D. Similar expressions can be obtained for the light neutrino and charged lepton mass matrices from the lepton mixing matrix and its eigenvalues with x_ℓ replacing x_q.

The parameters x_q and x_ℓ control the choice of bases for the quark and lepton mass matrices, respectively. The up quark mass matrix is diagonal for $x_q = 0$, while the down quark mass matrix is diagonal for $x_q = 1$; likewise, the light neutrino mass matrix is diagonal for $x_\ell = 0$, while the charged lepton mass matrix is diagonal for $x_\ell = 1$.

4) Vary x_q and x_ℓ and the signs of the mass eigenvalues to search for simplicity in the SO(10) framework, i.e., pure 10 or pure 126 contributions for as many mass matrix elements as possible.

For this purpose we note that 10's contribute equally to the down quark and charged lepton mass matrices, while 126 contributions differ by a factor of -3; likewise for the up quark and Dirac neutrino mass matrices. In terms of the Yukawa couplings and the appropriate VEVs, the mass matrices are given by

$$\begin{aligned}
M^U &= \sum_i f^{(10_i)} v_{ui} + \sum_j f^{(126_j)} w_{uj} \\
M^D &= \sum_i f^{(10_i)} v_{di} + \sum_j f^{(126_j)} w_{dj} \\
M^{N_{Dirac}} &= \sum_i f^{(10_i)} v_{ui} - 3 \sum_j f^{(126_j)} w_{uj} \\
M^E &= \sum_i f^{(10_i)} v_{di} - 3 \sum_j f^{(126_j)} w_{dj}
\end{aligned} \tag{2}$$

5) For the best choice of x_q and x_ℓ which maximizes the simplicity, construct a simple model of the mass matrices with as many texture zeros as possible.

6) Evolve the mass eigenvalues and mixing matrices determined from the model at the SUSY GUT scale to the low scales and compare the results with the starting input data.

2   Application to Two Different Neutrino Mass and Mixing Scenarios

We now illustrate the technique by applying it to two different neutrino scenarios, both of which explain the solar neutrino depletion data with the non-adiabatic Mikheyev-Smirnov-Wolfenstein (MSW) effect,[4] where one includes the atmospheric neutrino depletion phenomenon[5] while the other accepts the cocktail model[6] interpretation of missing dark matter.

We take the same quark input data for both models. For the light quark masses, we shall adopt the values quoted by Gasser and Leutwyler[7] while the heavy quark masses are specified at their running mass scales:

$$
\begin{aligned}
m_u(1GeV) &= 5.1\ MeV, & m_d(1GeV) &= 8.9\ MeV \\
m_c(m_c) &= 1.27\ GeV, & m_s(1GeV) &= 175\ MeV \\
m_t(m_t) &= 150\ GeV, & m_b(m_b) &\simeq 4.25\ GeV
\end{aligned}
\tag{3a}
$$

For the Cabibbo-Kobayashi-Maskawa (CKM) mixing matrix, we adopt the following central values at the weak scale

$$
V_{CKM} = \begin{pmatrix}
0.9753 & 0.2210 & (-0.283 - 0.126i) \times 10^{-2} \\
-0.2206 & 0.9744 & 0.0430 \\
0.0112 - 0.0012i & -0.0412 - 0.0003i & 0.9991
\end{pmatrix}
\tag{3b}
$$

where we have assumed a value of 0.043 for V_{cb} and applied strict unitarity to determine V_{ub}, V_{td} and V_{ts}.

2.1   Lepton Scenario (A) involving the Non-adiabatic MSW Solar and Atmospheric Neutrino Depletion Effects

In this scenario, we single out the central points in the two neutrino mixing planes

$$
\begin{aligned}
\delta m_{12}^2 &= 5 \times 10^{-6}\ eV^2, & \sin^2 2\theta_{12} &= 8 \times 10^{-3} \\
\delta m_{23}^2 &= 2 \times 10^{-2}\ eV^2, & \sin^2 2\theta_{23} &= 0.5
\end{aligned}
\tag{4}
$$

With the neutrino masses assumed to be non-degenerate, we take for the lepton input

$$
\begin{aligned}
m_{\nu_e} &= 0.5 \times 10^{-6}\ eV, & m_e &= 0.511\ MeV \\
m_{\nu_\mu} &= 0.224 \times 10^{-2}\ eV, & m_\mu &= 105.3\ MeV \\
m_{\nu_\tau} &= 0.141\ eV, & m_\tau &= 1.777\ GeV
\end{aligned}
\tag{5a}
$$

and

$$
V_{LEPT}^{(A)} = \begin{pmatrix}
0.9990 & 0.0447 & (-0.690 - 0.310i) \times 10^{-2} \\
-0.0381 - 0.0010i & 0.9233 & 0.3821 \\
0.0223 - 0.0030i & -0.3814 & 0.9241
\end{pmatrix}
\tag{5b}
$$

where we have simply assumed a value for the electron-neutrino mass to which our analysis is not very sensitive and constructed the lepton mixing matrix by making use of the unitarity conditions with the same phase in (5b) as in (3b).

We evolve the masses and mixing matrices to the SUSY GUT scale and use the extended Kusenko[3] procedure to construct the mass matrices numerically. The simplest SO(10) structure for the mass matrices is found with $x_q = 0$ and $x_\ell = 0.88$, in which case the matrices have the following SO(10) transformation properties:

$$M^U \sim M^{N_{Dirac}} \sim diag(126;\ 126;\ 10) \tag{6a}$$

$$M^D \sim M^E \sim \begin{pmatrix} 10', 126 & 10', 126' & 10' \\ 10', 126' & 126 & 10' \\ 10' & 10' & 10 \end{pmatrix} \tag{6b}$$

Note that the same 10 is assumed to contribute to the 33 elements of the above mass matrices, with Yukawa coupling unification achieved for $\tan\beta \simeq 48.9$.

In this scenario we are able to construct a simple SO(10) model with just nine independent parameters for the following four matrices, such that

$$M^U = diag(F',\ E',\ C') \qquad M^{N_{Dirac}} = diag(-3F',\ -3E',\ C')$$

$$M^D = \begin{pmatrix} 0 & A & D \\ A & E & B \\ D & B & C \end{pmatrix} \qquad M^E = \begin{pmatrix} F & 0 & D \\ 0 & -3E & B \\ D & B & C \end{pmatrix} \tag{7a}$$

where only D is complex and

$$C'/C = v_u/v_d, \qquad E'/E = -4F'/F = w_u/w_d \tag{7b}$$

in terms of the ratios of the 10 and of the 126 vacuum expectation values associated with the diagonal Yukawa couplings. In this model for scenario (A), two 10's and two 126's are required as indicated in (6a,b), while four texture zeros appear in the two matrices for M^U and M^D and for $M^{N_{Dirac}}$ and M^E.

With

$$F' = -1.098 \times 10^{-3}, \qquad E' = 0.314, \qquad C' = 120.3 \tag{8a}$$

$$\begin{aligned} C &= 2.4607, & so \quad & v_u/v_d = \tan\beta = 48.9 \\ E &= -0.3830 \times 10^{-1}, & hence \quad & w_u/w_d = -8.20 \\ F &= -0.5357 \times 10^{-3}, & B &= 0.8500 \times 10^{-1} \\ A &= -0.9700 \times 10^{-2}, & D &= (0.4200 + 0.4285i) \times 10^{-2} \end{aligned} \tag{8b}$$

in GeV, the masses and mixing matrices are calculated at the GUT scale by use of the projection operator technique of Jarlskog[8] and then evolved to the low scales. The following low-scale results emerge for the quarks:

$$\begin{aligned} m_u(1GeV) &= 5.10\ MeV, & m_d(1GeV) &= 9.33\ MeV \\ m_c(m_c) &= 1.27\ GeV, & m_s(1GeV) &= 181\ MeV \\ m_t(m_t) &= 150\ GeV, & m_b(m_b) &= 4.09\ GeV \end{aligned} \tag{9a}$$

$$V_{CKM} = \begin{pmatrix} 0.9753 & 0.2210 & (0.2089 - 0.2242i) \times 10^{-2} \\ -0.2209 & 0.9747 & 0.0444 \\ 0.0078 - 0.0022i & -0.0438 - 0.0005i & 0.9994 \end{pmatrix} \tag{9b}$$

which are to be compared with the input starting data given in (3a) and (3b).

In the absence of any VEVs coupling the left-handed neutrino fields together, we observe that the heavy righthanded Majorana neutrino mass matrix can be computed at the GUT scale from the seesaw mass formula

$$M^R = -M^{N_{Dirac}}(M^{N_{eff}})^{-1}M^{N_{Dirac}} \tag{10a}$$

which can be well approximated by the nearly geometric form

$$M^R = \begin{pmatrix} F'' & -\frac{2}{3}\sqrt{F''E''} & -\frac{1}{3}\sqrt{F''C''}e^{i\phi_{D''}} \\ -\frac{2}{3}\sqrt{F''E''} & E'' & -\frac{2}{3}\sqrt{E''C''}e^{i\phi_{B''}} \\ -\frac{1}{3}\sqrt{F''C''}e^{i\phi_{D''}} & -\frac{2}{3}\sqrt{E''C''}e^{i\phi_{B''}} & C'' \end{pmatrix} \tag{10b}$$

where $E'' = \frac{2}{3}\sqrt{F''C''}$ and $\phi_{B''} = -\phi_{D''}/3$. In terms of the three additional parameters $C'' = 0.6077 \times 10^{15}$, $F'' = 0.1745 \times 10^{10}$ and $\phi_{D''} = 45^o$, the resulting heavy Majorana neutrino masses are determined to be

$$\begin{aligned} M_{R_1} &= 0.249 \times 10^9 \; GeV \\ M_{R_2} &= 0.451 \times 10^{12} \; GeV \\ M_{R_3} &= 0.608 \times 10^{15} \; GeV \end{aligned} \tag{10c}$$

From the model parameters, the seesaw formula and Jarlskog's projection operator technique,[8] the light lepton masses and their mixing matrix can be constructed at the GUT scale and then evolved downward to the low scales where we find

$$\begin{aligned} m_{\nu_e} &= 0.534 \times 10^{-5} \; eV, & m_e &= 0.504 \; MeV \\ m_{\nu_\mu} &= 0.181 \times 10^{-2} \; eV, & m_\mu &= 105.2 \; MeV \\ m_{\nu_\tau} &= 0.135 \; eV, & m_\tau &= 1.777 \; GeV \end{aligned} \tag{11a}$$

and

$$V_{LEPT} = \begin{pmatrix} 0.9990 & 0.0451 & (-0.029 - 0.227i) \times 10^{-2} \\ -0.0422 & 0.9361 & 0.3803 \\ 0.0174 - 0.0024i & -0.3799 - 0.0001i & 0.9371 \end{pmatrix} \tag{11b}$$

The agreement with the initial input values in (5a,b) is excellent.

2.2 Lepton Scenario (B) involving the Non-adiabatic MSW Solar Depletion Effect and the Cocktail Model for Mixed Dark Matter

In this scenario, the parameters of the 23 mixing plane given in (4) are replaced by

$$\delta m_{23}^2 = 49 \; eV^2, \qquad \sin^2 2\theta_{23} = 10^{-3} \tag{12a}$$

with the tau-neutrino now assumed to account for the 30% hot dark matter component of the cocktail model[6] for mixed dark matter with a mass of $m_{\nu_\tau} = 7.0$ eV. The mixing angle has been set close to the present upper bound from accelerator experiments, so the mixing matrix is now given by

$$V_{LEPT}^{(B)} = \begin{pmatrix} 0.9990 & 0.0447 & (-0.289 - 0.129i) \times 10^{-2} \\ -0.0446 & 0.9989 & 0.0158 \\ 0.0036 - 0.0013i & -0.0157 & 0.9998 \end{pmatrix} \tag{12b}$$

Following the same general procedure as applied for scenario (A), we find the simplest SO(10) structure is obtained for $x_q = 0.5$ and $x_\ell = 0.0$, but only one texture zero is then present. Instead we make use of the bases where $x_q = 0$ and $x_\ell = 0.3$ as in scenario (A). The quark mass matrices are then exactly the same as before, while the SO(10) structures are more complicated as seen from

$$M^U \sim M^{N_{Dirac}} \sim diag(10, 126;\ 126;\ 10) \tag{13a}$$

$$M^D \sim M^E \sim \begin{pmatrix} 10, 10', 126 & 10', 126' & 10' \\ 10', 126' & 126 & 10', 126' \\ 10' & 10', 126' & 10 \end{pmatrix} \tag{13b}$$

Without repeating all the details, we simply write down the form of the model matrices derived in this scenario and find

$$M^U = diag(F',\ E',\ C') \qquad M^{N_{Dirac}} = diag(-2.5F',\ -3E',\ C')$$

$$M^D = \begin{pmatrix} 0 & A & D \\ A & E & B \\ D & B & C \end{pmatrix} \qquad M^E = \begin{pmatrix} \frac{5}{6}F & -\frac{1}{3}A & D \\ -\frac{1}{3}A & -3E & \frac{1}{3}B \\ D & \frac{1}{3}B & C \end{pmatrix} \tag{14}$$

where again only D is complex. In this model for scenario (B), two 10's and two 126's are required as indicated in (13a,b), while four texture zeros appear in the two matrices for M^U and M^D but only three texture zeros appear for $M^{N_{Dirac}}$ and M^E. Since the quark mass matrices must lead to the same numerical results as in scenario (A), the values for the parameters introduced above are those given in (8a) and (8b).

The heavy right-handed Majorana mass matrix can be approximated by the nearly geometric form

$$M^R = \begin{pmatrix} F'' & -2\sqrt{F''E''} & -\sqrt{F''C''}e^{i\phi_{D''}} \\ -2\sqrt{F''E''} & -E'' & 2\sqrt{E''C''}e^{i\phi_{B''}} \\ -\sqrt{F''C''}e^{i\phi_{D''}} & 2\sqrt{E''C''}e^{i\phi_{B''}} & C'' \end{pmatrix} \tag{15a}$$

where $E'' = \frac{1}{8}\sqrt{F''C''}$, aside from an overall sign. With $C'' = 0.2323 \times 10^{14}$, $F'' = 0.1096 \times 10^{11}$ and $\phi_{D''} = 18.7^o$, $\phi_{B''} = 23.0^o$ and $\phi_{C''} = 41.8^o$, we find the resulting heavy Majorana neutrino masses for this case are determined to be

$$\begin{aligned} M_{R_1} &= 0.841 \times 10^9\ GeV \\ M_{R_2} &= 0.312 \times 10^{12}\ GeV \\ M_{R_3} &= 0.235 \times 10^{14}\ GeV \end{aligned} \tag{15b}$$

By again making use of the simplified matrices at the GUT scale first to compute the lepton masses and mixing matrix V_{LEPT} by the projection operator technique of Jarlskog and then to evolve the results to the low scales, we find at the low scales for the (B) scenario

$$\begin{aligned} m_{\nu_e} &= 0.544 \times 10^{-6}\ eV, & m_e &= 0.511\ MeV \\ m_{\nu_\mu} &= 0.242 \times 10^{-2}\ eV, & m_\mu &= 107.9\ MeV \\ m_{\nu_\tau} &= 6.99\ eV, & m_\tau &= 1.776\ GeV \end{aligned} \tag{16a}$$

and

$$V_{LEPT} = \begin{pmatrix} 0.9992 & 0.0410 & (0.150 - 0.107i) \times 10^{-2} \\ -0.0411 & 0.9991 & 0.0113 \\ -0.0010 - 0.0011i & -0.0123 & 0.9999 \end{pmatrix} \qquad (16b)$$

to be compared with the initial low scale input in (5a) and (12).

3 Summary

In this talk we have sketched a procedure which enables one to construct fermion mass matrices at the GUT scale which yield the low energy data taken as input. The models constructed for the two neutrino scenarios work well in the SO(10) SUSY GUT framework with relatively few parameters, but the structure exhibited for scenario (B) with a 7 eV tau-neutrino is not as simple as that for scenario (A) based on the observed muon-neutrino atmospheric depletion effect. Discrete family symmetries giving rise to these models are now under investigation. Our general procedure to construct mass matrices can be applied to other symmetry-based frameworks as well.

The research of CHA was supported in part by Grant No. PHY-9207696 from the National Science Foundation, while that of SN was supported in part by the U.S. Department of Energy, Grant No. DE-FG05-85ER 40215.

References

[1] C. H. Albright and S. Nandi, Fermilab-PUB-93/316-T and Fermilab-PUB-94/061-T and OSU Preprints 282 and 286, submitted for publication.

[2] S. Naculich, Phys. Rev. D **48**, 5293 (1993).

[3] A. Kusenko, Phys. Lett. B **284**, 390 (1992).

[4] S. P. Mikheyev and A. Yu Smirnov, Yad Fiz. **42**, 1441 (1985) [Sov. J. Nucl. Phys. **42**, 913 (1986)]; Zh. Eksp. Teor. Fiz. **91**, 7 (1986) [Sov. Phys. JETP **64**, 4 (1986)]; Nuovo Cimento **9C**, 17 (1986); L. Wolfenstein, Phys. Rev. D **17**, 2369 (1978); **20**, 2634 (1979).

[5] K. S. Hirata et al., Phys. Lett. B **205**,416 (1988); D. Casper et al., Phys. Rev. Lett. **66**, 2561 (1991); W. W. M Allison et al., Argonne preprint ANL-HEP-CP-93-34.

[6] Q. Shafi and F. Stecker, Phys. Rev. Lett. **53**, 1292 (1984); Ap. J. **347**, 575 (1989); G. G. Raffelt, Max Planck Institute preprint, MPI-Ph/93-81; J. Ellis and P. Sikivie, CERN preprint, CERN-TH.6992/93.

[7] J. Gasser and H. Leutwyler, Phys. Rep. C **87**, 77 (1982).

[8] C. Jarlskog, Phys. Rev. D **35**, 1685 (1987); **36**, 2138 (1987); C. Jarlskog and A. Kleppe, Nucl. Phys. **B286**, 245 (1987).

Predicting the Top Quark, Higgs and Sparticle Masses in Supersymmetric Grand Unified Models

B. ANANTHANARAYAN
INSTITUT DE PHYSIQUE THÉORIQUE,
UNIVERSITÉ DE LAUSANNE
CH 1015, LAUSANNE, SWITZERLAND
AND
Q. SHAFI
BARTOL RESEARCH INSTITUTE,
UNIVERSITY OF DELAWARE
NEWARK, DE 19716, USA

Abstract: We estimate the top quark, sparticle and scalar higgs masses within a supersymmetric grand unified framework in the two limits $\tan\beta \approx 1$ and $\tan\beta \simeq m_t^0/m_b^0$ and the electroweak symmetry is radiatively broken. The requirement that the calculated b quark mass lies close to its measured value, together with the cosmological constraint $\Omega_{LSP} \approx 1$, fixes the top quark mass to be $m_t(m_t) \lesssim 170\ GeV$ in the former case, and $m_t(m_t) \approx 140 - 185\ GeV$ in the latter. Where as in the former case the sparticle spectrum can be light with the sparticle zoo accessible at LEP200 energies, in the latter case they are heavier with the LSP (of bino purity $\gtrsim 98\%$) for instance, having a mass $\sim 200 - 350\ GeV$. The higgs spectrum however, in the former case remains heavy characterized by $m_A \gtrsim 300\ GeV$ whereas in the latter case $m_A \lesssim 220\ GeV$. In the former case the lightest CP-even higgs is expected to be no heavier than $\sim 100\ GeV$, and no heavier than $140\ GeV$ in the latter.

The lower bounds on the masses of the top quark and the scalar higgs of the standard model now stand at $113\ GeV$[1] and $60\ GeV$[2] respectively. Estimates of the top quark mass based on a global analyses of the electroweak data suggest the value $150^{+19+15}_{-24-20}\ GeV$[3]. This approach does not, however, provide any precise indication of where the higgs mass may lie, and an answer almost certainly requires an extension of the standard model.

Recent improvements in the determination of the three standard model gauge couplings has renewed interest in the exciting possibility that they do indeed converge to the same value at some superheavy scale $\sim 10^{16}\ GeV$[4]. Within the framework of standard grand unification (GUT), compatibility with the available data requires an intermediate mass scale, on the order of $10^{12\pm1}\ GeV$ in some schemes. Perhaps a somewhat more elegant possibility is provided by supersymmetric grand unification (SUSY GUT), with SUSY broken at a scale $M_S \sim 100$ GeV to a few TeV. Among other things, this latter approach offers the prospects of resolving the formidable gauge hierarchy problem[5].

Presumably supersymmetry is the way to go, but the minimal supersymmetric standard model (MSSM)[6] introduces two new parameters in the scalar higgs sector, usually parametrized as m_A (the tree level mass of the CP-odd scalar boson) and $\tan\beta$ (the ratio of the two vacuum expectation values (vevs), needed to provide masses to up-type and down-type quarks). Knowing m_A and $\tan\beta$ allows one to estimate the tree level masses of the scalar higgs sector of MSSM. An important challenge to the model builder therefore is to find a framework in which something definite can be said about these as well as a number of other undetermined parameters of MSSM. One could also hope that SUSY GUTs can shed light on at least some of the many parameters appearing in the minimal supersymmetric $SU(3) \times SU(2) \times U(1)$ model (MSSM). A particularly important example is offered by the well known asymptotic relation $m_b^0 \approx m_\tau^0$[7]. It has become clear in recent years[8] that the determination of the b-quark mass from this relation depends in a rather interesting way on $m_t(m_t)$, $\alpha_S(M_Z)$ and $\tan\beta$. In particular, and we verify this for $\alpha_S(M_Z)$ near 0.12 and $m_t(m_t) \lesssim 170\,GeV$, the two regions, $\tan\beta$ close to unity $(1 \lesssim tan\beta \lesssim 1.6)$, and $\tan\beta \gg 1$ are singled out.

With $\tan\beta$ near unity and a top mass in the range of 150 - 170 GeV (favored by precision electroweak measurements), the top–quark Yukawa coupling h_t is near the infrared quasi–fixed point $h_t(m_t) \simeq 1.1$ [9]. This may be an appealing feature in that a wide range of initial values $h_t(M_X)$ near the GUT scale will be mapped to essentially the same value of $h_t(m_t)$ thereby making it insensitive to the initial conditions. Furthermore, with $\tan\beta$ close to unity, the weak-scale radiative corrections to the b mass can be quite small $(\lesssim 1 - 2\%)$.

Recently investigations were made with the possibility that $\tan\beta \approx 1$[10]. Related investigations have appeared in which $\tan\beta$ close to unity has been considered, although not exclusively so [11]. It turns out that the low $\tan\beta$ case is considerably less restrictive as far as the allowed parameter space is concerned. In particular, some of the sparticles including a stau and stop may well be accessible to LEP 200. With $m_t(m_t) \lesssim 170\,GeV$ the lightest CP even scalar higgs mass is expected to be $\lesssim 100\,GeV$.

It has been emphasized[12, 13, 14] that embedding the MSSM in grand unified theories (GUTs) based on gauge groups such as $SO(10)$ or $SU(3)_c \times SU(3)_L \times SU(3)_R$ (but not $SU(5)$!) can constrain $\tan\beta$ to a region close to $m_t^0/m_b^0 \gg 1$ (where the superscript stands for the values of the masses computed within this framework before inclusion of the weak-scale radiative corrections that we will discuss later) which follows from the requirement that at M_X $h_t = h_b = h_\tau$. With $m_A \gtrsim m_Z$, as suggested by the decay $b \to s\gamma$, one readily understands in this scheme why no higgs scalar has been found at LEPI. The tree level mass of the lighter CP-even scalar mass is approximately M_Z, and radiative corrections can increase it further. In order to estimate the latter one needs a reliable estimate of the top quark mass, a quantity closely related to the important issue of radiative electroweak breaking[15].

It was first pointed out in[12] that with $\tan\beta \approx m_t^0/m_b^0$, the requirement that $m_b(m_b) = 4.25 \pm 0.10 GeV$ [16] also helps constrain the top mass within a relatively narrow range. This has subsequently been refined by several authors[17], so that we expect $m_t(m_t)$ to be in a range consistent with the numbers expected from the global radiative analyses[3]. As we will see, the radiative breaking scenario coupled with the constraint Ω_{LSP}[18, 19] also imposes a fairly stringent upper bound on the top mass, namely $m_t(m_t) \lesssim 185\,GeV$.

The starting point of our computations are the two-loop renormalization group (RG) equations for the gauge and Yukawa couplings [20]. For simplicity, we will only consider two possible values for the (common) supersymmetry threshold M_S, a 'low' value comparable to m_t, and a 'high value' of order a TeV. Using $\alpha(M_Z) = 1/128$ and $\sin^2\theta_W(M_Z) = 0.233$ as inputs, we estimate the unification scale M_X to be about 2×10^{16} GeV, while the unified gauge coupling $\alpha_G(M_X) \approx 1/26$.

The running b quark mass has been estimated to be $m_b(m_b) = 4.25 \pm 0.1$ GeV [16]. Considering the various hadronic uncertainties that are inherent in this determination, we will allow for a more generous 2σ error and take $m_b(m_b)$ to lie in the mass range $4.05 - 4.45$ GeV. The upper end will turn out to be crucial in constraining $\tan\beta$. It corresponds to a physical mass (pole mass) $M_b = 5.2$ GeV (5.1 GeV) for $\alpha_s(M_Z) = 0.12$ (0.115). A value of M_b much larger than this seems unlikely.

The simplest (minimal) GUTs based on $SU(5)$ or $SO(10)$ predict the asymptotic relation $m_b^0 = m_\tau^0$, which holds above the unification scale $M_X \simeq 2 \times 10^{16}$ GeV. Deviation from the minimal GUT schemes, it appears, are necessary in order to obtain realistic quark and lepton masses. Even then, the above relation can hold to a good approximation. For instance, implementation of the Fritzsch ansatz in GUTs requires a non-minimal Higgs system [21]. From the asymptotic relation $Tr(M_\ell) = Tr(M_d)$,

$$m_d^0 - m_s^0 + m_b^0 = m_e^0 - m_\mu^0 + m_\tau^0 \tag{1}$$

Equation (6) leads to $m_b^0 = m_\tau^0(1 - \epsilon)$, where $\epsilon \approx (2/3)m_\mu^0/m_\tau^0 \approx 0.04$. Here $m_\mu^0 \simeq 3m_s^0$ has been used.

Another possible source for fermion masses corresponds to the non renormalizable interactions with contributions of order

$$M_W \left(\frac{M_X}{M_{Pl}}\right)^n \lesssim \mathcal{O}(100 \ MeV), \tag{2}$$

with the $n = 1$ contribution being of order the muon mass. One expects the relation $m_b^0 = m_\tau^0$ to hold at the 5% level in this case.

Consequently, in addition to allowing a 2σ error in $m_b(m_b)$, we also allow for deviation of 5% in the asymptotic relation $m_b^0 = m_\tau^0$, i.e.,

$$m_b^0 = m_\tau^0[1 \pm 0.05] \tag{3}$$

In Fig. 1, we plot $m_b(m_b)$ versus $\tan\beta$ with $m_t = 150 \ GeV$, $\alpha_s(M_Z) = 0.12$ and $M_S = 1 \ TeV$. We have employed the appropriate two-loop renormalization group equation for the running of the gauge and Yukawa couplings between M_X and M_Z, while below M_Z we have used the 3-loop QCD[22] and one–loop QED β functions in the evolution of α_s, α and the mass parameters. The solid line in Fig. 1 corresponds to $m_b^0 = m_\tau^0$, while the dot-dashed lines account for the $\pm 5\%$ deviation from the exact equality. It can be seen that there are two acceptable solutions for $\tan\beta$, corresponding to a given value of $m_b(m_b)$. The lower value is quite close to unity ($\tan\beta \simeq 1.1$ in this case), while the larger values exceed

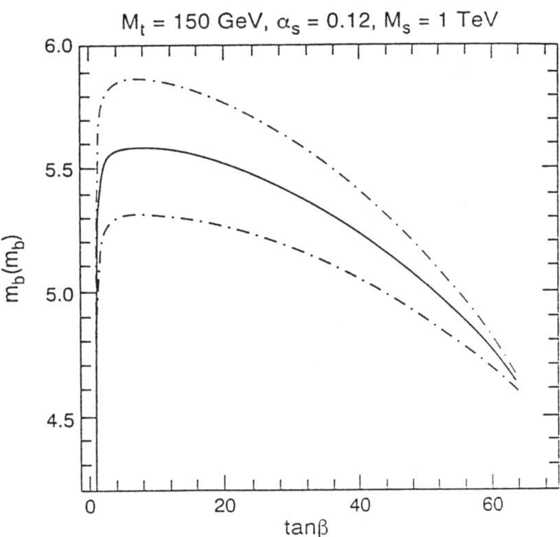

Figure 1.   Example of a plot of $m_b(m_b)$ vs. $\tan\beta$.

$\tan\beta \doteq m_t/m_b$. Barring radiative corrections all intermediate values of $\tan\beta$ are disfavored! If we require $m_b(m_b) \lesssim 4.45\,GeV$, then $\tan\beta \lesssim 1.1$ or $\tan\beta \gtrsim 60$ are singled out.

Of course, all this assumes that radiative corrections to the b and τ masses can be ignored. It has been shown in *[23]* that this very much depends on a combination of parameters which include $\tan\beta$, the universal gaugino mass and the supersymmetric higgsino mass. In the case that we are interested in, to wit, $\tan\beta$ close to unity, we can a posteriori verify that the radiative correction can be kept small ($\stackrel{<}{\sim}$ few %), without imposing stringent new constraints on the available parameter space.

The predictions for $m_b(m_b)$ are rather sensitive to the input values chosen for $\alpha_s(M_Z)$, m_t and M_S. Based on variations of all these parameters it was concluded*[10]* that the two regions $\tan\beta$ close to unity, or $\tan\beta \gg 1$ are singled out.

It is an interesting coincidence that the asymptotic mass relation $m_b^0 \simeq m_\tau^0$ prefers the two extreme values of $\tan\beta$. Since the parameter space is narrow near the end points, a certain amount of fine–tuning is needed to obtain the desired values of $\tan\beta$. To see this explicitly, let us analyze the tree–level (neutral) higgs potential of the MSSM:

$$V_0 = \mu_1^2|H_1^0|^2 + \mu_2^2|H_2^0|^2 + \mu_3^2(H_1^0 H_2^0 + h.c.)$$
$$+\frac{g_1^2 + g_2^2}{8}(|H_1^0|^2 - |H_2^0|^2)^2 \tag{4}$$

where the mass parameters μ_i^2 are defined as

$$\mu_1^2 = m_{H_1}^2 + \mu^2; \quad \mu_2^2 = m_{H_2}^2 + \mu^2; \quad \mu_3^2 = B\mu \ . \tag{5}$$

Here μ is the supersymmetric higgsino mass, and we let v_1, v_2 (both positive) denote the two vevs, such that $v^2 = v_1^2 + v_2^2$ and $\tan\beta = v_2/v_1$. Minimization of Eq. (4) yields

$$sin2\beta = \frac{2\mu_3^2}{\mu_1^2 + \mu_2^2}$$

$$v^2 = \frac{2}{g_1^2 + g_2^2} \left[\frac{\mu_2^2 - \mu_1^2}{cos2\beta} - (\mu_1^2 + \mu_2^2) \right] \quad . \tag{6}$$

The large $\tan\beta$ solution would require $\mu_3^2 \ll (\mu_1^2 + \mu_2^2)$. The naturalness of this scenario has recently been discussed[24].

The scheme where $\tan\beta$ is close to unity requires a different fine–tuning condition. Let us take the limit of strict equality, $\tan\beta = 1$. In this case,

$$\mu_1^2 + \mu_2^2 - 2\mu_3^2 = 0 \tag{7}$$

from Eq. (6). Furthermore, $\mu_1^2 = \mu_2^2$ is needed to remedy the divergent behavior in v^2 (see Eq. (6)). In this limit, the Higgs potential has a custodial $SU(2)$ symmetry (under which H_1 and H_2 rotate into each other). This symmetry, of course, is broken badly in the Yukawa sector where $h_t \gg h_b$, as well as by the hypercharge gauge interactions. These effects show up in the Higgs sector only after one–loop radiative corrections are taken into account.

The $\tan\beta = 1$ solution preserves the custodial $SU(2)$ symmetry (at tree level), while the solution $\tan\beta \gg 1$ breaks this symmetry maximally. In the former case, the Yukawa interactions explicitly do not respect the custodial symmetry, whereas in the latter case they do. The hypercharge gauge interactions break this custodial symmetry in either case.

As a consequence of Eq. (6), the $\tan\beta = 1$ solution has a zero mass scalar at tree–level which can be seen as follows. The Higgs potential as a function of v^2, after minimizing with respect to β, is given by

$$V(v^2)|_{tan\beta=1} = v^2 \left[\frac{1}{2}(\mu_1^2 + \mu_2^2) - \mu_3^2 \right] \quad . \tag{8}$$

The right-hand side of Eq. (8) vanishes as a consequence of Eq. (7), leaving v^2 undetermined. The resulting ambiguity along the radial direction gives rise to a (pseudo) Goldstone mode. The relation (7) will be modified by the radiative corrections and is to be regarded as the starting point of the SUSY analogue of the Coleman-Weinberg mechanism[25]. This has been studied in some detail in the existing literature. The zero mass higgs mode, on account of the large top quark Yukawa coupling, as well as the existence of the scalar stops, can acquire a substantial mass through radiative corrections. Within the MSSM, with the large number of free parameters, the value of this mass can exceed $100 \ GeV$. However, by embedding within a supergravity/GUT approach, we find this to be not the case.

It is important to ask if $\tan\beta \approx 1$ can be realized in a natural manner. One approach would be to replace the supersymmetric higgsino mass term by $\lambda N(H_1 H_2 - M^2)$, where N denotes a singlet superfield. The superpotential in this case possesses an R-symmetry under which all matter superfields as well as the superpotential change sign. Now the $SU(2) \times U(1)$ gauge symmetry can be broken at tree level whilst preserving supersymmetry. It is then "natural" to have $\tan\beta \simeq 1$, since we can now take the limit where all the SUSY breaking mass parameters are small compared to M^2. Although intriguing, we will not pursue this possibility any further here.

Our analysis of radiative electroweak breaking will be based on the renormalization group improved tree level scalar potential given in Eq. (4). The various parameters evolve according to the well known one loop RGE[15] from M_X to the low energy scale Q_0. A judicious choice of Q_0[26] will yield a fairly reliable estimate of the parameters, including bounds on the sparticle mass spectrum.

A technical note on what allows us to solve for such vacua is in order. The fact that the one-loop gauge coupling evolution equations can be integrated independent of the knowledge of all other parameters[27] is the corner stone of this program (this of course yields the scale M_X, the unified coupling constant at M_X, α_G and the low energy prediction for $\sin^2\theta$). Then the knowledge of the Yukawa and gauge couplings alone is sufficient to integrate the one-loop evolution equations for the Yukawa couplings, albeit coupled to gauge coupling evolution. Then the knowledge of the gaugino masses, scalar masses and tri linear couplings in addition to the knowledge of the Yukawa and gauge couplings is sufficient to integrate the one-loop equations for the parameters that have entered the program at the last stage. It is in this manner that the Yukawa couplings hold the key to the origin of the masses of the sparticles of the heaviest generation in addition to those of the standard model fermions (whereas for the two lighter generations the Yukawa couplings remain too small to influence the mass spectrum, now determined by the gauge couplings alone).

The spontaneous breaking of the electroweak gauge symmetry imposes the constraint

$$\mu_1^2 \mu_2^2 \leq \mu_3^4 \tag{9}$$

while the boundedness of the scalar potential yields

$$\mu_1^2 + \mu_2^2 \geq 2|\mu_3|^2 \ . \tag{10}$$

The previous considerations of $m_b(m_b)$ help us pin down the range of values at M_X for the Yukawa couplings h_b, h_τ and h_t. The parameter $\tan\beta$ at scale $Q_0 \sim M_S$ lies between 1 and 1.6. In addition, we have the parameters $M_{\frac{1}{2}}$, m_0, A at M_X, which denote the common gaugino mass, the common scalar mass and the common trilinear scalar coupling. Once these are specified, the one loop beta functions are integrated to yield the gaugino and scalar mass parameters, as well as the trilinear couplings at Q_0.
Following Ref. [26], we impose the constraint

$$|A(M_X)| < 3 \tag{11}$$

Note that the structure of the one loop beta functions does not require us to know the values of the parameters μ (the SUSY higgsino mass) and the bilinear coupling parameter $B(\equiv \mu_3^2/\mu)$ at this stage of the computation. Indeed, once $\tan\beta$ is known from the considerations described earlier, we obtain from the minimization conditions:

$$\mu^2(Q_0) = \left[\frac{1}{2}M_Z^2(\tan^2\beta - 1) - (m_{H_1}^2 - \tan^2\beta\, m_{H_2}^2)\right]/(1 - \tan^2\beta) \ . \tag{12}$$

Further it can be seen that

$$\mu_3^2(Q_0) = \tan\beta\left[-m_{H_2}^2 - \mu^2 - \frac{1}{2}M_Z^2\left(\frac{\tan^2\beta - 1}{\tan^2\beta + 1}\right)\right] \ . \tag{13}$$

It may also be noted that the two parameters that had remained unknown at the beginning of the analysis, $\mu(M_X)$ and $B(M_X)$ can now be evalued to one loop order by integrating the coupled system of renormalization group equations back to M_X since $\mu(Q_0)$ and $B(Q_0)$ are now known. It is therefore possible to isolate certain families of solutions where all the parameters of the theory are required to obey certain boundary conditions. For instance, it is now possible to find those classes of vacua such that at M_X [28]

$$\mu_3^2 = \mu_2^2 = \mu_1^2. \tag{14}$$

There is a two fold ambiguity in the physical spectrum for a given point in the parameter space due to the fact that μ can be of either sign. This is found to affect the physical spectrum significantly because of the structure of the terms in the sfermion, neutralino and chargino mixing matrices when the electro-weak symmetry is broken. We find that the dependence of the spectrum on A is minimal and therefore stick to the choice $A = 0$ for the most part. The beta functions are such that the physical spectrum remains unaltered under the interchange $(-M_{\frac{1}{2}}, A)$ with $(M_{\frac{1}{2}}, -A)$. Thus, with $A = 0$ the physical spectrum shows no dependence on the sign of M.

Armed with this a systematic search is performed to find the acceptable region of the parameter space and the associated sparticle spectrum. It may be seen that with all other parameters fixed, as $\tan\beta$ approaches 1, the parameter μ rises. An unusually large $|\mu|$ may be unattractive, and so we impose a somewhat ad hoc upper bound of $1\,TeV$ on $|\mu|$ (radiative corrections to the b-mass can become significant if $|\mu|$ is excessively large). This establishes a lower bound on $\tan\beta$ (separately for the choice of positive or negative μ). For a given of $\tan\beta$ there is a unique value of the common gaugino and scalar mass for which the lightest neutralino is the LSP (albeit degenerate with the lighter stau and stop). This follows from the following observations. As m_0 is increased, with all other parameters fixed, μ increases (thus an upper bound on m_0 emerges naturally). For a sufficiently small $\tan\beta$, as m_0 is increased, the mass of the lighter scalar top falls, due to the mixing term increasing in importance. However, a sufficiently large m_0 is required to ensure that the lighter scalar tau is heavier than the lightest neutralino (the would be LSP). As m_0 increases, the bino-purity of the lightest neutralino is also found to increase. We impose a

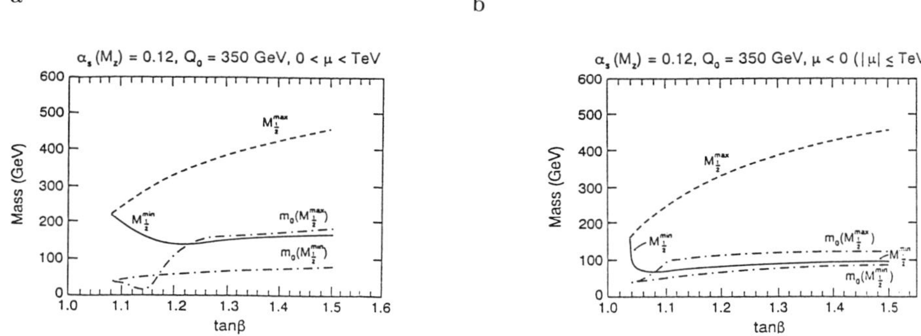

Figure 2. a. Plot of mass parameters $M_{\frac{1}{2}}$ and m_0 vs. $\tan\beta$ for $\mu > 0$. b. Same as Fig. 2a with $\mu < 0$.

constraint that the bino-purity of the lightest neutralino be $\gtrsim 95\%$[19] in order to ensure that the LSP is a good candidate for cold dark matter.

The above considerations are used in determining the upper and lower bounds on the two primary soft SUSY parameters $M_{\frac{1}{2}}$ and m_0. It is found that as $\tan\beta$ migrates away from its lower bound the parameter space opens up somewhat. This is because as $\tan\beta$ increases, even with fairly large values of $M_{\frac{1}{2}}$ and m_0, we meet all of the consistency conditions, with $|\mu|$ staying considerably smaller than a TeV.

We illustrate in the figures the behavior of the parameters of the theory. In Figs. 2a, 2b we exhibit the upper and lower bounds on $M_{\frac{1}{2}}$ (and the corresponding m_0 values) as functions of $\tan\beta$. [Note that without the restriction $|\mu| \lesssim 1\,TeV$, the upper bound on $M_{\frac{1}{2}} \approx 800\,GeV$, corresponding to a bino mass of $\approx 350\,GeV$.] Due to the fact that for a given point in much of the parameter space the squark mixing is less important when μ is negative, the lower bounds emerge smaller. In Figs. 3a, 3b we show, for a particular choice of some of the parameters, the variations of the lighter stau and stop masses as m_0 is varied. In Fig. 4 we show the behavior for a larger value of $\tan\beta$. Comparing Figs. 3 and 4 we find that the parameter space is much more constrained in the former case.

To explain the shapes a brief discussion is in order. At the smallest value of $\tan\beta$ realizable, we meet the conditions that $|\mu| \approx 1\,TeV$ and that the lighter stop and stau are degenerate with the lightest neutralino. As $\tan\beta$ increases, we can find smaller values of the unified gaugino mass such that the lighter stop and stau continue to be degenerate with the lightest neutralino, with $|\mu| < 1\,TeV$. However, with increasing $\tan\beta$ the lightest neutralino begins to mix strongly with one of the higgsinos. We require that the bino purity remain in excess of, and restrict the parameter range further by requiring that the $< 95\%$ lighter stau mass be no larger than three times the LSP mass. The maximum value of $M_{\frac{1}{2}}$ for a given $\tan\beta$ is reached when m_0 is chosen such that the lightest neutralino is the LSP, and simultaneously the upper bound on $|\mu|$ is saturated. In Fig. 3, 4, the shapes of the stau and stop mass contours are

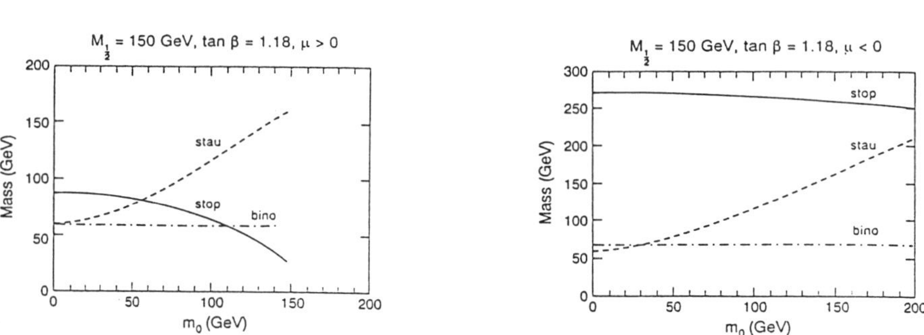

Figure 3. a. Plot of bino, stop and stau mass vs. m_0 for a choice of parameters with $\mu > 0$. b. Same as 3a with $\mu < 0$.

easily understood. Since the mixing term is negligible for the stau, increasing m_0 leads to increasing values of the stau. In the stop sector, however, as m_0 increases, $|\mu|$ increases leading to appreciable mixing and the diminishing of the smaller eigenvalue of the stop mixing matrix, i.e., the mass of the lighter physical stop. Indeed as discussed earlier, as $\tan\beta$ falls, this mixing becomes even more pronounced for a given point in the parameter space, and thus the parameter space becomes more constrained since the stop must not remain lighter than the lightest neutralino.

It has been argued and found to be true by comparing our computations with other related studies that the tree level potential is consistent to within 10% with the complete one loop scalar potential. This is achieved by the selection of Q_0 to be of the order of the geometric mean of the scalar top masses. We also study the variations in the spectrum for Q_0 between the two extreme limits of M_Z and $1\,TeV$.

As far as the sparticles are concerned it is not inconceivable that the lighter stau and stop will be found at LEP 200. Equally likely, however, is the possibility that the sparticles are all heavy, accessible perhaps only at the LHC.

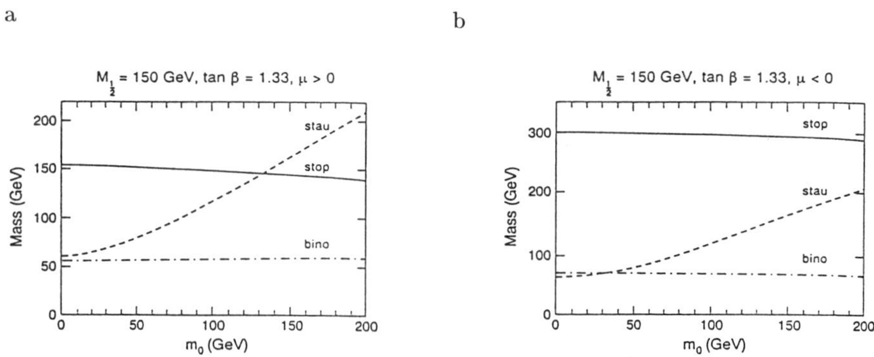

Figure 4. a. Same as Fig. 3a with a larger $\tan\beta$ value. b. Same as Fig. 3b with a larger $\tan\beta$ value.

This is the case where the bino LSP has mass $\sim 350\ GeV$, one of the staus is the lightest charged sparticle, while the colored sparticles approach and even exceed the TeV range.

As far as the higgs sector is concerned, with $m_t(m_t) \lesssim 170\ GeV$, $|\mu| < 1\ TeV$, stop masses $\lesssim 1-2\ TeV$, and $A_t(Q_0) \lesssim 1\ TeV$, the lightest (CP even) higgs mass is expected to be between $60\ GeV$ and $100\ GeV$[29]. It offers the best prospects for a truly remarkable discovery at LEP 200. The CP odd scalar mass exceeds $300\ GeV$ which implies that the remaining scalar higgs will not be accessible for quite some time.

The analysis of the higgs potential as discussed in the earlier sections is further simplified by the fact that there is a self-consistent determination of $\tan\beta$ with $h_t = h_b = h_\tau = h$. Solving for μ and B for a given point in the parameter space $M_{\frac{1}{2}}$, m_0, A and h we can now proceed to discuss the features of the sparticle spectrum and the higgs spectrum as well[13, 14].

Indeed, the neutralino, chargino and sfermion mass matrices[6] display some dependence on the sign of μ, it turns out to be not very significant. We will therefore assume that $\mu > 0$. Furthermore, our analysis shows that deviations of $A(M_X)$ from zero can never be large, so that the bounds depend on $A(M_X)$ only in a minor way. With $A(M_X) \approx 0$, the sparticle spectrum is unchanged if $M_{\frac{1}{2}} \to -M_{\frac{1}{2}}$. It is important to point out that even with $A(M_X) \approx 0$, the quantities A_t and A_b, when evaluated at the scale Q_0, lie in the TeV range.

Let us now see how the bounds on $m_t(m_t)$ are obtained. The first thing to note is that for each h (the unified Yukawa coupling at M_X), there is a lower bound on $M_{\frac{1}{2}}$ for a satisfactory realization of the scenario. We call this lower bound $M_{\frac{1}{2}}^{min}|_{m_A=0}$, where $m_A^2 = \mu_1^2 + \mu_2^2$ is the mass squared of the CP odd boson. Once this is found, consider some fixed $M_{\frac{1}{2}} > M_{\frac{1}{2}}^{min}|_{m_A=0}$, and find the value of m_0 above which m_A^2 turns negative, keeping the other parameters fixed.

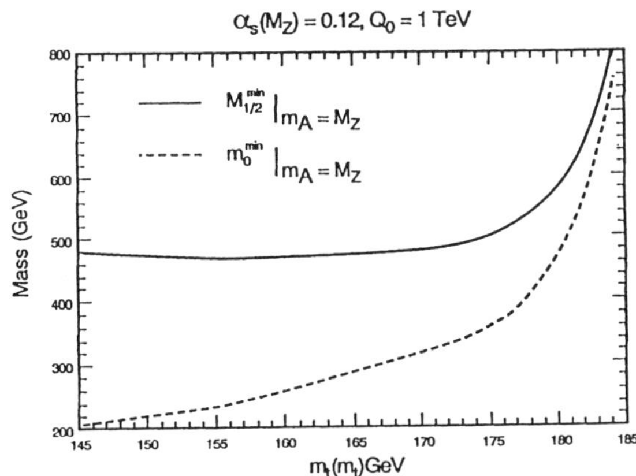

Figure 5. Contours of $M_{\frac{1}{2}}^{min}$ and m_0^{min} for $m_A = M_Z$ with large $\tan\beta$.

As m_0 is decreased below this value m_A increases. Thus, for fixed $M_{\frac{1}{2}}$ and h, there is a maximum value, m_A^{max}, which can be obtained by lowering m_0. For fixed h, for smaller values of $M_{\frac{1}{2}}$, the value of of m_A^{max} is smaller. It is possible, therefore, for fixed h and a given m_A to find $M_{\frac{1}{2}}^{min}|_{m_A}$ (and a corresponding $m_0^{min}|_{m_A}$). Varying h yields a profile for $M_{\frac{1}{2}}^{min}|_{m_A}$ (and a corresponding one for $m_0^{min}|_{m_A}$), which can be plotted against $m_t(m_t)$ (corresponding to h). An example is shown in Fig. 5, corresponding to $m_A = m_Z$. In most of the region of interest, with larger values of h and fixed m_A, the quantity $M_{\frac{1}{2}}^{min}|_{m_A}$ increases. We can see from the shape of this profile that with increasing $m_t(m_t)$, to obtain larger m_A requires increasing values of $M_{\frac{1}{2}}$. In other words, if one plots, for given m_A, contours of $M_{\frac{1}{2}}^{min}|_{m_A}$ (and a corresponding $m_0^{min}|_{m_A}$) as functions of $m_t(m_t)$, then for given m_{A_1}, m_{A_2} with $m_{A_1} > m_{A_2}$, contours corresponding to m_{A_1} lie above those corresponding to m_{A_2}. Note that in this work we will take $\alpha_s(M_Z)$ in the range 0.11 to 0.12.

Next we discuss the role played by cosmological considerations in restricting the value of $m_t(m_t)$. As argued above, as h increases the value of $M_{\frac{1}{2}}^{min}|_{m_A}$ required for obtaining a satisfactory scenario also rises. It turns out that for $m_t(m_t) \gtrsim 145\ GeV$, the LSP essentially consists of the bino with a small ($\lesssim 2\%$) higgsino component. According to Ref. [19], the bino mass should be below $350\ GeV$ in order that Ω_{LSP} does not exceed unity, which means that the common gaugino mass at M_X should not exceed $\sim 800\ GeV$. Using fig. (1), this translates into an upper bound $m_t(m_t) \lesssim 185\ GeV$ (for $\alpha_S(M_Z) = 0.12$). [With $\alpha_S(M_Z) = 0.11, Q_0 = 1\ TeV$, the upper bound on $m_t(m_t)$ is reduced by a few percent.]

In summary, with $\tan\beta \approx m_t^0/m_b^0$, the radiative electroweak breaking scenario, combined with cosmological considerations of the LSP, predict that the top quark is to be found in the mass range $145\ GeV \lesssim m_t(m_t) \lesssim 185\ GeV$. Furthermore, the LSP (with bino purity $\gtrsim 98\%$) mass is estimated to be $\sim 200 - 350\ GeV$. Some comments about the scale-dependence of the results are in order. By varying $\alpha_s(M_Z)$ between 0.11 and 0.125 and Q_0 between 0.5 and TeV, we find that $M_{\frac{1}{2}}^{min}$ and m_0^{min} vary by $\sim 5 - 10\%$. This alters the upper bound on $m_t(m_t)$ by about $\pm 2\ GeV$. (Similar considerations later apply to the scalar higgs masses.)

As indicated earlier, with $\tan\beta \approx m_t^0/m_b^0$, an independent estimate of $m_t(m_t)$ is obtained by requiring that the calculated $m_b(m_b)$ is close to the value $4.25 \pm 0.10\ GeV$ given in [16]. Some recent estimates[17] yield $m_t(m_t) \approx 155 - 200\ GeV$. One cannot, however, ignore the potentially large radiative corrections to the b quark mass from one-loop diagrams involving supersymmetric particles, particularly the gluino and the b squarks[23]. For the case at hand these corrections turn out to be of order 30% and must be chosen negative in order to ensure that $m_b(m_b)$ comes out in the right range. The net effect of this is to reduce the top mass prediction by about $15 - 20\ GeV$[14]. Note that $\tan\beta \approx m_t^0/m_b^0$ is somewhat smaller than m_t/m_b, where the mass parameters m_t and m_b are the values obtained including the corrections mentioned above.

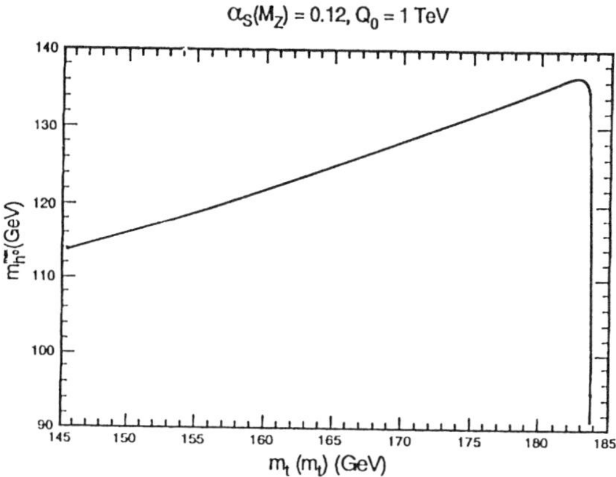

Figure 6. Plot of $m_{h^0}^{max}$ as a function of
$m_t(m_t)$.

We can now combine the two independent etimates of $m_t(m_t)$, from considerations of $m_b(m_b)$, and from radiative electroweak breaking coupled with the requirement that $\Omega_{LSP} \approx 1$, to yield the overlap range of $140 - 185\ GeV$ for $m_t(m_t)$.

Let us now consider the important issue of the scalar higgs masses of the MSSM. In the class of models under discussion, the parameter m_A is constrained by the approximate upper bound of $3\ M_W$, corresponding to $\alpha_S(M_Z) = 0.12$. [With $\alpha_S(M_Z) = 0.11$ this is reduced by $\approx 10\%$.] Thus, we have an unusually well specified version of the MSSM. Following the second paper in [30], we have estimated the radiative corrections to the scalar higgs masses and some of the results are presented in Figs. 6 and 7. In Fig. 6 we display the dependence of $m_{h^\circ}^{max}$ on $m_t(m_t)$ which yields the absolute upper bound on the former of $\approx 140\ GeV$. In Fig. 7 the dependence on $m_t(m_t)$ of the scalar masses m_{H°, m_A and $m_{H^\pm}$ are displayed. The results are $m_A^{max} \approx m_{H^\circ}^{max} \lesssim 220\ GeV$ and $m_{H^\pm}(\equiv \sqrt{m_A^2 + m_W^2}) \lesssim 240\ GeV$ (for $\alpha_S(M_Z) = 0.12$). Somewhat lower values are obtained with $\alpha_S(M_Z) = 0.11$. Note that the upper bound on m_A gets somewhat relaxed after the radiative corrections indicated above are included[14].

In order to explain the shape of the curve determining $m_{h^\circ}^{max}$ (Fig. 6), we first draw attention to the fact that a sufficiently large value of m_A is needed to ensure that the lighter (CP-even) higgs h° receives a substantial part of the radiative corrections to the tree-level relations in the presence of large Yukawa couplings. [Recall that the lightest higgs receives substantial corrections only in the "Weinberg-Salam" limit, $m_A >> M_Z$.] This explains, for 'sufficiently small' $m_t(m_t)$, the near linear rise in $m_{h^\circ}^{max}$ (Fig. 6). However, for $m_t(m_t) \gtrsim 180\ GeV$, the parameter space rapidly shrinks and m_A^{max} declines sharply (Fig. 7). As a result, the radiative corrections to the mass of the lightest higgs begin to diminish in importance, and finally $m_{h^\circ}^{max}$ becomes nearly degenerate with

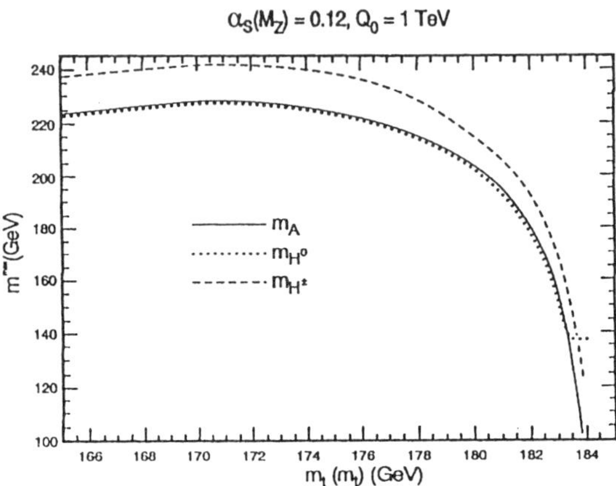

Figure 7. Plot showing maximum allowed
values of the higgs masses, m_A^{max}, $m_{H^0}^{max}$ and
$m_{H^\pm}^{max}$.

m_A^{max}. This is seen in Fig. 6 in the initial linear rise of $m_{h^\circ}^{max}$, followed by a
slow down and culminating in a sharp fall. Meanwhile, as $m_t(m_t)$ increases,
the maximum tree-level mass of the heavier CP-even neutral higgs H° becomes
nearly degenerate with M_Z (whereas it is nearly degenerate with m_A^{max} for the
lower range of allowed $m_t(m_t)$) and receives the larger share of the radiative
corrections. Thus, $m_{H^\circ}^{max}$ reaches a minimum as $m_t(m_t)$ increases and then rises
(Fig. 7).

In order to bound m_{h° from below we note that a lower bound on m_A more
stringent than from the LEP data is obtained from considerations of the decay
$b \to s\gamma$. The CLEO collaboration[31] has recently reported the first observation
of this process at a rate consistent with the standard model estimates. For a
top quark mass of 150 GeV, theoretical estimates [32] (prior to the discovery)
suggested that $m_A \gtrsim M_Z$ (perhaps even $\gtrsim$ 130 GeV!). Combining this with
the above considerations, we are led to an estimate of the h° mass in the
range 92 − 140 GeV. The model therefore "explains" why the higgs scalar
h° has not been seen at LEPI. It also may elude an upgraded LEPII unless
m_A happens to lie close to M_Z. [For instance, with $m_A \approx$ 100 GeV and
$m_t(m_t) = 170\ GeV$, $m_{h^\circ} \approx 100\ GeV$.]

To conclude, the MSSM raises many more questions than it answers. We
have argued that a supergravity/grand unified approach in which $\tan\beta$ is close
to m_t^0/m_b^0 provides an attractive and perhaps even the most predictive extension. Indeed, we are able to shed light on a number of the most fundamental parameters of MSSM. This includes a rather precise determination of the
top quark mass, (140 − 185 GeV), a narrow allowed range for the LSP mass
($\sim$ 200 − 350 GeV), and a stringent upper bound ($\approx$ 220 GeV) on m_A which
forces the scalar higges to all have masses below about 250 GeV. [Note, however,
that by relaxing the asymptotic relation $h_t = h_b(= h_\tau)$ by a small amount, say
10% with $h_b(= h_\tau) \lesssim h_t$, the bounds we obtained can be significantly changed.

For instance, we then obtain a bino as light as 120 GeV.] The sparticle mass spectrum is also quite constrained. The lightest charged sparticles include a stau ($m_{LSP} \lesssim m_{\tilde{\tau}_1} \lesssim 580\ GeV$) and a chargino ($370\ GeV \lesssim m_{\chi_1^{\pm}} \lesssim 670\ GeV$). The heaviest states include the gluino ($1 TeV \lesssim m_{\tilde{g}} \lesssim 1.9\ TeV$) and the squarks of the first two families' ($1.2\ TeV \lesssim m_{\tilde{u},\tilde{d}} \lesssim 2.1\ TeV$). The discovery of the top quark will narrow the allowed range and is therefore eagerly awaited.

ACKNOWLEDGEMENTS.   B. A. thanks the Swiss National Science Foundation for support during the course of this work. We thank the Organizing Committee and Pierres Ramond and Sikivie for their hospitality at the Second I. F. T. Workshop on Yukawa Couplings and the Origin of Mass. We thank K. S. Babu, G. Lazarides and X. M. Wang for their collaboration during various stages of the investigations summarized here.

References

[1]   CDF Collaboration, Fermilab, Conf. - 93/212-E, 1993; Also see S. Abachi et al., Phys. Rev. Lett. 72 (1994) 2138.

[2]   M. Felcini, CERN - PPE/92-208, Dec. 1992.

[3]   For a recent review, see P. Langacker, "Precision tests of the standard model," University of Pennsylvania preprint, UPR-0555T, 1993; See also J. Ellis, G.L. Fogli and E. Lisi, Nucl. Phys. B393 (1993) 3.

[4]   U. Amaldi, W. de Boer and H. Fürstenau, Phys. Lett. B260 (1991) 817; J. Ellis, S. Kelley and D. V. Nanopoulos, Phys. Lett. B260 (1991) 131; P. Langacker and M. X. Luo, Phys. Rev. D44 (1991) 817; F. Anselmo, L. Cifarelli, A. Peterman and A. Zichichi, N. C. 104A (1991) 817.

[5]   For recent discussions see G. Dvali and Q. Shafi, Phys. Lett. B326 (1994) 258; G. Lazarides and C. Panagiotakopoulos, Univ. Thessaloniki preprint, Dec. 1993; K. S. Babu and S. M. Barr, Phys. Rev. D48 (1993) 5354; G. Dvali, Phys. Lett. B324 (1994) 59.

[6]   For reviews see, H. P. Nilles, Phys. Rep. 110 (1984) 1; H. E. Haber and G. L. Kane, Phys. Rep. 117 (1985) 75; F. Zwirner, Lectures delivered at the Trieste Summer School in High Energy Physics and Cosmology, 1991.

[7]   M. Chanowitz, J. Ellis and M. K. Gaillard, Nucl. Phys. B128 (1977) 506.

[8]   For some discussions and further references see H. Arason et. al., Phys. Rev. Lett. 67 (1991) 2933; B. Ananthanarayan and Q. Shafi, Proc. PASCOS Conference, March 1991; S. Dimopoulos, L. Hall and S. Raby, Phys. Rev. Lett. 68 (1992) 68; V. Barger, M. S. Berger and P. Ohmann, Phys. Rev. D47 (1993) 1093; W. A. Bardeen et. al., Phys. Lett. B320 (1994) 110; P. Langacker and N. Polonsky, Phys. Rev. D49 (1994) 1454.

[9]   B. Pendleton and G. G. Ross, Phys. Lett. 98B (1981) 291; C. T. Hill, Phys. Rev. D24 (1981) 691; C. T. Hill, C. N. Leung and S. Rao, Nucl. Phys. B262 (1985) 517; M. Carena et. al., Nucl. Phys. B369 (1992) 33; N. Polonsky, Neighborhood Workshop, Univ. Delaware, Oct. 1993.

[10]   B. Ananthanarayan, K. S. Babu and Q. Shafi, Nucl. Phys. B428 (1994) 19.

[11]   V. Barger, M. S. Berger and P. Ohmann, Madison preprint MAD/PH/801 (1993); M. Carena et. al., CERN preprint TH.7060/93 (1993); G. Kane et. al., Univ. Michigan preprint, UM-TH-93/24 (1993); P. Langacker and N. Polonsky, Univ. Pennsylvania preprint UPR-0594T (1994).

[12] B. Ananthanarayan, G. Lazarides and Q. Shafi, Phys. Rev. D44 (1991) 1613; Q. Shafi and B. Ananthanarayan, "Will LEPII narrowly miss the Weinberg-Salam-Higgs Boson?" *1991 Summer School in High Energy Physics and Cosmology*, pp. 233, E Gava, *et al.* eds., World Scientific, Singapore, 1992.

[13] B. Ananthanarayan, G. Lazarides and Q. Shafi, Phys. Lett. 300B (1993) 245.

[14] B. Ananthanarayan, Q. Shafi and X. M. Wang, Phys. Rev. D50 (1994) 5980; M. Olechowski and S. Pokorski, Nucl. Phys. B404 (1993) 590; M. Carena, M. Olechowski, S. Pokorski and C. E. M. Wagner, Nucl. Phys. B426 (1994) 269.

[15] K. Inoue et. al.; Prog. Theo. Phys., 68 (1983) 927; *ibid* 71 (1984) 413; L. E. Ibàñez and C. López, Nucl. Phys. B233 (1984) 511; L. E. Ibàñez, C. Lòpez and C. Muñoz, Nucl. Phys. B256 (1985) 218; M. Drees and M. M. Nojiri, Nucl. Phys. B369 (1992) 54 and references therein.

[16] J. Gasser and H. Leutwyler, Phys. Rep. 87 (1992) 77.

[17] L.J. Hall, R. Rattazzi and U. Sarid, LBL preprint 33997 (1993); G. Anderson et. al., Phys. Rev. D49 (1994) 3660; See also V. Barger et al, Univ. of Madison preprint MAD/PH/781 (1993); P. Langacker and N. Polonsky, Phys. Rev. D49 (1994) 1454.

[18] Q. Shafi and R.K. Schaefer, Nature 359 (1992) 199 and references therein.

[19] K. Olive and M. Srednicki, Phys. Lett. 230B (1989) 78; K. Griest, M. Kamionkowski and M. S. Turner, Phys. Rev. D41 (1990), 3565; M. Drees and M. M. Nojiri, Phys. Rev. D47 (1993) 376; J. McDonald, K. Olive and M. Srednicki, Phys. Lett. B283 (1992) 80; L. Roszkowski, Univ. Michigan preprint UM-TH-93-06 (1993).

[20] See for example V. Barger, M. S. Berger and P. Ohmann, Phys. Rev. D47 (1993) 1093.

[21] H. Fritzsch, Phys. Lett. 70B (1977) 436; Nucl. Phys. B155 (1979) 189; K. S. Babu and Q. Shafi, Phys. Rev. D47 (1993) 5004.

[22] S. G. Gorishny et. al., Mod. Phys. Lett. A5 (1990) 2703.

[23] See L. J. Hall, R. Rattazi and U. Sarid in Ref. [17]; G. Anderson et al. in Ref. [17]; M. Carena et al. in Ref. [14], B. Ananthanarayan, Q. Shafi and X. M. Wang in Ref. [14].

[24] See L. J. Hall, R. Rattazi and U. Sarid in Ref. [17]; A. Nelson and L. Randall, Phys. Lett. 316B (1993) 516.

[25] M. Diaz and H. Haber, Phys. Rev. D46 (1992) 3086.

[26] G. Gamberini, G. Ridolfi and F. Zwirner, Nucl. Phys. B331 (1990) 331.

[27] In the non-susy case see H. Georgi, H. R. Quinn and S. Weinberg, Phys. Rev. Lett. 33 (1974) 451; In the susy case see S. Dimopoulos, S. Raby and F. Wilczek, Phys. Rev. D 24 (1981) 1681.

[28] B. Ananthanarayan, G. Dvali and Q. Shafi, in preparation; For an interesting scheme to solve the doublet-triplet problem that requires the parameters of the theory to obey specific relations see R. Barbieri, G. Dvali and M. Moretti, Phys. Lett. B312 (1993) 137.

[29] J. Kodaira, Y. Yasui and K. Sasaki, High energy physics bulletin board, hep-ph 9311366 (1993).

[30] J. Ellis, G. Ridolfi and F. Zwirner, Phys. Lett. 257B (1991) 83; J. Ellis, G. Ridolfi and F. Zwirner, Phys. Lett. 262B (1991) 477; Y. Okada, M. Yamaguchi and T. Yanagida, Prog. Theor. Phys. Lett 85 (1991) 1; Y. Okada, M. Yamaguchi and T. Yanagida, Phys. Lett. 262B (1991) 54; R. Barbieri, M. Frigeni and F. Caravaglios, Phys. Lett. 258B (1991) 167; J. R. Espinosa and M. Quirós, Phys. Lett. 266B (1991) 389; H. E. Haber and R. Hempfling, Phys. Rev. Lett. 66 (1991) 1815; P. N. Pandita, "Radiative corrections to Higgs Bososn Masses in Supersymmetric Modes," Talk given at the X DAE High Energy Symposium and references therein.

[31] R. Ammar et al., Phys. Rev. Lett. 71 (1993) 674.

[32] J.L. Hewett, Phys. Rev. Lett. 70 (1993) 1045; V. Barger, M.S. Berger and R.J.N. Phillips, Phys. Rev. Lett. 70 (1993) 1368.

Missing (up) Mass, Accidental Anomalous Symmetries and the Strong CP Problem

TOM BANKS[1]
DEPARTMENT OF PHYSICS AND ASTRONOMY
RUTGERS UNIVERSITY,
PISCATAWAY, NJ 08855-0849, USA

YOSEF NIR[2]
PHYSICS DEPARTMENT
WEIZMANN INSTITUTE OF SCIENCE,
REHOVOT, 76100, ISRAEL

NATHAN SEIBERG[3]
DEPARTMENT OF PHYSICS AND ASTRONOMY
RUTGERS UNIVERSITY,
PISCATAWAY, NJ 08855-0849, USA

Abstract: We reconsider the massless up quark solution of the strong CP problem. We show that an anomaly free horizontal symmetry can naturally lead to a massless up quark and to a corresponding accidental anomalous symmetry. Reviewing the controversy about the phenomenological viability of $m_u = 0$ we conclude that this possiblity is still open and can solve the strong CP problem.

1 Introduction

In the present state of experimental particle physics, our strongest clues to the nature of physics beyond the standard model are the various fine tuning problems that are revealed by the standard model fit to experimental data. Surely the most pressing of these are the fine tuning of the cosmological constant, and the gauge hierarchy problem. Next in importance come the various small numbers associated with the fermion mass matrices, and among these, the strong CP problem is (at least numerically) the most striking. Several mechanisms, of varying degrees of plausibility, have been invented to account for the small value of the QCD vacuum angle that is required to explain the observed bounds on the neutron electric dipole moment. In particular, it is often suggested that if only a massless up quark were compatible with the spectrum of hadrons, it would also provide a neat and economical solution of the strong CP problem.

Recently a class of models which employed discrete symmetries and some modest dynamical assumptions were proposed *[1]* to explain the gross features of the quark and lepton mass matrices. The simplest of these models automatically incorporated a discrete symmetry which guaranteed that the up quark mass was zero to all orders in perturbation theory. The discrete symmetry

---

[1] banks@physics.rutgers.edu

[2] ftnir@weizmann.bitnet

[3] seiberg@physics.rutgers.edu

enforced an accidental anomalous $U(1)$ symmetry on all renormalizable terms in the Lagrangian. This symmetry remains unbroken in perturbation theory despite the fact that the discrete symmetry is spontaneously broken.

The real and imaginary parts of the determinant of the quark mass matrix are *a priori* independent parameters in the standard model. In the standard model, setting either or both of them to zero is unnatural. The argument that the theory is more symmetrical for $Re\ m_u = Im\ m_u = 0$, and that the fine tuning is therefore natural, is spurious. The axial $U(1)$ "symmetry" of the standard model with one massless quark is anomalous, and is really no symmetry at all. In the absence of further constraints on very high energy physics we should expect all relevant and marginally relevant operators that are forbidden only by this symmetry to appear in the standard model Lagrangian with coefficients of order one.

In some of the models of *[1]*, it is natural to set the real and imaginary parts of the up quark mass matrix to zero at high energy. The accidental anomalous $U(1)$ symmetry of these models guarantees that both the real and imaginary parts are generated only by nonperturbative QCD processes. In section II of the paper we review the details of the mechanism which leads to the accidental anomalous $U(1)$ symmetry, and argue that this is essentially the unique way to generate such an accidental symmetry.

To evaluate the viability of these models, we were led to reexamine the controversy which has surrounded the question of whether a massless up quark is consistent with low-energy hadron phenomenology. In section III of this paper we review the literature on this subject. Fitting low energy data to a first order chiral Lagrangian, one can extract the "low-energy quark masses," μ_i. These should be distinguished from the quark mass parameters, m_i of the QCD Lagrangian at high scale (say 1 TeV) *[2, 3]* . The distinction between them is second order in the m's. In particular, μ_u receives an additive contribution of order $\frac{m_d m_s}{\Lambda_{\chi SB}}$ where $\Lambda_{\chi SB} \sim 1\ GeV$ is some characteristic QCD energy scale, perhaps the fundamental scale $4\pi f_\pi$. Even if $m_u = 0$, the parameter μ_u can be nonzero. Its value is $\mu_u = \beta \frac{m_d m_s}{\Lambda_{\chi SB}}$ where the dimensionless coefficient, β, is estimated to be a number of order one. Unfortunately, with present technology we are unable to calculate it reliably starting from QCD. The low-energy data is compatible with $m_u = 0$ provided $\beta \approx 2$ which is of order one. Therefore, $m_u = 0$ seems to be a viable possibility. Kaplan and Manohar *[4]* tried to avoid a calculation of the parameter β by extending the low-energy analysis to second order in the m's. This led them to find an ambiguity in the parametrization of the second order Lagrangian which prevented them from determining the value of m_u. Several authors *[5, 6, 7, 8, 9]* tried to resolve this ambiguity and to determine m_u by adding more physical input. Our understanding is that this ambiguity cannot be resolved solely on the basis of low energy data. Therefore, we conclude that the possibility that $m_u = 0$ is still open.

In section IV we take up the question of whether a vanishing up quark mass really solves the strong CP problem. In models with an accidental anomalous $U(1)$ symmetry, both real and imaginary parts of the up quark mass are generated by nonperturbative QCD dynamics. The message of section III was that the real part so generated is, within our ability to calculate, compatible with low energy data. Is the same true of the imaginary part? This is a particular case of a general question first addressed in 1979 by Ellis and Gaillard [10], and since studied by a number of authors [11, 12]: if the QCD vacuum angle is set to zero at some large scale Λ_0, what will its low-energy value be? We argue that in the present context there is a very general operator analysis of this question. The analysis shows that CP violation in the high-energy theory can induce a low-energy θ_{QCD} only via the action of a CP violating irrelevant operator $\mathcal{O}$. In this case, the same operator will make a direct perturbative contribution to the neutron electric dipole moment of order $\frac{\Lambda_{\chi SB}}{\mu_u}\theta_{QCD}$ where μ_u is the low energy up quark mass described above and $\Lambda_{\chi SB}$ again denotes a typical QCD scale. In other words, if both the real and the imaginary part of m_u vanish at high-energy, nonperturbative strong CP violation will be smaller than perturbative contributions to the neutron electric dipole moment. We conclude that the models of [1] provide a phenomenologically viable and relatively economical solution to the strong CP problem, in a framework which may resolve many of the other puzzles of the fermion mass matrix in the standard model.

In an appendix we present a warmup exercise for the calculation of direct contributions to the neutron electric dipole moment in models with an accidental anomalous $U(1)$ symmetry.

2 $m_u = 0$, Naturally

CP violation in the QCD Lagrangian arises from the terms:

$$\mathcal{L}_{CP} = \theta\,\frac{g^2}{32\pi^2}\,G^{\mu\nu}\tilde{G}_{\mu\nu} - i\bar{q}(Im\,M_q)\gamma_5 q \tag{1}$$

where M_q is the quark mass matrix. There is only one independent CP violating parameter in 1,

$$\bar{\theta} = \theta + \arg\det M. \tag{2}$$

Strong interactions, through their $\bar{\theta}$-dependence, lead to an electric dipole moment for the neutron. An estimate of this contribution, using current algebra, gives [13]

$$\mathcal{D}_n = +3.6 \times 10^{-16}\,\bar{\theta}\,e\,cm. \tag{3}$$

The experimental bound on $\mathcal{D}_n$ [14],

$$\mathcal{D}_n < 1.2 \times 10^{-25}\,e\,cm \tag{4}$$

implies

$$\bar{\theta} < 10^{-9}. \tag{5}$$

In pure QCD the fine tuning required to satisfy 5 is natural. A new symmetry, CP, is obtained for $\bar{\theta} = 0$. However, within the Standard Model, this fine tuning is not natural, since the standard model lagrangian with nonzero CKM phase δ is not CP invariant even for $\bar{\theta} = 0$. The experimental observation of CP violation in the neutral K system requires that δ be nonzero.

The above parametrization gives the erroneous impression that one fine tuning, $m_u = 0$, sets the up quark mass to zero and solves the strong CP problem simultaneously. In fact, if we choose to absorb the vacuum angle into the quark mass matrix, we can instead parametrize QCD by the real positive values of $N_f - 1$ quark masses, and the real and imaginary parts of the up quark mass[4]. The bound on $\bar\theta$ can be reexpressed as

$$Im\ m_u\ <\ 5 \times 10^{-12}\ GeV, \qquad (6)$$

which is about fourteen orders of magnitude below the electroweak breaking scale. Given this bound, the imaginary part of the up quark mass makes a completely negligible contribution to hadron masses, and chiral lagrangian fits to pseudoGoldstone boson masses are in fact constraints only on the independent, CP conserving, real part of the up quark mass. The latter is only constrained to be about five orders of magnitude below the electroweak scale, and we will see that the data are consistent with it being equal to zero.

In order to render the vanishing of both the real and the imaginary parts of m_u natural (with all other quarks massive), a continuous $U(1)$ symmetry

$$\bar u \to e^{i\alpha}\bar u \qquad (7)$$

must be imposed. It guarantees that the left handed field $\bar u$ couples only to the gauge fields. Like the Peccei Quinn symmetries of axion models, this $U(1)$ symmetry is anomalous and hence not an exact symmetry of the full theory. Therefore, it is unnatural to impose it on a Lagrangian at a given energy scale, unless we can provide a reason why symmetry violating physics at much larger energy scales does not violate all conclusions based on the anomalous $U(1)$. We conclude that, the value of neither the real nor the imaginary part of the up quark mass is natural in the standard model, although the imaginary part is certainly the worst offender.

We should remark that in SUSY theories, the non-renormalization theorems allow us to set the coupling of the up quark to the Higgs to zero. In such theories a massless up quark might appear to be technically natural. (Clearly, the symmetry 7 acts then also on the $\bar u$ squark.) However, unless we arbitrarily impose the $U(1)$ symmetry on the soft SUSY breaking terms (in particular a coupling of the $\bar u$ squark to a squark and a Higgs breaks the symmetry) radiative corrections would lead to couplings of $\bar u$ to the Higgs and to an up quark mass. Since we do not expect the full theory to respect the anomalous $U(1)$, the non-renormalization theorems do not help.

The only natural way to ensure an anomalous $U(1)$ symmetry is to make it an accidental symmetry. That is, it follows from another symmetry plus renormalizability. All terms which violate it are then irrelevant in the renormalization group sense, and will have negligible effect if the physics responsible for them is pushed off to a sufficiently high energy scale. Thus our strategy

---

[4]In order for the eigenvalues of $Re\ M$ to have the conventional meaning of quark masses, it is necessary to have $Im\ M$ proportional to the unit matrix. The difference between Re M and the quark masses in our convention is $\mathcal{O}(\bar\theta^2)$ and therefore negligible.

is as follows: We impose an anomaly free symmetry H, which can be continuous or discrete, and require that the most general renormalizable H invariant Lagrangian exhibits the accidental $U(1)$ symmetry 7.

It is clear that in order to achieve this state of affairs, the symmetry H must act differently on $\bar{u}$ and on $\bar{c}$ and $\bar{t}$. (It can also act on the other light fields.) It therefore satisfies the definition of a *horizontal symmetry*. As such we can immediately apply some of the general analysis of horizontal symmetries presented in [1].

If H acts only on $\bar{u}$, it is anomalous. To cancel this anomaly other fields in the theory must be in a complex representation of H. Since they are massive, H must be spontaneously broken.

We now examine how H can be spontaneously broken. The expectation value of the single Higgs in the standard model cannot break H [1]. More precisely, a subgroup of $U(1)_Y \times H$ isomorphic to H is always unbroken ($U(1)_Y$ is hypercharge). Similarly, in a two Higgs version of the standard model with natural flavor conservation (such as the supersymmetric standard model) H cannot be broken by the expectation values of the two Higgs fields ϕ_u and ϕ_d [1]. To show that, note that in the absence of a $\phi_u\phi_d$ term in the superpotential and in the soft breaking terms, the renormalizable theory is invariant under the anomaly free $U(1)$ symmetry

$$
\begin{aligned}
\bar{u} &\to e^{-3i\alpha}\bar{u} \\
\bar{d} &\to e^{i\alpha}\bar{d} \\
\bar{s} &\to e^{i\alpha}\bar{s} \\
\bar{b} &\to e^{i\alpha}\bar{b} \\
\phi_d &\to e^{-i\alpha}\phi_d
\end{aligned}
\tag{8}
$$

(all other fields are invariant) which guarantees the accidental $U(1)$ 7. However, its breaking by $\langle\phi_d\rangle$ leads to an unacceptable Goldstone boson. With the $\phi_u\phi_d$ term in the Lagrangian we can again use $U(1)_Y$ to make both Higgs fields invariant. Therefore, we must add more fields to the theory.

The simplest way to organize the analysis is to integrate out the extra fields, which we assume to be heavier than the weak scale, and to consider non-renormalizable terms added to the standard model Lagrangian. Typical non-renormalizable terms which could avoid the previous no-go theorem are

$$
Q\phi_d\bar{d}\left(\frac{\phi_u\phi_d}{M^2}\right)^n + Q\phi_u\bar{u}\left(\frac{\phi_u\phi_d}{M^2}\right)^n
\tag{9}
$$

where M is the scale of the fields which have been integrated out. Such terms explicitly break the anomaly free continuous symmetry 8 to a discrete subgroup in the absence of a $\phi_u\phi_d$ term in the Lagrangian. We can also add a field, S, which is invariant under the standard model gauge group, and whose vacuum expectation value breaks H. Then, terms like

$$Q\phi_d \bar{d} \left(\frac{S}{M}\right)^n + Q\phi_u \bar{u} \left(\frac{S}{M}\right)^n \tag{10}$$

can lead to the entries in the mass matrix. Rather low values of $\langle S \rangle$ and M are consistent with the constraints on flavor changing processes [1]. Terms of the form 910 can be generated by integrating out massive fermions of masses of order M [15] and lead to interesting mass matrices for the quarks.

We conclude that the framework of [1] provides the unique natural way of enforcing the accidental $U(1)$ symmetry, and hence $m_u = 0$, to all orders in perturbation theory. We must now examine the question of whether a theory with such an accidental anomalous symmetry is phenomenologically viable.

3 Microscopic and Macroscopic Mass Matrices

In this section we will review the controversy that has revolved around the "experimentally determined" value of the up quark mass. Early estimates [16] of the pion and kaon masses in terms of quark masses in lowest order χPT led to the conclusion that the up quark mass could not be much smaller than half the down quark mass.

In quantum field theory, the parameters in an effective Lagrangian are scale dependent. A microscopic theory more fundamental than the standard model fixes them at some high-energy, e.g. of order $1 \ TeV$. In order to relate these to low-energy parameters we must understand how the parameters renormalize in QCD.

To lowest order in chiral perturbation theory, the renormalization of quark masses is multiplicative and flavor independent. This is what enabled Weinberg [16] to obtain renormalization invariant predictions for high-energy quark mass ratios in terms of low-energy meson masses. Once we go beyond lowest order χPT however, this is no longer true. In particular, the up quark mass receives an additive renormalization proportional to $m_d^* m_s^*$ via the mechanism of [2, 3]. Their result may be phrased as the statement that a renormalization group transformation from a scale λ to $\lambda - \Delta\lambda$ (for $\lambda \gg \Lambda_{\chi SB} \sim 1 \ GeV$ where the theory is weakly coupled) leads to

$$m_i(\lambda - \Delta\lambda) = m_i(\lambda)(1 + \mathcal{O}(g^2(\lambda), m_j^2))$$

$$+ \frac{\det M_q^\dagger}{m_i^*} \times \left(\int_{\lambda^{-1}}^{(\lambda - \Delta\lambda)^{-1}} d\rho \frac{1}{\rho^{N_f - 3}} e^{-\frac{8\pi^2}{g^2(\rho)}} g^n(\rho)\beta(1 + \mathcal{O}(g^2(\rho), m_j^2)) \right) \tag{11}$$

where the mass matrix M_q has complex eigenvalues m_i (the vacuum angle θ has been rotated into M_q). The second term arises from small instantons and the dimensionless constant β depends on the number of colors N and flavors N_f.

This term also depends on the regularization scheme through our choice of integrating over the instanton scale size ρ from λ^{-1} to $(\lambda - \Delta\lambda)^{-1}$. This ambiguity is very small. It is of order $\exp(-\frac{8\pi^2}{g^2(\lambda)}) \ll 1$. Therefore, even though $m_i(\lambda)$ is ambiguous, the limit $\lim_{\lambda\to\infty} m_i(\lambda)$ is well defined and M_q at high energies is not ambiguous.

In renormalizing between two high-energy scales (e.g. 1 TeV and 100 GeV) the additive renormalization is accurately calculated in the dilute instanton gas approximation, and is very small because it is nonperturbative in the small high-energy value of the QCD coupling. When the QCD coupling becomes strong this is no longer the case and the best we can do is to estimate the additive contribution to the low energy quark mass as $\frac{m_d m_s}{\Lambda_{\chi SB}}$. Therefore, if $m_u = 0$, $\mu_u \sim \frac{m_d m_s}{\Lambda_{\chi SB}}$. This estimate is compatible with the low-energy quark mass ratios extracted from lowest order chiral perturbation theory. A more detailed comparison requires a nonperturbative solution of QCD, and might in principle (though probably not in practice since the effect pointed out in [2] and [3] is invisible in the quenched approximation) be extracted from a lattice calculation.

To conclude this section, we emphasize that the high energy values of the quark masses are, for all practical purposes, unambiguously determined by potentially observable QCD Green's functions. For example, in the QCD sum rule approach to hadron phenomenology, it is precisely these high energy values which are related to low energy hadron parameters[5]. Unfortunately, it is difficult to estimate the uncontrollable systematic errors in sum rule calculations. As far as we have been able to determine, all such calculations are compatible with a vanishing value for the high energy up quark mass.

3.1 First order analysis

To make these considerations more precise, let us outline the procedure for actually extracting the high-energy quark masses from low-energy data. The first step is to fit low-energy data to the low-energy chiral Lagrangian (we use the notations of reference [17]

$$\mathcal{L} = \frac{F_\pi^2}{4}\operatorname{tr}\left\{\partial_\mu U \partial^\mu U^\dagger + 2B(\mathcal{M}U^\dagger + \mathcal{M}^\dagger U)\right\}. \tag{12}$$

Using $SU(3)_L \times SU(3)_R$, $\mathcal{M}$, which is in the $(3,\bar{3})$ representation, can be brought to the form

$$\mathcal{M} = \begin{pmatrix} \mu_u e^{i\theta} & & \\ & \mu_d & \\ & & \mu_s \end{pmatrix} \tag{13}$$

where all the parameters are real. The mass parameters μ_i can be interpreted as the low-energy values of the high-energy quark masses m_i. The result of the fit is

$$\theta < 10^{-9}, \quad \frac{\mu_u}{\mu_d} \sim 0.57, \quad \frac{\mu_d}{\mu_s} \sim 0.05. \tag{14}$$

---

[5]This implies that if our assertions about the inequality of low energy and high energy quark mass ratios are correct, then values for the quark masses which are extracted by combining sum rules with chiral lagrangian results for mass ratios are incorrect in principle.

Now let us expand the μ's in the short distance m's. Using the $SU(3)_L \times SU(3)_R$ transformation laws of the m's we find that for three flavors

$$\mu_u = \beta_1 m_u + \beta_2 \frac{m_d^* m_s^*}{\Lambda_{\chi SB}} + \mathcal{O}(m^3)$$

$$\mu_d = \beta_1 m_d + \beta_2 \frac{m_u^* m_s^*}{\Lambda_{\chi SB}} + \mathcal{O}(m^3) \tag{15}$$

$$\mu_s = \beta_1 m_s + \beta_2 \frac{m_u^* m_d^*}{\Lambda_{\chi SB}} + \mathcal{O}(m^3)$$

where β_1 and β_2 are dimensionless coefficients. The μ's in 15 are complex and can be brought to the form 13 using an $SU(3)_L \times SU(3)_R$ transformation. The additive renormalization proportional to β_2 is the strong coupling version of the instanton term in 11 and it represents the breaking of the axial $U(1)$ by the anomaly. In terms of the matrix $\mathcal{M}$ and the underlying quark mass matrix M_q, equation 15 can be written as

$$\mathcal{M} = \beta_1 M_q + \beta_2 \frac{1}{\Lambda_{\chi SB}} \det M_q^\dagger \frac{1}{M_q M_q^\dagger} M_q + \mathcal{O}(M_q^3). \tag{16}$$

The dimensionless coefficients β_1 and β_2 cannot be found without a strong coupling calculation in QCD. On general grounds we expect them to be of order one. They both suffer from ambiguities resulting from the regularization scheme. The only sense in which any of them is small is in the large N approximation, where β_2 is of order $\frac{1}{N}$.

Assuming $\frac{m_u}{m_d} \ll 1$, 15 leads for real m's to

$$\frac{m_u}{m_d} = \frac{\mu_u}{\mu_d} - \frac{\beta_2}{\beta_1^2} \frac{\mu_s}{\Lambda_{\chi SB}} + \mathcal{O}\left((\frac{\mu_i}{\Lambda_{\chi SB}})^2 \right). \tag{17}$$

Therefore, $\frac{m_u}{m_d}$ cannot be determined without knowledge of $\frac{\beta_2}{\beta_1^2}$. If the dimensionless ratio $\frac{\beta_2}{\beta_1^2}$ is near 2, the up quark can be massless. Even if this ratio is not near 2, the ambiguity in $\frac{m_u}{m_d}$ is significant.

3.2 Second order analysis

The previous discussion can be criticized on the basis that it is a first order analysis in the μ's but it includes some second order contribution in the m's. To make the analysis consistent one should include all second order terms in the m's and hence, go to second order in the μ's. The most general potential of second order in M_q can be written, using $\chi \equiv 2BM_q$, as

$$V(U) = \left\{ \frac{F_\pi^2}{4} \operatorname{tr} \chi U^\dagger + r_1 [\operatorname{tr} \chi^\dagger U \chi^\dagger U - (\operatorname{tr} \chi^\dagger U)^2] \right\} + h.c.$$
$$+ r_2 [\operatorname{tr} \chi^\dagger U \chi^\dagger U + (\operatorname{tr} \chi^\dagger U)^2] + h.c. \tag{18}$$
$$+ r_3 (\operatorname{tr} \chi^\dagger U)(\operatorname{tr} \chi U^\dagger) + \mathcal{O}(\mathcal{M}^3).$$

(The dimensionless coefficients r_1, r_2 and r_3 are linear combinations of L_6, L_7 and L_8 of [17].) Remembering that M_q is in the $(3, \bar{3})$ representation of the $SU(3)_L \times SU(3)_R$ flavor symmetry, we learn that the terms multiplying r_1, r_2 and r_3 are in the representations $(3, \bar{3})$, $(\bar{6}, 6)$ and $(8, 8)$, respectively. Since the two terms in the curly brackets in 18 transform as one irreducible representation, there must be a redundancy in the parametrization. Indeed, using the identity of 3×3 matrices

$$\det A - \frac{1}{2}A[(\operatorname{tr} A)^2 - \operatorname{tr} A^2] + A^2 \operatorname{tr} A - A^3 = 0 \tag{19}$$

applied to $A = \chi^\dagger U$ we find

$$\det \chi^\dagger \operatorname{tr} \frac{1}{\chi^\dagger} U^\dagger + \frac{1}{2}\operatorname{tr} \chi^\dagger U \chi^\dagger U - \frac{1}{2}(\operatorname{tr} \chi^\dagger U)^2 = 0. \tag{20}$$

Therefore, the shifts

$$\chi \to \chi + \frac{8a}{F_\pi^2}(\det \chi^\dagger)\frac{1}{\chi^\dagger} \tag{21}$$
$$r_1 \to r_1 + a$$

with an arbitrary complex number a change the potential 18 by terms of order M_q^3, which are neglected.

Because of this redundancy in the parametrization of the effective Lagrangian, the mass matrix M_q cannot be determined by fitting the experimental data to 18. At best, we can determine M_q up to an arbitrary shift by $\frac{aB}{F_\pi^2}(\det M_q^\dagger)\frac{1}{M_q^\dagger}$.

The ambiguity 21 is reminiscent of the expression for $\mathcal{M}$ in terms of M_q 16. The functional similarity stems from the fact that the axial $U(1)$ is broken preserving only the $SU(3)_L \times SU(3)_R$ symmetry. However, these two effects are different. Equation 16 relates the short distance mass M_q to the long distance mass $\mathcal{M}$ defined by a fit to a first order Lagrangian. The non-linear term in this equation expresses a renormalization effect. The ambiguity 21 expresses a redundancy of the second order low-energy description which prevents us from using only the low-energy chiral Lagrangian to determine M_q.

The distinction between these two effects is more clear in the "trivial" two flavor case. The identity analogous to 20 in this case is

$$\operatorname{tr} \psi^\dagger U = \det \psi^\dagger \operatorname{tr} \frac{1}{\psi^\dagger} U^\dagger \tag{22}$$

where ψ is an arbitrary 2×2 matrix. It shows that the matrix χ in the potential $\operatorname{tr} \chi^\dagger U + c.c$ is ambiguous. First, it is clear that only the eigenvalues of χ are physical. Second, using equation 22 the transformation

$$\chi \to \chi + \psi - \det \psi^\dagger \frac{1}{\psi^\dagger} \tag{23}$$

leaves the potential invariant and allows us to set one of the eigenvalues of χ to zero[6]. The non-perturbative contribution to $\mathcal{M}$ arises at first order in M_q. It is the second term in

$$\mathcal{M} = \beta_1 M_q + \beta_2 \det M_q^\dagger \frac{1}{M_q M_q^\dagger} M_q + \mathcal{O}(M_q^2). \tag{24}$$

Clearly, the redundancy in the parametrization has nothing to do with the change in the strength of this effect (changing β_2). The only common thing about the ambiguity 23 and the additive renormalization 24 is that they both allow for $m_u = 0$ to be compatible with low energy data.

Some authors [5, 6, 7] have suggested the use of more input about low energy hadron physics to resolve this ambiguity and determine M_q. We would like to make clear what bothers us about these attempts. These authors fix the ambiguity by insisting that the coefficients χ in $Tr\chi U^\dagger$ in the chiral Lagrangian are proportional to the bare quark masses, i.e. by setting what we have called $\mu_u \propto m_u$. They then attempt to compute the other coefficients in the Lagrangian (implicitly) with this choice of Kaplan-Manohar "gauge." However, their computations make no reference to this particular choice. We could equally well consider them to be determinations of the couplings in some other "gauge." And since their computations involve relatively low energy hadronic physics there is no reason to think that they have anything to do with the special parametrization in which the low energy and high energy up quark masses are proportional to each other. Other approaches of resolving the ambiguity use the large N expansion. While this may be a valid approach, one must beware of considerations that are justified only in this limit. The whole issue is of higher order in $\frac{1}{N}$ and therefore cannot be settled at the leading order in this expansion.

Finally, we note that the argument is sometimes made that the consistency of the quark masses extracted from first order chiral perturbation theory analysis of baryons and mesons, suggests that the scale dependent additive renormalization which we have discussed, is small. That is, the additive renormalization should give an effective upquark mass proportional to $m_d m_s$ in both the baryon and meson effective lagrangians, but the coefficients would not be the same. We have examined the extraction of quark masses from baryon masses [18, 6] and believe that it suffers from large uncertainties. The results are sensitive to nonanalytic corrections ($\propto m_q^{3/2}$), and to model dependent calculations of electromagnetic contributions. They are not precise enough to rule out the possibility that $m_u = 0$.

We conclude that at the level of precision (order of magnitude) of nonperturbative QCD calculations available to us at present, low-energy phenomenology is completely compatible with a vanishing value of the high-energy up quark mass.

Only a nonperturbative calculation in QCD can prove or disprove the phenomenological viability of $m_u = 0$. Therefore, in view of the recent progress in numerical methods in lattice gauge theory, we would like to encourage a

---

[6]Note that therefore, in the two flavor case there cannot be any CP violation in the first order chiral Lagrangian.

detailed analysis of the possibility of a massless up quark by these methods. We emphasize however that the additive renormalization 11, vanishes in the quenched approximation.

4   Does $m_u = 0$ Solve the Strong CP Problem?

As we have seen in section II, models of the type studied by *[1]* can incorporate discrete symmetries which guarantee that both the real and imaginary parts of the up quark mass are zero (at 1 TeV) while CP violation in the Cabibbo-Kobayashi-Maskawa matrix is allowed. The discrete symmetries enforce an accidental anomalous $U(1)$ symmetry on the Lagrangian. If all physics not explicitly included in the Lagrangian comes from energy scales substantially higher than 1 TeV then this accidental symmetry guarantees that the only nonvanishing contributions to the up quark mass come from nonperturbative QCD effects.

We have argued above that in our present state of impotence with regard to precise nonperturbative calculations in QCD, the real part of the up quark mass generated in this manner appears to be consistent with hadron phenomenology. The purpose of the following discussion is to determine whether an analogous conclusion can be made for the imaginary part of the up quark mass.

This task is made easier by the observation that violations of the anomalous $U(1)$ symmetry are only significant at scales below 1 GeV. Thus if we construct an effective field theory for scales just above 1 GeV, it must obey the $U(1)$ symmetry. As a consequence, there will be no CP violating terms in the renormalizable part of this effective Lagrangian. The imaginary part of the low-energy up quark mass will thus be generated by a combination of nonperturbative QCD effects and CP violating irrelevant operators in the effective field theory. Let $\mathcal{O}_d$ be such an operator, of dimension d. It will give a contribution to the imaginary part of the up quark mass of order $C \frac{m_d m_s}{\Lambda_{\chi SB}} (\frac{\Lambda_{\chi SB}}{M})^{d-4}$, where $\frac{C}{M^{d-4}}$ is the coefficient of $\mathcal{O}_d$ in the effective Lagrangian, M is the heavy scale at which CP violating effects originate (e.g. the mass of the W boson), and $\Lambda_{\chi SB}$ is, as always, a typical QCD scale of order 1 GeV. The consequent contribution to the neutron electric dipole moment is of order $C \frac{m_d m_s}{\Lambda_{\chi SB}^3} (\frac{\Lambda_{\chi SB}}{M})^{d-4}$.

The operator $\mathcal{O}_d$ also makes a direct contribution to the neutron electric dipole moment of order $C \frac{1}{\Lambda_{\chi SB}} (\frac{\Lambda_{\chi SB}}{M})^{d-4}$, which is larger than that coming from the "induced strong CP violation" by a factor of order $\frac{\Lambda_{\chi SB}^2}{m_d m_s}$. *Thus, if the direct contributions to the neutron electric dipole moment coming from perturbatively induced CP violating irrelevant operators are within experimental bounds, there will be no strong CP problem.* Perturbative CP violating effects might put significant constraints on particular models with new physics at the TeV scale. We will not study these direct contributions in detail here (except for a short discussion of the standard model in the appendix). What we have shown is that if these constraints are satisfied, we need not worry about strong CP violation.

ACKNOWLEDGEMENTS.   It is a pleasure to thank A. Cohen, A. Dabholkar, M. Dine, K. Intriligator, D. Kaplan, A. Nelson, J. Polchinski, S. Shenker and E. Witten for useful discussions. Y.N. wishes to thank the Rutgers group for its hospitality. This work was supported in part by DOE grant DE-FG05-90ER40559. YN is an incumbent of the Ruth E. Recu Career Development chair, and is supported in part by the Israel Commission for Basic Research, by the United States – Israel Binational Science Foundation (BSF), and by the Minerva Foundation.

5 Appendix A: Direct Contributions to the Neutron Electric Dipole Moment

Here we present an operator analysis of direct perturbative contributions to the neutron electric dipole moment in the standard model with an accidental $U(1)$ symmetry. This is a warmup for a full calculation in models of the sort studied in *[1]*.

The lowest dimension CP violating operators that might be relevant here are the dimension five chromoelectric dipole moments of the light quarks. In models with an accidental $U(1)$ symmetry, no dipole moments involving an up quark are allowed in the effective Lagrangian above the QCD scale. It is also worth reiterating *[19]* that all dipole moments are suppressed by factors of light quark masses because they violate the nonabelian chiral symmetries of massless QCD. As a consequence their contributions to the neutron electric dipole moment are *a priori* smaller than chirally allowed dimension six operators, including the three gluon operator *[20]*, and the *axial polyp [19]* operators:

$$
\mathcal{L}_u^3 = \Delta_3^u \, \mathcal{P}_{3-}^u,
$$
$$
\mathcal{P}_{3-}^u = i\bar{u}\gamma^\mu \{G_{\mu\nu}, \overrightarrow{D}^\nu\}\gamma_5 u,
$$

$$(25)$$

There is a similar term, $\mathcal{L}_u^1$ with a photon field $F_{\mu\nu}$ replacing the gluon field $G_{\mu\nu}$ in 25. In the standard model, the largest contributions to quark electric dipole moments at a given scale actually come from "fusing" together one of these polyp operators and a quark mass term via the exchange of a gluon right at the infrared cutoff scale. In many previous analyses of CP violation, the electric dipole moment operators were considered to be more important than, or on equal footing with, the polyp operators. We believe that such arguments rest on the dubious procedure of sticking a "constituent" quark mass into an effective lagrangian. The purpose of an effective lagrangian calculation is to make a clean separation between physics at different scales, and in particular between weakly coupled high energy physics, and strongly interacting QCD. We do not believe that it is consistent with the "rules of the game" to calculate the coefficients of an effective lagrangian (without the use of a computer) at scales at which QCD is strongly coupled. Thus, constituent quark masses should not appear as coefficients in an honest effective lagrangian. Rather, we should work with operators normalized at a scale where QCD is still weakly coupled and estimate their hadronic matrix elements by dimensional analysis or some more sophisticated hadronic "model." In the framework of a such a philosophy, the direct contributions of CP violating polyps are more important than chromoelectric dipole moments of light quarks.

At the order of magnitude level then, the calculation of the neutron electric dipole moment in the models of *[1]* probably reduces to the calculation of Δ_3^q for the various quarks. As a warmup exercise for this calculation we will present an estimate for the Δ_3^u coefficient within the Standard Model.

As a single W-propagator cannot introduce CP violation, we are led to consider the diagrams of order $g_s\alpha_w^2$ shown in Fig. 1 - 3. The dependence of such diagrams on the quark sector parameters is of the form

$$\sum_{i,k=1}^{3}\sum_{j=1}^{3} f_1(m_i)f_2(m_j)f_3(m_k)Im[V_{ui}V_{ij}^{\dagger}V_{jk}V_{ku}^{\dagger}]$$

$$= \sum_{i,k=1}^{3}\sum_{j=1}^{3} f_1(m_i)f_2(m_j)f_3(m_k)\, J \sum_{m,n=1}^{3} \epsilon_{1jm}\epsilon_{ikn}. \tag{26}$$

J is the CP violating invariant measure of Jarlskog *[21]*. As only left handed fields have charged current interactions, there can be no odd number of helicity flips in the loops, so the f_i functions are functions of quark squared-masses, m_q^2. Assume that we had replaced $f_2(m_j)$ by a function independent of the quark masses. Since $\sum_j \epsilon_{1jm} = 0$, such a quantity would not contribute to the coefficient of CP violating operators. Similarly, if either of $f_1(m_i)$ and $f_3(m_k)$ were constant, there would be no contribution. Therefore, we can rewrite the above quantity by adding and subtracting contributions which do not contribute to CP violating terms. The result is:

$$J\{[f_1(m_s^2) - f_1(m_b^2)][f_2(m_c^2) - f_2(m_t^2)][f_3(m_d^2) - f_3(m_b^2)] - (m_s^2 \leftrightarrow m_d^2)\}. \tag{27}$$

Next, let us concentrate on the inner loop of diagrams 1(a) and 1(b). We assume that the momentum p entering this loop is small on the scale of m_W.

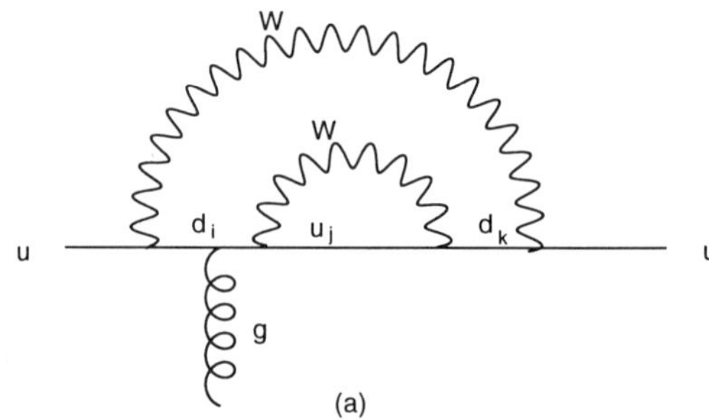

Figure 1. First Feynman diagram for induced θ

(This assumption will be justified below.) Then such a loop contributes

$$
\begin{aligned}
L_1 &\sim \int d^4 l \frac{1}{(l+p)^2 + m_W^2} \left(\frac{\slashed{l}}{l^2 + m_c^2} - \frac{\slashed{p}}{l^2 + m_t^2} \right) \\
&= (m_t^2 - m_c^2) \int d^4 l \frac{\slashed{l}}{((l+p)^2 + m_W^2)(l^2 + m_c^2)(l^2 + m_t^2)}.
\end{aligned}
\tag{28}
$$

We next expand in powers of p. The zeroth order term does not contribute because of Lorentz invariance. The first order term is

$$
L_1 \sim (m_t^2 - m_c^2)\slashed{p} \int \frac{d^4 l \ l^2}{(l^2 + m_W^2)^2 (l^2 + m_c^2)(l^2 + m_t^2)}.
$$

We can scale the integral by m_t^2, assumed to be of order m_W^2. The integral is finite, and consequently $L_1 \sim \slashed{p}$. If m_t had been smaller than m_W, there would have been an additional suppression factor of order $\frac{m_t^2}{m_W^2}$.

Thus we can replace the inner loop with an insertion of a $\slashed{p}$-operator, where p is the momentum entering this loop. In this approximation, the external loop of the diagram with the gluon on the left (Fig. 1) gives

$$
\begin{aligned}
L_2 &\sim \int d^4 l [(\slashed{l} + \slashed{q})\gamma_\mu \slashed{l}\slashed{l}] \frac{1}{((l-k)^2 + m_W^2)} \left\{ \left[\frac{1}{(l^2 + m_s^2)} - \frac{1}{(l^2 + m_b^2)} \right] \right. \\
&\quad \times \left[\frac{1}{((l+q)^2 + m_d^2)(l^2 + m_d^2)} - \frac{1}{((l+q)^2 + m_b^2)(l^2 + m_b^2)} \right] \\
&\quad \left. -(m_s^2 \leftrightarrow m_d^2) \right\}
\end{aligned}
\tag{29}
$$

where q is the incoming gluon momentum and k is the incoming u-quark momentum. The diagram with the gluon on the right (Fig. 2) gives a similar contribution in which the coefficient of $\frac{(\slashed{l} + \slashed{q})\gamma_\mu \slashed{l}}{(l-k)^2 + m_W^2}$ is modified by simultaneously exchanging m_d with m_s and l with $l + q$. If we now add these diagrams and perform the explicit subtraction of terms with $m_d \leftrightarrow m_s$, we obtain:

$$
\begin{aligned}
L_2 &\sim (m_s^2 - m_d^2)(m_b^2 - m_s^2)(m_b^2 - m_d^2) \\
&\quad \int \frac{d^4 l \ [l^2 - (l+q)^2](\slashed{l} + \slashed{q})\gamma_\mu \slashed{l}}{((l-k)^2 + m_W^2)((l+q)^2 + m_b^2)(l^2 + m_b^2)} \\
&\quad \times \frac{1}{((l+q)^2 + m_s^2)(l^2 + m_s^2)((l+q)^2 + m_d^2)(l^2 + m_d^2)}.
\end{aligned}
\tag{30}
$$

We are interested in the terms linear in both k and q. Thus we expand to first order in k and q. As L_2 is a highly convergent integral even when we

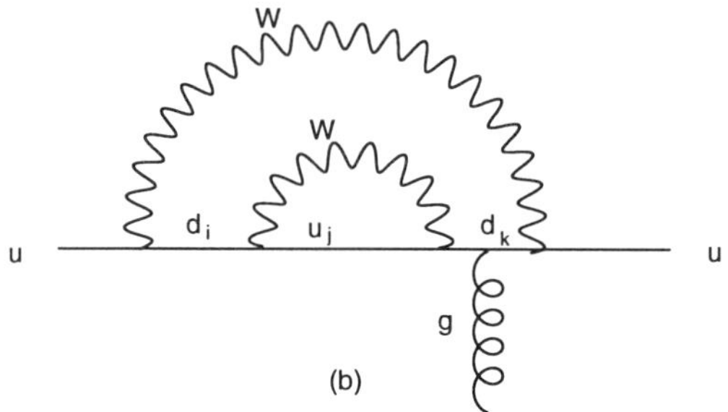

Figure 2. Second Feynman diagram for induced θ

neglect the momentum dependence of the W propagator, we were justified in assuming that the momentum entering the internal loop was small. We obtain a factor m_W^{-4}, multiplying a finite integral. Scaling the loop momentum by m_b and expanding in $(\frac{m_d}{m_s})^2$ and $(\frac{m_s}{m_b})^2$ we get

$$L_2 \sim \frac{m_s^2}{m_W^4} \int \frac{d^4l \; l \cdot k l \cdot q \slashed{l} \gamma_\mu \slashed{l}}{(l^2 + 1)^2 (l^2 + (\frac{m_d}{m_b})^2)^2 (l^2 + (\frac{m_s}{m_b})^2)^2}. \tag{31}$$

This leads to[7]

$$\Delta_3^u \sim \left(\frac{\alpha_w}{\pi}\right)^2 J \frac{m_s^2}{m_W^4}. \tag{32}$$

Using this result, it is easy to estimate the induced θ. We can "fuse" the polyp operator with the up quark mass (which in our case is induced dynamically) to find

$$Im \; m_u \sim (Re \; m_u) \Lambda_{\chi SB}^2 \Delta_3^u \sim J(\frac{\alpha_w}{\pi})^2 (Re \; m_u) \frac{\Lambda_{\chi SB}^2 m_s^2}{m_W^4} \tag{33}$$

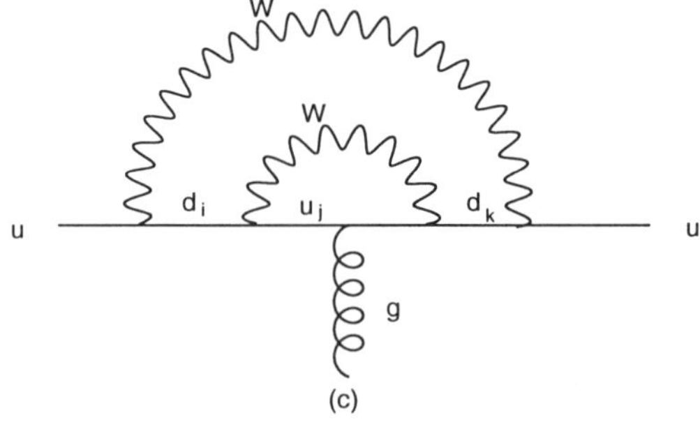

Figure 3. Third Feynman diagram for induced θ

---

[7]For the diagram in Fig. 3, the inner loop can be replaced with a gluon vertex $g_s \gamma_\mu$. It contributes to Δ_3^u with the same order of magnitude as diagrams (a) and (b).

which agrees with the estimate of *[10]*. Note that with the extra suppression factor coming from $L_1 \sim \frac{m_t^2}{m_W^2}\not{p}$ when $m_t < m_W$, our estimate 33 would agree with that of Shabalin *[11]*.

References

[1]   M. Leurer, Y. Nir and N. Seiberg, Nucl. Phys. B398 (1993) 319; hep-ph/9310320, RU-93-43, WIS-93/93/Oct-PH, Nucl. Phys. in press.

[2]   H. Georgi and I.N. McArthur, Harvard preprint HUTP-81/A011 (1981).

[3]   K. Choi, C.W. Kim and W.K. Sze, Phys. Rev. Lett. 61 (1988) 794.

[4]   D.B. Kaplan and A.V. Manohar, Phys. Rev. Lett. 56 (1986) 2004.

[5]   H. Leutwyler, Nucl. Phys. B337 (1990) 108.

[6]   J.F. Donoghue and D. Wyler, Phys. Rev. D45 (1992) 892.

[7]   J.F. Donoghue, B.R. Holstein and D. Wyler, Phys. Rev. Lett. 69 (1992) 3444.

[8]   K. Choi, Nucl. Phys. B383 (1992) 58; Phys. Lett. B292 (1992) 159.

[9]   M. A. Luty and R. Sundrum, Phys. Lett. B312 (1993) 205; R. F. Lebed and M. A. Luty, LBL-34779, UCB-PTH-93/28, hep-ph/9401232.

[10]   J. Ellis and M.K. Gaillard, Nucl. Phys. B150 (1979) 141.

[11]   E.P. Shabalin, Sov. J. Nucl. Phys. 28 (1979) 75; 31 (1980) 864.

[12]   M. Dugan, B. Grinstein and L. Hall, Nucl. Phys. B255 (1985) 413; H. Georgi and L. Randall, Nucl. Phys. B276 (1986) 241.

[13]   V. Baluni, Phys. Rev. D19 (1979) 2227; R.J. Crewther, P. Di Vecchia, G. Veneziano and E. Witten, Phys. Lett. 88B (1979) 123.

[14]   K.F. Smith *et al.*, Phys. Lett. B234 (1990) 191; I.S. Altarev *et al.*, JETP Lett. 44 (1986) 460.

[15]   C.D. Froggatt and H.B. Nielsen, Nucl. Phys. B147 (1979) 277.

[16]   S. Weinberg, in *A Festschrift for I.I. Rabi*, ed. L. Motz (NY Acad. Sci., 1977) p. 185.

[17]   H. Leutwyler, in *Perspectives in the Standard Model*, eds. R.K. Ellis, C.T. Hill and J.D. Lykken (World Scientific, 1992), p. 97.

[18]   J. Gasser and H. Leutwyler, Phys. Rep. 87 (1982) 77.

[19]   A. De Rújula, M.B. Gavela, O. Pène and F.J. Vegas, Nucl. Phys. B357 (1991) 311.

[20]   S. Weinberg, Phys. Rev. Lett. 63 (1989) 2333.

[21]   C. Jarlskog, Phys. Rev. Lett. 55 (1985) 1039.

Supergravity Solutions in the Low-$\tan\beta$ Fixed Point Region

V. BARGER, M. S. BERGER, AND P. OHMANN[1]
PHYSICS DEPARTMENT,
UNIVERSITY OF WISCONSIN,
MADISON, WI 53706, USA

Abstract: There has been much discussion in the literature about applying the radiative electroweak symmetry breaking (EWSB) requirement to GUT models with supergravity. We motivate and discuss the application of the EWSB requirement to the low $\tan\beta$ fixed-point region and describe the solutions we find.

1   Introduction

Improvements in LEP data over the past few years have generated significant excitement at the prospect of grand unification within the Minimal Supersymmetric Standard Model (MSSM) [1]. In addition to the gauge coupling unification suggested by LEP, Yukawa unification – in particular $\lambda_b(M_G) = \lambda_\tau(M_G)$ [2] – has been extensively studied, both at the one-loop and two-loop levels[3]. Such a constraint places significant restrictions on the allowed parameter space, especially that of m_t and $\tan\beta$. For values of $m_b(m_b)$ within the range 4.25 ± 0.10 GeV [4], the resulting allowed parameter space lies almost exclusively within the fixed - point region, as defined by $\lambda_i^G\gtrsim1$ for $i = t, b$, and/or τ [3],[5]–[11].

If one makes only the additional assumption that $m_t(m_t)\lesssim175$ GeV (consistent with the recently released CDF measurement $m_t^{pole} = 174 \pm 10\,^{+13}_{-12}$ GeV[12] which corresponds to a running mass $m_t(m_t) \simeq 166 \pm 10 \pm 13$ GeV, then one is restricted to two very narrow regions in the m_t, $\tan\beta$ plane. One of these regions, the low $\tan\beta$ fixed-point region, has been the focus of our recent renormalization group analysis with supersymmetric grand unification[5], and we therefore examine whether these solutions satisfy the additional constraint imposed by Radiative Electroweak Symmetry Breaking (EWSB).

2   Fixed Points and $\lambda_b = \lambda_\tau$ Unification

Fixed-points arise naturally from imposing $\lambda_b = \lambda_\tau$ unification at the GUT scale along with the typically allowed range for the bottom quark mass 4.25 ± 0.10 GeV. Figure 1 shows the allowed parameter space for $\lambda_b = \lambda_\tau$ unification, and Figure 2 shows contours of Yukawa couplings (at M_{GUT}). Note that Figure 1 is a subset of Figure 2 (allowing for the small $\sim$ 5-10 GeV difference between $m_t(m_t)$ and m_t^{pole}); in fact, imposing this m_b mass constraint ensures the fixed-point nature of the solutions. Figure 3 shows the typical evolution of λ_t for these solutions.

As described by Figures 1 and 2, the allowed m_t-$\tan\beta$ parameter space can be divided into three distinct regions:

1) $\tan\beta\lesssim2$ (λ_t fixed point)

---

[1]Talk presented by P. Ohmann

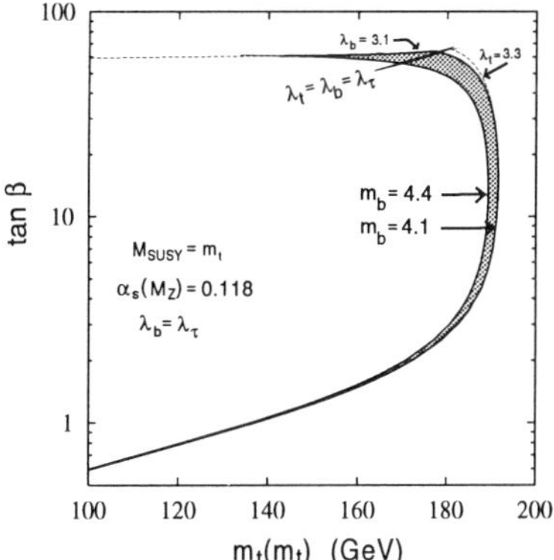

Figure 1. Contours of constant $m_b(m_b)$ in the $m_t(m_t)$, tan β plane (from Ref. [3]).

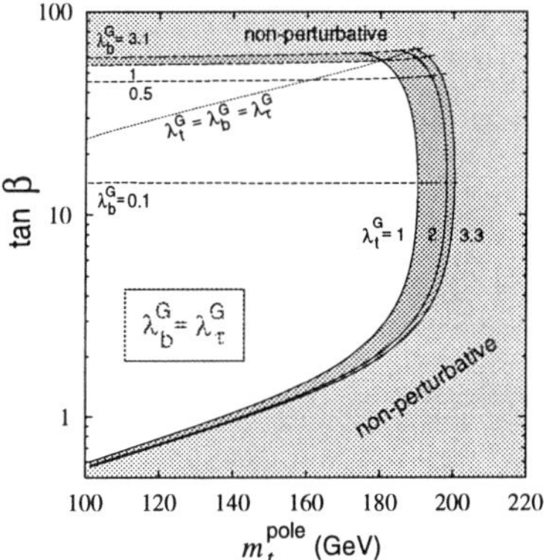

Figure 2. The fixed-point regions are given by Yukawa couplings at the GUT scale being larger than about 1 ($\lambda_i^G \gtrsim 1$). Even larger values of the Yukawa couplings results in a breakdown of perturbation theory.

2) $\tan\beta \gtrsim 50$ ($\lambda_b = \lambda_\tau$ fixed point)

3) $2 \lesssim \tan\beta \lesssim 50$ (λ_t fixed point)

It should be noted that threshold corrections, if large, may either enforce or mitigate the fixed-point nature of some of the solutions [8]–[10],[13]–[17]. If $m_t \lesssim 175$ GeV, then only the first two regions remain. While both solution sets may still be viable, there are criteria which seem to favor the first region: namely, the large tan β region typically results in large threshold corrections and in large enhancements to flavor changing neutral currents in processes like $b \to s\gamma$ and $B\overline{B}$ mixing and to proton decay[18]–[19]. However, it is possible these may be successfully eliminated by assuming certain symmetries[9].

One of the most important aspects of the large top mass is that it makes possible an understanding of the radiative breaking of the electroweak symme-

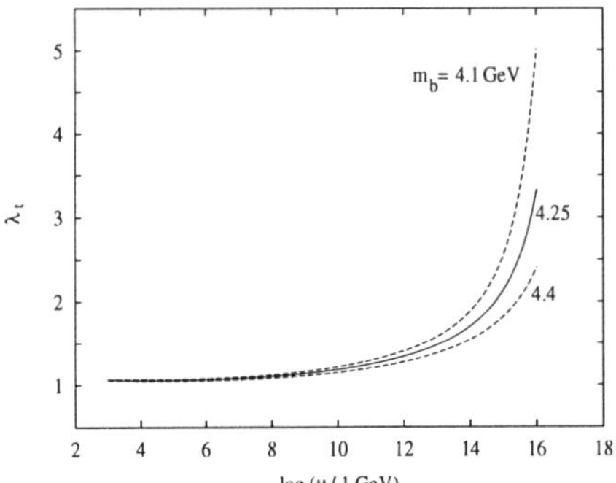

Figure 3. If λ_t is large at M_G, then the renormalization group equation causes $\lambda_t(Q)$ to evolve rapidly towards an infrared fixed point as $Q \to m_t$ (from Ref. *[3]*).

try; the large top quark Yukawa drives a Higgs mass-squared negative. However, when both the top and bottom quark Yukawas are the same size at the GUT scale, one must rely on the difference in their hypercharges to effect the symmetry breakdown [20].

The low $\tan \beta$ λ_t fixed-point region can be well described by the following relation between the top quark mass and $\tan \beta$.

$$\lambda_t(m_t) = \frac{\sqrt{2}m_t(m_t)}{v \sin \beta} \approx 1.1 \;\Rightarrow\; m_t(m_t) \approx \frac{v}{\sqrt{2}} \sin \beta = (192 GeV) \sin \beta \quad (1)$$

Converting this relation to the top quark pole mass yields[3, 5]

$$m_t^{pole} \approx (200 GeV) \sin \beta . \tag{2}$$

3 Electroweak Symmetry Breaking

There has been much discussion in the literature about imposing a Radiative Electroweak Symmetry Breaking (EWSB) constraint on GUT models [21]–[33]; in particular we address this issue with regard to the low $\tan \beta$ λ_t fixed-point region.

The EWSB constraint is enforced by minimizing the effective Higgs potential; at tree-level this is given by:

$$V_0 = (m_{H_1}^2 + \mu^2)|H_1|^2 + (m_{H_2}^2 + \mu^2)|H_2|^2 + m_3^2(\epsilon_{ij} H_1{}^i H_2{}^j + h.c.)$$
$$+ \frac{1}{8}(g^2 + g'^2)\left[|H_1|^2 - |H_2|^2\right]^2 + \frac{1}{2}g^2|H_1^{i*} H_2^i|^2 , \tag{3}$$

where $m_{H_1}^2$, $m_{H_2}^2$, and $m_3^2 = B\mu$ are soft-supersymmetry breaking parameters, ϵ_{ij} is the antisymmetric tensor, and H_1 and H_2 are complex doublets given by

$$H_1 = \begin{pmatrix} \frac{1}{\sqrt{2}}(\psi_1 + v_1 + i\phi_1) \\ H_1^- \end{pmatrix} ,$$

$$H_2 = \begin{pmatrix} H_2^+ \\ \frac{1}{\sqrt{2}}(\psi_2 + v_2 + i\phi_2) \end{pmatrix} .$$

$$(4)$$

Minimizing this potential with respect to the two real components of the neutral Higgs fields ψ_1 and ψ_2 yields the tree-level EWSB minimization conditions:

$$\frac{1}{2}M_Z^2 = \frac{m_{H_1}^2 - m_{H_2}^2 \tan^2\beta}{\tan^2\beta - 1} - \mu^2 , \tag{5}$$

$$-B\mu = \frac{1}{2}(m_{H_1}^2 + m_{H_2}^2 + 2\mu^2)\sin 2\beta . \tag{6}$$

The masses in these equations are running masses that depend on the scale Q in the RGEs that describe their evolution. Hence the solutions obtained are functions of the scale Q. Equations (5) and (6) are particularly convenient since the gauge couplings dependence (the D-terms in the language of supersymmetry) is isolated in Eq. (5). In addition, these minimization equations are readily solvable (even at the one-loop level) with the ambidextrous approach, which we describe in the next section.

The minimization equations also clearly show the fine-tuning problem that may be present in the radiative breaking of the electroweak symmetry. For large values of $|\mu|$, there must be a cancellation between large terms on the right hand side of equation (5) to obtain the correct experimentally measured M_Z (or equivalently the electroweak scale). For $\tan\beta$ near one, a cancellation of large terms must occur.

A heavy top quark produces large corrections to the Higgs potential of the MSSM[34]. Gamberini, Ridolfi, and Zwirner showed[23] that the tree-level Higgs potential is inadequate for the purpose of analyzing radiative breaking of the electroweak symmetry because the tree-level Higgs vacuum expectation values v_1 and v_2 are very sensitive to the scale at which the renormalization group equations are evaluated. The one-loop contribution to the effective potential is given by

$$\Delta V_1 = \frac{1}{64\pi^2} Str \left[\mathcal{M}^4 \left(\ln\frac{\mathcal{M}^2}{Q^2} - \frac{3}{2} \right) \right] , \tag{7}$$

where ΔV_1 is given in the dimensional reduction $(\overline{DR})$ renormalization scheme [35]. The supertrace is defined as $Str f(\mathcal{M}^2) = \sum_i C_i(-1)^{2s_i}(2s_i + 1)f(m_i^2)$ where C_i is the color degrees of freedom and s_i is the spin of the ith particle.

The one-loop corrections to the Higgs potential effectively moderates this sensitivity to the scale Q. The one-loop corrections are conveniently calculated using the tadpole method[32],[36],[37]. The one-loop corrected minimization conditions can then be used to generate a complete supersymmetric particle spectrum which satisfies EWSB. Including only the leading contribution coming from the top quark loop (and neglecting the D-term contributions to the squark masses) one obtains the expressions

$$\frac{1}{2}M_Z^2 = \frac{m_{H_1}^2 - m_{H_2}^2 \tan^2\beta}{\tan^2\beta - 1} - \mu^2 - \frac{3g^2 m_t^2}{32\pi^2 M_W^2 \cos 2\beta}\left[2f(m_t^2) - f(m_{\tilde{t}_1}^2) - f(m_{\tilde{t}_2}^2)\right.$$

$$\left. + \frac{f(m_{\tilde{t}_1}^2) - f(m_{\tilde{t}_2}^2)}{m_{\tilde{t}_1}^2 - m_{\tilde{t}_2}^2}\left((\mu\cot\beta)^2 - A_t^2\right)\right] ,$$

(8)

$$-B\mu = \frac{1}{2}(m_{H_1}^2 + m_{H_2}^2 + 2\mu^2)\sin 2\beta - \frac{3g^2 m_t^2 \cot\beta}{32\pi^2 M_W^2}\left[2f(m_t^2) - f(m_{\tilde{t}_1}^2) - f(m_{\tilde{t}_2}^2)\right.$$

$$\left. - \frac{f(m_{\tilde{t}_1}^2) - f(m_{\tilde{t}_2}^2)}{m_{\tilde{t}_1}^2 - m_{\tilde{t}_2}^2}(A_t + \mu\cot\beta)(A_t + \mu\tan\beta)\right] ,$$

(9)

where

$$f(m^2) = m^2\left(\ln\frac{m^2}{Q^2} - 1\right) .$$

(10)

The extra one-loop contribution included above renders the solution less sensitive to the scale Q[26]–[29], [32],[27], as can be shown explicitly by examining the relevant renormalization group equations for the parameters that enter into the minimization conditions. The complete expressions for the one-loop contributions can be found in Ref. [32]. The fine-tuning problem is alleviated somewhat, but not entirely, by the inclusion of one-loop corrections to the Higgs potential. As our naturalness criterion we require

$$|\mu(m_t)| < 500 \ GeV .$$

(11)

4 Ambidextrous Approach

Other RGE studies of the supersymmetric particle spectrum have evolved from inputs at the GUT scale (the top-down method[38]) or from inputs at the electroweak scale (the bottom-up approach[28]). The ambidextrous approach [39] incorporates some boundary conditions at both electroweak and GUT scales. We specify m_t and $\tan\beta$ at the electroweak scale (along with M_Z and M_W) and the common gaugino mass $m_{\frac{1}{2}}$, scalar mass m_0, and trilinear coupling A^G at the GUT scale. The soft supersymmetry breaking parameters are evolved from the GUT scale to the electroweak scale and then $\mu(M_Z)$ and $B(M_Z)$ (or $\mu(m_t)$ and $B(m_t)$) are determined by the one-loop minimization equations.

This strategy is effective because the RGEs for the soft-supersymmetry breaking parameters do not depend on μ and B. This method has two powerful advantages: First, any point in the $m_t - \tan\beta$ plane can be readily investigated in specific supergravity models since m_t and $\tan\beta$ are taken as inputs. Second, the minimization equations are easy to solve in the ambidextrous approach: equation (8) can be solved iteratively for $\mu(M_Z)$ (to within a sign), and then equation (9) explicitly gives $B(M_Z)$. We stress the numerical simplicity: no derivatives need be calculated and no functions need to be numerically minimized.

5   Low $\tan\beta$ Fixed Point Solutions

We now describe our numerical approach in more detail. Starting with our low-energy choices for m_t, $\tan\beta$, α_3, and m_b (and using the experimentally determined values for α_1, α_2 and m_τ[40]), we integrate the MSSM RGEs from m_t to M_G with M_G taken to be the scale Q at which $\alpha_1(Q) = \alpha_2(Q)$. We then specify $m_{\frac{1}{2}}$, m_0, and A at M_G, and integrate back down to m_t where we solve the full one-loop minimization equations (see Ref. [32]) for $\mu(m_t)$ and $B(m_t)$. We can then integrate the RGEs back to M_G to obtain $\mu(M_G)$ and $B(M_G)$.

In particular, we choose values of m_t, $\tan\beta$, α_3, and m_b representative of the low $\tan\beta$ λ_t fixed point region. In addition to requiring EWSB to be satisfied, we impose the following experimental bounds:

Table 1. Approximate experimental bounds.

Particle	Experimental Limit (GeV)
gluino	120
squark, slepton	45
chargino	45
neutralino	20
light higgs	60

Together with our naturalness criteria $|\mu(m_t)| < 500$ GeV, these bounds give the allowed region in the $m_0, m_{1/2}$ plane shown as the shaded areas in Fig. 4.

Hence there are solutions in the low-$\tan\beta$ λ_t fixed-point region which satisfy EWSB constraints (as well as our naturalness criterion) at the one-loop level. Note that the $\mu < 0$ solutions have more allowed parameter space than do the $\mu > 0$ solutions. A few additional remarks are pertinent: the prediction for m_h in the low-$\tan\beta$ region is particularly sensitive to higher order corrections[16],[41],[42]. Hence the precise location of the $m_h = 60$ GeV contour is somewhat uncertain. Also, the dark matter line in Figure 4 should be regarded as semi-quantitative only since the contributions of s-channel poles that can enhance the annihilation rate have been neglected [43].

6   Conclusions

Given only two reasonable assumptions
  * unification of couplings and $\lambda_b(M_G) = \lambda_\tau(M_G)$ at the GUT scale.
  * $m_b(m_b) = 4.25 \pm 0.10$ GeV

we are restricted to the fixed-point region of $m_t - \tan\beta$ parameter space. With only one additional assumption, $m_t(m_t) \lesssim 175$ GeV, we are restricted to either
  1) $\tan\beta \lesssim 2$ (λ_t fixed point), or
  2) $\tan\beta \gtrsim 50$ ($\lambda_b = \lambda_\tau$ fixed point). The small $\tan\beta$ solution is favored by proton decay and flavor changing neutral current constraints. We investigated the additional constraint imposed by radiative electroweak symmetry breaking upon this first region, and found solutions which are both experimentally viable and meet the naturalness criterion $|\mu(M_Z)| \simeq |\mu(m_t)| < 500$ GeV.

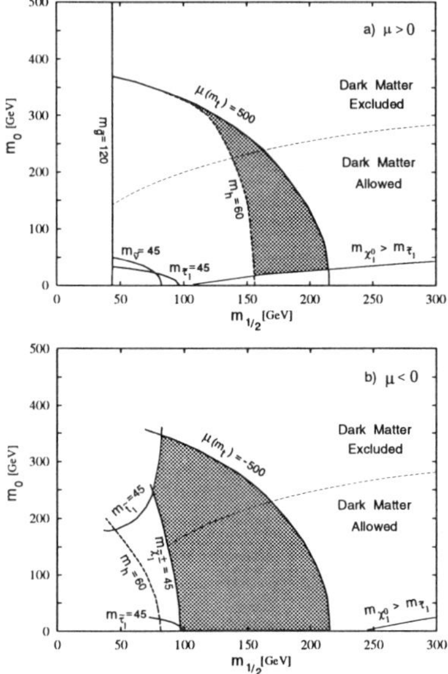

Figure 4. Allowed regions of parameter space for $m_t(m_t) = 160$ GeV, $\tan\beta = 1.47$ (a low-$\tan\beta$ λ_t fixed-point solution) (from Ref. *[32]*).

7 Acknowledgements

We would like to thank Roger Phillips for collaboration. This research was supported in part by the University of Wisconsin Research Committee with funds granted by the Wisconsin Alumni Research Foundation, in part by the U.S. Department of Energy under contract no. DE-AC02-76ER00881, and in part by the Texas National Laboratory Research Commission under grant nos. RGFY93-221 and FCFY9302. MSB was supported in part by an SSC Fellowship. PO was supported in part by an NSF Graduate Fellowship.

References

[1] U. Amaldi, W. de Boer, and H. Furstenau, Phys. Lett. **B260** (1991) 447; J. Ellis, S. Kelley and D. V. Nanopoulos, Phys. Lett. **B260** 131 (1991); P. Langacker and M. Luo, Phys. Rev. **D44** 817 (1991).

[2] M. Chanowitz, J. Ellis, and M. K. Gaillard, Nucl. Phys. **B128**, 506 (1977).

[3] V. Barger, M.S. Berger, and P. Ohmann, Phys. Rev. **D47**, 1093 (1993).

[4] J. Gasser and H. Leutwyler, Phys. Rep. **87**, 77 (1982).

[5] V. Barger, M. S. Berger, P. Ohmann, and R. J. N. Phillips, Phys. Lett. **B314**, 351 (1993), V. Barger, M. S. Berger, T. Han, and M. Zralek, Phys. Rev. Lett. **68**, 3394 (1992).

[6] B. Pendleton and G. G. Ross, Phys. Lett. **98B**, 291 (1981); C. T. Hill, Phys. Rev. **D24**, 691 (1981).

[7] C. D. Froggatt, I. G. Knowles, and R. G. Moorhouse, Phys. Lett. **B249**, 273 (1990); **B298**, 356 (1993).

[8] M. Carena, S. Pokorski, and C. E. M. Wagner, Nucl. Phys. **B406**, 59 (1993); W. Bardeen, M. Carena, S. Pokorski, and C. E. M. Wagner, Phys. Lett. **B320**, 110 (1994).

[9] L. J. Hall, R. Rattazzi, and U. Sarid, Lawrence Berkeley Lab preprint LBL-33997 (1993).

[10] P. Langacker and N. Polonsky, Phys. Rev. **D49**, 1454 (1994); N. Polonsky, Talk presented at the XVI Kazimierz Meeting on Elementary Particle Physics, Penn U. preprint UPR-0588-T, Kazimierz, Poland (1993), hep-ph 9310292.

[11] B. C. Allanach and S. F. King, University of Southampton preprint SHEP-93/94-15 (1994), hep-ph 9403212.

[12] CDF Collaboration, Fermilab preprint, April 1994.

[13] R. Barbieri and L. J. Hall, Phys. Rev. Lett. **68**, 752 (1992).

[14] L. J. Hall and U. Sarid, Phys. Rev. Lett. **70**, 2673 (1993).

[15] R. Hempfling, DESY preprint DESY-93-092 (1993).

[16] P. Langacker and N. Polonsky, Univ. of Pennsylvania preprint UPR-0594T (1994).

[17] B. D. Wright, University of Wisconsin preprint MAD/PH/812 (1994).

[18] R. Arnowitt and P. Nath, Phys. Rev. **70**, 3696 (1993).

[19] J. Hisano, H. Murayama and T. Yanagida, Phys. Rev. Lett. **69**, 1014 (1992); Nucl. Phys. **B402**, 46 (1993).

[20] M. Carena, M. Olechowski, S. Pokorski, and C. E. M. Wagner, Max Planck Institute preprint MPI-Ph/93-103(1994), hep-ph 9402253.

[21] For a review see P. Nath, R. Arnowitt, and A. H. Chamseddine, "Applied $N = 1$ Supergravity," World Scientific (1984).

[22] L. E. Ibañez and G. G. Ross, Phys. Lett. **B110**, 215 (1982); L. Alvarez-Gaumé, M. Claudson and M. B. Wise, Nucl. Phys. **207**, 96 (1982); J. Ellis, D. V. Nanopoulos, and K. Tamvakis, Phys. Lett. **B121**, 123 (1983); L. E. Ibañez, Nucl. Phys. **218**, 514 (1983); L. Alvarez-Gaumé, J. Polchinski and M. B. Wise, Nucl. Phys. **221**, 495 (1983); J. Ellis, J. S. Hagelin, D. V. Nanopoulos, and K. Tamvakis, Phys. Lett. **B125**, 275 (1983); L. E. Ibañez and C. Lopez, Phys. Lett. **B126**, 54 (1983); Nucl. Phys. **B233**, 511 (1984); C. Kounnas, A. B. Lahanas, D. V. Nanopoulos, and M. Quiros, Nucl. Phys. **B236**, 438 (1984); U. Ellwanger, Nucl. Phys. **B238**, 665 (1984); L. E. Ibañez, C. Lopez, and C. Muñoz, Nucl. Phys. **B256**, 218 (1985); A. Bouquet, J. Kaplan, and C. A. Savoy, Nucl. Phys. **B262**, 299 (1985); H. J. Kappen, Phys. Rev. **D35**, 3923 (1987); M. Drees, Phys. Rev. **D38**, 718 (1988); H. J. Kappen, Phys. Rev. **D38**, 721 (1988); B. Gato, Z. Phys. **C37**, 407 (1988); G. F. Giudice and G. Ridolfi, Z. Phys. **C41**, 447 (1988); S. P. Li and H. Navelet, Nucl. Phys. **B319**, 239 (1989).

[23] G. Gamberini, G. Ridolfi, and F. Zwirner, Nucl. Phys. **B331**, 331 (1990).

[24] R. Arnowitt and P. Nath, Phys. Rev. **69**, 725 (1992); Phys. Lett. **B289**, 368 (1992).

[25] G. G. Ross and R. G. Roberts, Nucl. Phys. **B377**, 571 (1992).

[26] S. Kelley, J. L. Lopez, D. V. Nanopoulos, H. Pois, and K. Yuan, Nucl. Phys. **B398**, 3 (1993).

[27] B. de Carlos and J. A. Casas, Phys. Lett. **B309**,320 (1993); B. de Carlos, CERN preprint CERN-TH-7024-93, Talk given at the *XVI International Warsaw Meeting on Elementary Particle Physics* (1993), hep-ph 9310232.

[28] M. Olechowski and S. Pokorski, Nucl. Phys. **B404**, 590 (1993).

[29] P. Ramond, Institute for Fundamental Theory Preprint UFIFT-HEP-93-13 (1993), hep-ph 9306311; D. J. Castaño, E. J. Piard, and P. Ramond, Institute for Fundamental Theory Preprint UFIFT-HEP-93-18 (1993), hep-ph 9308335.

[30] W. de Boer, R. Ehret, and D. I. Kazakov, Inst. für Experimentelle Kernphysik preprint IEKP-KA/93-13, Contribution to the International Symposium on Lepton Photon Interactions, Ithaca, NY (1993), hep-ph 9308238.

[31] M. Carena, M. Olechowski, S. Pokorski, and C. E. M. Wagner, CERN preprint CERN-TH.7060/93 (1993), hep-ph 9311222.

[32] V. Barger, M. S. Berger, and P. Ohmann, University of Wisconsin preprint MAD/PH/801 (1993), hep-ph 9311269.

[33] G. Kane, C. Kolda, L. Roszkowski, and J. D. Wells, University of Michigan preprint UM-TH-93-24 (1993), hep-ph 9312272.

[34] M. S. Berger, Phys. Rev. **D41**, 225 (1990).

[35] W. Siegel, Phys. Lett. **B84**, 193 (1979); D. M. Capper D. R. T. Jones, and P. van Nieuwenhuizen, Nucl. Phys. **B167**, 479, (1980).

[36] S. Weinberg, Phys. Rev. **D7**, 2887 (1973); S. Y. Lee and A. M. Sciaccaluga, Nucl. Phys. **B96**, 435 (1975); K. T. Mahanthappa and M. A. Sher, Phys. Rev. **D22**, 1711 (1980).

[37] M. Sher, Phys. Rep. **C179**, 273 (1989).

[38] R. Hempfling, DESY preprint DESY 94-057.

[39] S. Kelley, J. L. Lopez, D. V. Nanopoulos, H. Pois, K. Yuan, Phys. Lett. **B273**, 423 (1991).

[40] Particle Data Book, Phys. Rev. **D45** (1992).

[41] J. Kodaira, Y. Yasui, and K. Sasaki, Hiroshima preprint HUPD-9316 (1993).

[42] R. Hempfling and A. H. Hoang, DESY preprint DESY-93-162 (1993).

[43] M. Drees and M. M. Nojiri, Phys. Rev. **D47**, 376 (1993).

Generic Properties of Yukawa Couplings in String Models

P. BINÉTRUY
LPTHE, UNIVERSITÉ PARIS-SUD
F-91405 ORSAY CEDEX, FRANCE

1 Introduction

The question of Yukawa couplings in string models seems to have been receiving renewed attention in the recent months. Obviously, as candidates for the theory underlying the Standard Model, superstrings should yield some clue on the diversity of masses observed in the low energy world. It is a natural question to raise in a theory with such big claims as strings advertize. It seems also that, in view of the difficulties encountered to solve the fundamental problem of supersymmetry breaking and the generation of the hierarchy M_W/M_{Pl}, it is a relief to address a problem which does not rely too heavily upon this question. In what follows, I will give the generic features of Yukawa couplings in string models and the different lines of attack chosen to tackle the problem of hierarchies among the Yukawa couplings.

2 Generic Value and the Infrared Fixed Point

It is a general property of superstring models that if a Yukawa coupling λ_Y is non-zero, it is typically of order of a gauge coupling g

$$\lambda_Y \sim g. \tag{1}$$

Such a relation should be understood at the string scale and, as we will discuss below, gets renormalized at lower scales. The relation (1) is best understood in the context of 10-dimensional string models compactified to 4 dimensions. Indeed the gauge supermultiplet of the 10-dimensional theory which consists of gauge fields $A_M, M = 0, \cdots, 9$ (8 physical degrees of freedom) and gauginos $\chi_A, A = 1, \cdots, 8$ gives rise, through compactification on a 6-dimensional compact manifold $\mathcal{M}$, to:

(i) gauge supermultiplets (A_μ, λ_α) in 4-dimensional Minkowski space ($\mu = 0, \cdots, 3$);

(ii) chiral supermultiplets (A_I, ψ_α) in 4 dimensions: $I = 4, \cdots, 9$ is a compact index, therefore A_I is a scalar field ϕ in Minkowski space.

Thus a Yukawa term of the form $\phi\bar{\psi}\psi$ originates from a term $A_M\bar{\chi}\chi$ in the underlying 10-dimensional theory, *i.e.* the minimal coupling of gaugino to gauge fields. The coupling constant is therefore a gauge coupling constant:

$$\mathcal{L}_{Yukawa} = \int_{\mathcal{M}} d^6y \; g f_{abc} A_I^a \psi_\eta^b \Gamma_{\eta\zeta}^I \psi_\zeta^c \tag{2}$$

where Γ^I is a gamma function of the 10-dimensional theory with compact index I.

This naturally leads to a topological formula for Yukawa couplings in the case of Calabi-Yau manifolds [1]. It is best described in terms of a superpotential

$$W = \lambda_{klm}\phi_k\phi_l\phi_m \tag{3}$$

Indeed since (2) reads in the case of a Calabi-Yau manifold

$$\mathcal{L}_{Yukawa} = \int_{\mathcal{M}} d^3z d^3\bar{z} g f_{abc} A^a_{\bar{\imath}} \psi^b_{\bar{\jmath}} \psi^c_{\bar{k}} \epsilon_{\bar{\imath}\bar{\jmath}\bar{k}} \tag{4}$$

where the complex coordinates $z^i, \bar{z}^{\bar{\imath}}, i, \bar{\imath} = 1, 2, 3$ parametrize the complex Calabi-Yau manifold and $\epsilon_{\bar{\imath}\bar{\jmath}\bar{k}}$ is the tensor completely antisymmetric in its antiholomorphic indices. The spinorial indices have been suppressed.

This leads to a "topological" formula for the Yukawa coupling λ_{klm}:

$$\lambda_{klm} = \int_{\mathcal{M}} d^3z d^3\bar{z} g f_{abc} u^a_k \wedge u^b_l \wedge u^c_m \wedge \Omega, \tag{5}$$

where the u^a_k are (0,1)-forms [1] and Ω is the single (3,0)-form characteristic of Calabi-Yau manifolds.

3   Generic Yukawa Couplings and Infrared Fixed Points

Following the previous section, let us consider a coupling with the generic value (1):

$$\lambda_Y \sim g. \tag{6}$$

As stressed above, this relation is valid at the string unification scale M_S and should therefore be renormalized down to low energies. The standard procedure makes use of the renormalisation group equations. Such equations have quasi-infrared fixed points [2], which play an important role for the type of couplings that we consider here.

Indeed, let us consider as an example the third family and let us suppose for the sake of the argument that $\lambda_t \gg \lambda_b, \lambda_\tau$. Then the quasi-infrared fixed point is reached for

$$\lambda_t^2 \sim \frac{8}{9} g_3^2, \tag{7}$$

where g_3 is the $SU(3)$ gauge coupling. Because this fixed point lies close to the generic value (6) at the string scale, it is most probable that it will be reached at low energies. As is well-known, it corresponds to a typical value for the top quark mass of $m_t \sim m_0 \sin\beta$ where m_0 is around 190 to 200 GeV and β is the parameter familiar in the minimal supersymmetric model. The boundary condition (6) therefore yields a natural explanation as to why the top is heavy.

In the same line of thought, this tends to show that any quark or lepton which satisfies (6) will have a similar mass, or at least a similar Yukawa coupling. Thus, for example, if all Yukawa couplings of the third family obey the boundary condition (6), *i.e.*

$$\lambda_t(M_S) = \lambda_b(M_S) = \lambda_\tau(M_S) = g(M_S) \tag{8}$$

then we expect similar couplings for the top and bottom quark at low energies, which leads to a large $\tan\beta$ scenario. This is a reflection of the well-known fact noted some years ago by Froggatt and Nielsen in a pioneering work [3]:

---

[1] The fields with antiholomorphic indices decompose as follows: $A^a_{\bar{\imath}} = \sum_k \phi_k(x) u^a_{k\bar{\imath}}(y)$.

evolution according to the renormalisation group equations fails to generate large hierarchies among Yukawa couplings.

Phenomenological low energy analyses based on the knowledge of the tau, bottom and top masses tend nowadays to favor two scenarios:

(i) a scenario with equal Yukawa couplings at the string scale [4] as in (8). The previous discussion shows that it can easily be implemented in string models. However, because it needs a large value of $\tan\beta$, it has problems of its own which might or might not be fatal [5].

(ii) a scenario with a large Yukawa coupling for the top; in this case, low energy phenomenology seems to favor an approximate unification of the smaller bottom and tau couplings [6], i.e.

$$\lambda_t(M_S) \gg \lambda_b(M_S) \sim \lambda_\tau(M_S). \tag{9}$$

This is compatible with values of $\tan\beta$ of order 1. It remains to be seen how the smaller couplings λ_b or λ_τ may be generated in string models. I will discuss this issue in the next section. Note also that the last section will describe a mechanism –proposed by Nambu in a different context [7]– that automatically generates a hierarchy of the type (9), that is one coupling much larger than all the other ones. But before we proceed to these questions, let us see an example of a model with a large top mass due to a generic value of the corresponding Yukawa coupling.

Let us consider a flipped $SU(5) \times U(1)$ model [8] where the low energy fermions are in representations $(10, 1/2)$ noted F, $(\bar{5}, -3/2)$ noted $\bar{f}$ and the singlet representation noted L^c. The superpotential reads

$$W = g \sum_{ijk} \alpha_{ijk} F_i F_j h_k + \beta_{ijk} F_i \bar{f}_j h_k + \gamma_{ijk} \bar{f}_i L^c_j h_k, \tag{10}$$

where i, j, k are family indices and $\alpha_{ijk}, \beta_{ijk}, \gamma_{ijk}$ are numbers equal to 0 or 1. At this stage, massive fermions such as the top are obtained from a mixing of massive states (of mass $g < h >$) and massless states [9]. Hence

$$\lambda_t \sim g\cos\theta \tag{11}$$

where θ is some mixing angle. Thus $\lambda_t < g$ and one recovers bounds for the top mass of the order of the infrared fixed point discussed above. One should note that in this model one also finds $\lambda_b = \lambda_\tau = g\cos\theta'$ where θ' is some other mixing angle. This allows to obtain at low energies $m_b \sim 3m_\tau \sim 5\,GeV$. The hierarchy is thus realized through some mixing to the massless states which must be such that $\cos\theta \gg \cos\theta'$.

4   Why Should Yukawas be Small ($\lambda_Y \ll g$)?

Let me recall how one obtains Yukawa couplings in string theories: they are given by the correlation functions of the corresponding vertex operators V computed on the 2-dimensional world-sheet, described by the string in its motion through spacetime. In other words,

$$\mathcal{L}_{Yukawa} = < V(\phi)V(\bar{\psi})V(\psi)V(T_i) \cdots V(\tilde{\phi}_j) \cdots > . \tag{12}$$

In this formula extra fields appear. Indeed Yukawa couplings depend in general on the vacuum expectation values (vevs) of fields such as:

(i) moduli fields T_i, known as Kähler moduli, whose vevs describe radii and shapes of the six-dimensional compact manifold.

(ii) extra singlet and non-singlet fields $\tilde{\phi}_j$: complex structure moduli, etc.

It is this dependence on extra fields which allows to generate some hierarchies among the Yukawa couplings. One can distinguish two different origins for this hierarchy.

4.1 Non-perturbative contributions

A simple non-renormalisation theorem states that the superpotential cannot depend on the moduli T [10]. The idea behind this result goes as follows.

In the case of a single (Kähler) modulus T, its real part can be interpreted as the radius-squared R^2 of the compact manifold $\mathcal{M}$ whereas its imaginary part is related to the antisymmetric tensor B_{MN} present in the 10-dimensional massless spectrum (M and N are here 6-dimensional compact indices). Similar interpretations apply to the other Kähler moduli. Because of its origin, $Im\ T$ behaves in a way very similar to an axion and only has derivative couplings. Hence the superpotential cannot depend on $Im\ T$, and being analytic in the fields, cannot depend on T as a whole.

This, however, is a perturbative statement. As a matter of fact, T can be interpreted as $1/\gamma^2$, where γ is the coupling constant of the 2-dimensional sigma-model that characterizes the string model considered on the world-sheet. And the above non-renormalisation theorem can be interpreted as stating the absence of quantum corrections to the superpotential, to all finite orders in the sigma model perturbation. However, non-perturbative contributions such as world-sheet instantons may appear: they are typically of order $\exp(-1/\gamma^2) \sim \exp(-T) \sim \exp(-R^2)$. Hence we may have Yukawa couplings with an exponential suppression [12]:

$$\lambda_Y = g e^{-R^2/\alpha'} \tag{13}$$

where the right dimensions have been restored by using the dimensionful string coupling α'. If $R \gg \sqrt{\alpha'}$, this generates small Yukawa couplings, as looked for.

As an example, we can take orbifold models. Orbifolds are constructed by identifying, in a manifold, points which are obtained from one another by a discrete group: in other words, they can be understood as the coset space of a manifold by a discrete group acting on this manifold [13]. They are not manifolds themselves since fixed points under the group correspond to a singularity of curvature (for example, a tetraedron can be thought of as an orbifold). Orbifolds are used as the 6-dimensional compact spaces necessary to compactify the string models don to 4 dimensions. Untwisted states correspond to strings closed in the original manifold whereas twisted modes correspond to "twisted" strings closed up to an operation of the discrete subgroup: the latter strings surround the fixed points.

It can be shown [14] that, whereas the superpotential couplings between untwisted states −or between states twisted around the same fixed point− are *of order g, the couplings between states twisted around different fixed points are of order $g \exp(-R^2/\alpha')$. To understand this, consider that a coupling is possible if the corresponding strings add up to a single string which can be deformed into a point. In order to do so in the latter case, one has to make at least one twisted string slide from one fixed point to another. The factor $\exp(-R^2/\alpha')$ is just the corresponding semi-classical factor ($S = R^2$ being the surface swept in the process).*

4.2 *Symmetries*

Another way to generate hierarchies which one can again trace back to the work of Froggatt and Nielsen *[3]*, consists in using approximate selection rules associated with spontaneously broken symmetries. Typically, a symmetry requires the fermion mass to vanish but a finite mass is generated at some order in a symmetry breaking interaction.

The symmetries used can be continuous or discrete. There is in fact a plethora of such symmetries in string models. But the freedom is not complete: continuous but also discrete symmetries *[15]* are severely constrained by the anomaly cancellation conditions.

Let me sketch how these symmetries are used on the simple example of a continuous $U(1)_{\tilde{Y}}$ symmetry. One first forbids large cubic couplings by making the symmetry generation-dependent. For example, a coupling TT^cH is allowed (*i.e.* $2\tilde{Y}_T+\tilde{Y}_H = 0$) whereas UU^cH and CC^cH are forbidden (*i.e.* $2\tilde{Y}_U+\tilde{Y}_H \neq 0$ and $2\tilde{Y}_C + \tilde{Y}_H \neq 0$). But higher nonrenormalisable couplings to other fields may be allowed by the symmetry: for example $UU^cH(\Phi/\Lambda)^n$ may be allowed if $2\tilde{Y}_U + \tilde{Y}_H + n\tilde{Y}_\Phi = 0$ (Λ is some large scale and Φ is one of these extra fields). Then if $< \Phi > \neq 0$ and $< \Phi > /\Lambda$ is small, one obtains a small Yukwa coupling:

$$\lambda_Y = g \left(\frac{< \Phi >}{\Lambda} \right)^n \ll g. \tag{14}$$

This game, which can be played in standard grand unified theories, has a new set of rules in string models, in particular because anomalies may be cancelled by the Green-Schwarz mechanism *[16]*.

4.3 *The string unification threshold*

Nonrenormalisation theorems can also be seen from the point of view of string perturbation theory *[11]*. One consequence is that the superpotential is not renormalized, to all orders of string perturbation theory (to be explicit, the one loop contribution in this expansion corresponds to world sheets with one handle, *i.e.* with the topology of a torus). At the field theory level, the string coupling is expressed in terms of the vev of the dilaton field ϕ:

$$g_{string} = e^{2\phi} \tag{15}$$

Through supersymmetry, ϕ is related to the antisymmetric tensor $B_{\mu\nu}$ (this time with 4-dimensional indices). Together with a fermion, the dilatino, they form what is known as a linear supermultiplet L, which is real. The superpotential, being analytic in the fields, cannot depend on L. This is related to the gauge invariance associated with the antisymmetric tensor. This in turn ensures that W cannot depend on ϕ and hence is not renormalized, to all orders of string perturbation theory.

This does not mean however that Yukawa couplings do not get renormalised. In fact, because there is wave function renormalisation, fields do get renormalized and so do the Yukawas, in order to cancel the renormalisation effect in the superpotential.

Take for example a one-loop diagram contributing to the renormalisation of a fermion field. At the string level, it consists of a torus with the two vertex operators of the external fermions (of say momentum Q) attached to it. Such a contribution involves two distinct classes of one-loop field theory diagrams: the ones involving massless string states and the ones involving massive string states. The first class of diagrams contributes, through infrared divergent logarithmic contributions $\log Q^2$, to the external fermion anomalous dimension. On the other hand, since the massive states have moduli-dependent (*i.e.* radii-dependent) masses, the second class gives some moduli-dependent contributions known as threshold corrections.

Indeed, these corrections make the boundary conditions (1) at the string scale moduli-dependent *[17]*. This is similar to what happens in the case of gauge couplings*[18]*, which is not surprising since we have seen earlier the strong connection between generic Yukawa couplings and gauge couplings. Indeed, in the case of the untwisted fields of an orbifold model, it has been shown recently *[17]* that, when one expresses the Yukawa couplings in terms of the string couplings, the moduli-dependent corrections of the two types of couplings cancel again each other:

$$\lambda_{klm}(M_S) = g(M_S)\lambda^0_{klm}(1 + g^2(M_S)\alpha_{klm})^{-1/2} \qquad (16)$$

where λ^0_{klm} and α_{klm} are moduli-independent constants.

5   Nambu Hierarchies

Finally, I would like to describe a new mechanism proposed by Nambu *[7]* for generating hierarchies among Yukawa couplings. It was originally devised in the context of the Standard Model *[19]* but remarkably adapts to the context of strings *[20]*.

Nambu's idea is to minimize the vacuum energy with respect to the Yukawa couplings *subject to a quadratic constraint*. Minimization in the absence of the constraints would lead to instabilities (the Yukawa couplings being either zero or infinite). The role of the constraint is precisely to constrain the Yukawas to a finite region of the parameter space. Minimization then tends to favor the boundary regions where one coupling is of order one whereas the others are exponentially smaller.

The quadratic constraint chosen by Nambu originates from the cancellation of the quadratic divergence, typically a Veltman condition *[21]* but we will see in what follows that other types of constraints may arise in string models *[22]*.

Nambu proposes a toy model with two Yukawa couplings in which the Veltman condition simply reads

$$\lambda_1^2 + \lambda_2^2 = a^2 \tag{17}$$

with a a constant. This constrains both Yukawas to the region [0,a]. Such a condition cancels the order Λ^2 contributions to the vacuum energy, where Λ is the cut-off. One is left in the scalar potential with the $O(\ln \Lambda)$ contributions which read:

$$< V_1 >= -A(\lambda_1^4 + \lambda_2^4) + B(\lambda_1^2 \ln \lambda_1^2 + \lambda_2^2 \ln \lambda_2^2). \tag{18}$$

This type of potential favors, in the case of $A > 0$, large hierarchies of couplings: $(\lambda_1, \lambda_2) = (a, 0)$ or $(0, a)$ when one disregards the logarithmic corrections proportional to B. The general effect of these logarithmic corrections is to generate a non-zero value $(\sim a \exp -(2aA/B))$ for the smaller Yukawa coupling.

The procedure of minimizing with respect to the Yukawas is certainly meaningful in the case of strings. Indeed we have seen that Yukawa couplings depend on the moduli fields. In this context, the minimization amounts to varying with respect to the underlying moduli. According to the non-renormalisation theorems discussed above, these moduli correspond to flat directions of the potential. The associated degeneracy may be lifted once supersymmetry is broken. For our approach to make sense, one needs to suppose that, at the low energies that we consider, some of the moduli still remain undetermined so that we can minimize with respect to them.

The question of the constraint may receive a different treatment in string models. If one considers the quadratic divergences, one is led to compute as in any supersymmetric theory the supertrace of the squared mass

$$STrM^2 = \sum_J (-)^J(2J + 1)m_J^2 \tag{19}$$

where the sum runs over all fields and all spins J. After supersymmetry breaking, the leading terms in $STrM^2$ are of order $m_{3/2}^2$, where $m_{3/2}$ is the gravitino mass. These terms must cancel since otherwise they would lead to instabilities by driving $m_{3/2}$ to zero or infinity [23]. The next order $O(m_{3/2}^4/M_{Pl}^2)$ should also cancel or at least be bounded when we consider the Yukawa couplings to be dynamical (and minimize with respect to them): the reason again is that otherwise they would lead to instabilities *i.e.* Yukawa couplings being zero or infinite. This gives a condition which is quadratic in the Yukawa couplings [20]. A similar condition could be obtained by minimising with respect to $m_{3/2}$ [24] but trying to obtain by minimisation the constraint that confines the Yukawas to a finite region might lead to instabilities. Finally, it is possible that the geometry of the moduli themselves leads to some constraints on the Yukawa couplings (constraints which, in this sense, would be more geometrical than dynamical, probably a welcome feature). The constraint in this case is generally multiplicative (i.e. $\lambda_1 \lambda_2 \cdots = a$) [22].

Once quadratic divergences are kept under control in the ways just described, the logarithmically divergent contributions naturally give in supersymmetric theories a vacuum energy of the form (18), with the right signs for A and B *[20]*.

A nice property of the Nambu mechanism is that in the case of more than two Yukawa couplings, one coupling is large whereas all the others are small (and of a similar magnitude).

6 Conclusion

Given the huge number of different string models, the task of describing the properties of Yukawa couplings might seem overwhelming. It turns out that, barring the technicalities and intricacies of the task of computing the Yukawa couplings in a specific model, the general features are remarkably simple and general.

Generically, the Yukawa couplings are of the order of the gauge couplings at the string unification. In order to obtain smaller couplings, one needs to consider extra fields, either moduli or extra gauge non-singlet fields associated with the breaking of an extra gauge symmetry which provides approximate selection rules. Finally, the scenario proposed by Nambu might provide some new ways of getting a hierarchy; it also invoves the underlying moduli structure.

Acknowledgements

I wish to thank Pierre Ramond and all the organizers for providing such a stimulating atmosphere during this workshop.

References

[1]   A. Strominger and E. Witten, *Comm. Math. Phys.* 101 (1985) 341.

[2]   B. Pendelton and G.G. Ross, *Phys. Lett.* 98B (1981) 291.

[3]   C.D. Froggatt and H.D. Nielsen, *Nucl. Phys.* B147 (1979) 277.

[4]   M. Carena, M. Olechowski, S. Pokorski and C.E.M. Wagner, preprint MPI-Ph/93-103, CERN-TH.7163/94.

[5]   S. Pokorski, *Proceedings of Supersymmetry-94*, Ann Arbor, May 1994; U. Sarid, these Proceedings.

[6]   M. Carena and C. Wagner, these Proceedings and references therein.

[7]   Y. Nambu, *Proceedings of the International Conference on Fluid Mechanics and Theoretical Physics in honor of Professor Pei-Yuan Chou's 90th anniversary*, Beijing, 1992; preprint EFI 92-37.

[8]   S.M. Barr, *Phys. Lett.* 112B (1982) 219; J.P. Derendinger, J.E. Kim and D.V. Nanopoulos, *Phys. Lett.* 139B (1984) 170; I. Antoniadis, J. Ellis, J. Hagelin and D.V. Nanopoulos, *Phys. Lett.* 194B (1987) 231.

[9]   I. Antoniadis, J. Ellis, J. Hagelin and D.V. Nanopoulos, *Phys. Lett.* 205B (1988) 459, 208B (1988) 209, 231B (1989) 65; G. Leontaris, J. Rizos and K. Tamvakis, *Phys. Lett.* 251B (1990) 83.

[10]   E. Witten, *Nucl. Phys.* B268 (1986) 79.

[11]   E. Martinec, *Phys. Lett.* 171B (1986) 189; M. Dine and N. Seiberg, *Phys. Rev. Lett.* 57 (1986) 2625.

[12] M. Dine, N. Seiberg, X.G. Wen and E. Witten, *Nucl. Phys.* B289 (1987) 319.

[13] L. Dixon, J.A. Harvey, C. Vafa and E. Witten, *Nucl. Phys.* B261 (1985) 678; *ibid.* B274 (1986) 285.

[14] L. Dixon, D. Friedan, E. Martinec and S. Shenker, *Nucl. Phys.* B282 (1987) 13.

[15] L. Ibáñez and G.G. Ross, *Nucl. Phys.* B368 (1992) 3.

[16] G.G. Ross, these Proceedings; L. Ibáñez and G.G. Ross, *Phys. Lett.* B332 (1994) 100.

[17] I. Antoniadis, E. Gava, K.S. Narain and T.R. Taylor, *Nucl. Phys.* B407 (1993) 706.

[18] V. Kaplunovsky, *Nucl. Phys.* B307 (1988) 145; L. Dixon, V. Kaplunovsky and J. Louis, *Nucl. Phys.* B355 (1991) 649.

[19] T. Gherghetta, these Proceedings.

[20] P. Binétruy and E. Dudas, *Phys. Lett.* B338 (1994) 23.

[21] M. Veltman, *Acta Phys. Pol.* B12 (1981) 437.

[22] P. Binétruy and E. Dudas, preprint LPTHE-Orsay 94/73.

[23] S. Ferrara, C. Kounnas and F. Zwirner, *Nucl. Phys.* B429 (1994) 589.

[24] C. Kounnas, I. Pavel and F. Zwirner, *Phys. Lett.* B335 (1994) 403.

The Infrared Fixed Point of the Top Quark Mass and its Implications within the MSSM[1]

M. CARENA AND C.E.M. WAGNER
THEORY DIVISION, CERN
CH 1211, GENEVA, SWITZERLAND

Abstract: We analyse the general features of the Higgs and Supersymmetric particle spectrum associated with the infrared fixed point solution of the top quark mass in the Minimal Supersymmetric Standard Model. We consider the constraints on the mass parameters which are derived from the condition of a proper radiative electroweak symmetry breaking. In the case of universal soft supersymmetry breaking parameters at the high energy scale, the radiative $SU(2)_L \times U(1)_Y$ breaking, together with the top quark Yukawa fixed point structure imply that, for any given value of the top quark mass, the Higgs and supersymmetric particle spectrum is fully determined as a function of only two supersymmetry breaking parameters. This result is of great interest since the infrared fixed point solution appears as a prediction in many different theoretical frameworks. In particular, in the context of the MSSM with unification of the gauge and bottom-tau Yukawa couplings, for small and moderate values of $\tan\beta$ the value of the top quark mass is very close to its infrared fixed point value. We show that, for the interesting range of top quark mass values $M_t \simeq 175 \pm 10$ GeV, both a light chargino and a light stop may be present in the spectrum. In addition, for a given top quark mass, the infrared fixed point solution of the top quark Yukawa coupling minimizes the value of the lightest CP-even Higgs mass m_h. The resulting upper bounds on m_h read $m_h \leq 90$ (105) (120) GeV for $M_t \leq 165$ (175) (185) GeV.

1   Introduction

In the present evidence of a heavy top quark, it is of interest to study in greater detail the phenomenological implications of the infrared fixed point predictions for the top quark mass. The low energy fixed point structure of the Renormalization Group (RG) equation of the top quark Yukawa coupling, determines the value of the top quark mass independent of the precise symmetry conditions at the high energy scale. This *quasi* infrared fixed point behaviour of the RG solution is present in the Standard Model (SM) *[1]* as well as in the Minimal Supersymmetric Standard Model (MSSM) *[2, 3]*, and it is associated with large values of the top quark Yukawa coupling, which, however, remain in the range of validity of perturbation theory. Within the MSSM, for a range of high energy values of the top quark Yukawa coupling, such that it can reach its perturbative limit at some scale $M_X = 10^{14} - 10^{19}$ GeV, the value of the physical top quark mass is focused to be

$$M_t = 190 - 210 \, GeV \sin\beta \,, \tag{1}$$

---

[1]Presented by M. Carena.

where $\tan\beta = v_2/v_1$ is the ratio of the two Higgs vacuum expectation values. The above variation in M_t is due to a variation in the value of the strong gauge coupling, $\alpha_3(M_Z) = 0.11$–0.13.

The infrared fixed point structure is independent of the supersymmetry breaking scheme under consideration. On the contrary, since the Yukawa couplings – especially if they are strong – affect the running of the mass parameters of the theory, once the infrared fixed point structure is present, it will play a decisive roll in the resulting (s)particle spectrum of the theory. In particular, in the low and moderate $\tan\beta$ regime, in which the effects of the bottom and tau Yukawa couplings are negligible, it is possible to determine the evolution of the soft supersymmetry breaking mass parameters of the model as a function of their boundary conditions at high energy scales and the ratio of the top quark Yukawa coupling h_t to its quasi infrared fixed point value h_f [4]-[7], giving definite predictions in the limit $h_t \rightarrow h_f$ [8].

In minimal supergravity grand unified models, the soft supersymmetry breaking mass parameters proceed from common given values at the high energy scale. In addition, to assure a proper breakdown of the electroweak symmetry, one needs to impose conditions on the low energy mass parameters appearing in the scalar potential. This yields interesting correlations among the free high energy mass parameters of the theory, which then translate into interesting predictions for the Supersymmetric (SUSY) spectrum [8]-[15].

In the above, we have emphasized the infrared fixed point structure, which determines the value of the top quark mass as a function of $\tan\beta$. There is a small dependence of the infrared fixed point prediction on the supersymmetric spectrum, which, however, comes mainly through the dependence on the spectrum of the running of the strong guage coupling. Moreover, considering the MSSM with unification of gauge couplings at a grand unification scale M_{GUT} [16], the value of the strong gauge coupling is determined as a function of the electroweak gauge couplings while its dependence on the SUSY spectrum can be characterized by a single effective threshold scale T_{SUSY} [17]-[18]. Thus, the stronger dependence of the infrared fixed point prediction on the SUSY spectrum can be parametrized through T_{SUSY}. (There is also an independent effect coming from supersymmetric threshold corrections to the Yukawa coupling, which, for supersymmetric particle masses smaller or of the order of 1 TeV, may change the top quark mass predictions in a few GEV, but without changing the physical picture [19]).

The infrared fixed point structure of the top quark mass is interesting by itself, due to the many interesting properties associated with its behaviour As we shall show below, it gives a highly predictive framework for the Higgs and supersymmetric particle spectrum. Moreover, it has been recently observed in the literature that the condition of bottom-tau Yukawa coupling unification in minimal supersymmetric grand unified theories requires large values of the top quark Yukawa coupling at the unification scale [17]-[18], [20]-[23]. Most appealing, exactly for values of the guage couplings compatible with recent predictions from LEP and for the experimentally allowed values of the bottom mass, the conditions of gauge and bottom–tau Yukawa coupling unification predict values of the top quark mass within 10% of its infrared fixed point results *[17, 24]*.

In this talk we shall consider approximate analytical solutions to the one loop RG equations of the low energy parameters, showing their dependence on the high energy soft supersymmetry breaking mass parameters and the top quark Yukawa coupling and analyzing the implications of the infrared fixed point solution. We shall then incorporate the radiative electroweak symmetry breaking condition, to derive approximate analytical correlations among the free, independent high energy parameters of the theory. The analytical results are extremely useful in understanding the properties derived from the full numerical study, in which a two loop RG evolution of the gauge and Yukawa couplings is considered. In the numerical analysis the evolution of the Higgs and supesymmetric mass parameters are considered at the one loop level, and the one loop radiative corrections to the Higgs quartic couplings are taken into account. We shall then concentrate on the infrared fixed point predictions for the Higgs and SUSY spectrum as a function of given values for the top quark mass. We shall also present an analysis of the results obtained in the context of gauge and bottom-tau Yukawa coupling unification, to show the proximity of the top quark mass predictions obtained in this framework to the infrared fixed point top quark mass values as a function of $\tan\beta$. We summarize our results in the last section.

2   Infrared Fixed Point and the Evolution of the Mass Parameters

In the Minimal Supersymmetric Standard Model, with unification of gauge couplings at some high energy scale $M_{GUT} \simeq 10^{16}$ GeV, the infrared fixed point structure of the top quark Yukawa coupling may be easily analyzed, in the low and moderate $\tan\beta$ regime, $1 \le \tan\beta < 10$, considering its analytical one loop RG solution. As we said before, for such values of $\tan\beta$ the effects of the bottom and tau Yukawa couplings are negligible. For large values of $\tan\beta$, instead, the bottom Yukawa coupling becomes large and, in general, a numerical study of the coupled equations for the couplings becomes necessary even at the one loop level. There are, however, particular cases for which, for sizeable effects from the bottom and top Yukawa couplings, approximate analytical expressions may still be obtained.

In terms of $Y_t = h_t^2/4\pi$, the one loop solution in the small and moderate $\tan\beta$ region reads, *[4, 6]*:

$$Y_t(t) = \frac{2\pi Y_t(0)E(t)}{2\pi + 3Y_t(0)F(t)},\tag{2}$$

with E and F being functions of the gauge couplings,

$$E = (1 + \beta_3 t)^{16/3b_3}(1 + \beta_2 t)^{3/b_2}(1 + \beta_1 t)^{13/9b_1}, \qquad F = \int_0^t E(t')dt', \tag{3}$$

where $\beta_i = \alpha_i(0)b_i/4\pi$, b_i is the beta function coefficient of the gauge coupling α_i and $t = 2\log(M_{GUT}/Q)^2$. As we mentioned above, the fixed point solution,

---

[2]The corresponding solution for the bottom and tau Yukawa couplings in this regime

$h_f(t)$, is obtained for values of the top quark Yukawa coupling that become large at the grand unification scale, that is, approximately,

$$Y_f(t) \simeq \frac{2\pi E(t)}{3F(t)},\tag{4}$$

where $Y_f = h_f^2/4\pi$. For values of the grand unification scale $M_{GUT} \simeq 10^{16}$ GeV, the fixed point value, Eq. (4), is given by $Y_f \simeq (8/9)\alpha_3(M_Z)$. Indeed, since $F(Q = M_Z) \simeq 300$, the infrared fixed point solution is rapidly reached for a wide range of values of $Y_t(0) \simeq 0.1 - 1$. This behaviour is shown in Fig. 2.1, in which the value of the running top quark mass, $m_t(t) = h_t(t)v_2 = h_t(t)v \sin\beta$, with $v^2 = v_1^2 + v_2^2$, is plotted as a function of the energy scale, for a moderate value of $\tan\beta = 5$. For a wide range of high energy values, the value of $h_t(m_t)$ tends to h_f implying that the running top quark mass tends to its infrared fixed point value,

$$m_t^{IR}(t) = h_f(t) \, v \, \sin\beta = m_t^{IRmax.}(t) \, \sin\beta\tag{5}$$

where for $\alpha_3(M_Z) = 0.11 - 0.13$, $m_t^{IRmax.}$ is approximately given by

$$m_t^{IRmax.}(M_t) \simeq 196GeV[1 + 2(\alpha_3(M_Z) - 0.12)].\tag{6}$$

One should remember that there is a significant quantitative difference between the running top quark mass, and the physical top quark mass M_t, defined as the location of the pole in its two point function. The main source of difference comes from the QCD corrections, which at the two loop level are given by

$$M_t = m_t(M_t) \left[1 + 4\alpha_3(M_t)/3\pi + 11(\alpha_3(M_t)/\pi)^2\right].\tag{7}$$

In Fig. 2.1 we present the result of a two loop RG analysis, showing the stability of the infrared fixed point under higher order loop contributions.

Moreover, using Eq. (4) it follows,

$$\frac{6Y_t(0)F(t)}{4\pi} = \frac{Y_t(t)/Y_f(t)}{1 - Y_t(t)/Y_f(t)},\tag{8}$$

with Y_t/Y_f the ratio of Yukawa couplings at low energies. The value of the top quark Yukawa coupling at M_{GUT}, $Y_t(0)$, appears in the RG solutions of the soft SUSY breaking parameters, and the above equation permits to express it as a function of the gauge couplings (through F) and the ratio Y_t/Y_f.

A similar analytical study can be done for the large $\tan\beta$ regime when the bottom and top Yukawa couplings are equal at the unification scale. Neglecting in a first approximation the effects of the tau Yukawa coupling and identifying the right-bottom and right-top hypercharges, the solution for $Y = Y_t \simeq Y_b$ reads,

$$Y(t) = \frac{4\pi Y(0)E(t)}{4\pi + 7Y(0)F(t)}\tag{9}$$

---

are: $Y_b(t) = Y_b(0)E(t)'/[1 + (3/2\pi)Y_t(0)F(t)]^{1/6}$ and $Y_\tau(t) = Y_\tau(0)\tilde{E}(t)$, where E' may be obtained from E by changing the exponent coefficient 13/9 by 7/9, and $\tilde{E}(t)$ can be obtained from $E(t)$ by changing the exponent coefficients 16/3, and 13/9 by 0 and 3, respectively. These expressions are useful only when b–τ Yukawa coupling unification is to be considered.

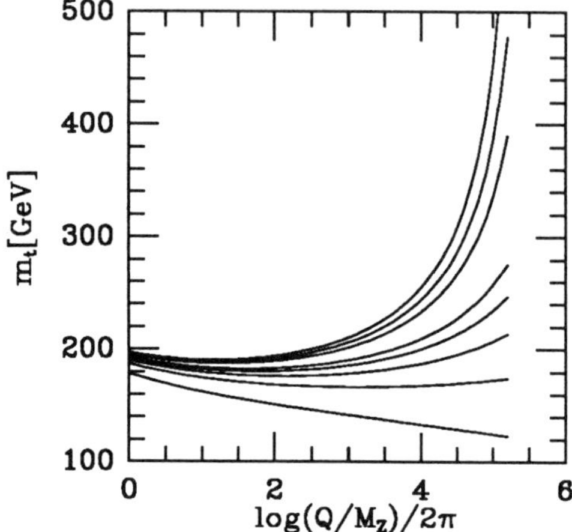

Figure 1. Running top quark Yukawa coupling evolution, normalized in order to get the running top quark mass at low energies, $h_t v_2$, for different boundary conditions at an energy scale $Q \simeq 10^{16}$ GeV.

Then, if the Yukawa coupling is large at the grand unification scale, at energies of the order of the top quark mass it will develop an infrared fixed point value approximately given by,

$$Y_f(t)^{(Y_t=Y_b)} \simeq \frac{4\pi E(t)}{7F(t)} \simeq \frac{6}{7}Y_f(t)^{(low\,\tan\beta)}. \tag{10}$$

Relaxing the unification condition of the bottom and top Yukawa couplings but still neglecting in a first approximation the effects of the tau Yukawa coupling and identifying the hypercharges, then, a general approximate analytical expression for Y_b and Y_t may be considered:

$$Y_{t,b}(t) = Y_{t,b}(0)E(t)/\left[1 + (3/2\pi)F(t)(Y_t(0) + Y_b(0))\right]^{1/6}$$
$$\left[1 + (3/2\pi)F(t)Y_{t,b}(0)\right]^{5/6}, \tag{11}$$

which goes to the correct limits for $Y_t \gg Y_b$ as well as for $Y_b \gg Y_t$, while for the case $Y_b \simeq Y_t$, gives the result for the top and bottom quark masses with an error of the order of 2 %. If both Yukawa couplings are large at the grand unification scale, unlike the two previous cases, their infrared fixed point expressions depend on the relative values of their boundary conditions at the high energy scale. In fact, the ratio of the top to bottom Yukawa couplings at the infrared fixed point depends on the ratio of their boundary conditions as follows,

$$\frac{Y_t^f}{Y_b^f} = \left(\frac{Y_t^f(0)}{Y_b^f(0)}\right)^{1/6}. \tag{12}$$

Using the above relation to replace the dependence of the general solutions, Eq. (11), on the boundary conditions of the Yukawa couplings, we obtain an infrared fixed point contour in the Y_t–Y_b plane,

$$\left[\left(Y_t^f\right)^6 + \left(Y_b^f\right)^6\right]^{1/6} = \frac{2\pi E}{3F} \tag{13}$$

In general, in the large $\tan\beta$ region the bottom quark Yukawa coupling becomes strong and plays an important role in the RG analysis. The are also possible large radiative corrections to the bottom mass coming from loops of supersymmetric particles, which are strongly dependent on the particular spectrum and are extremely important in the analysis if unification of bottom and tau Yukawa couplings is to be considered. Moreover, in some of the minimal models of grand unification, large $\tan\beta$ values are in conflict with proton decay constraints [25]. In the special case of tau-bottom-top Yukawa coupling unification the infrared fixed point solution for the top quark mass is not achievable unless a relaxation in the high energy boundary conditions of the mass parameters of the theory is arranged, and it is necessarily associated with a heavy supersymmetric spectrum. The large $\tan\beta$ regime will be analysed in detail in this workshop in the presentations of U. Sarid [26] and C. Wagner [27]. In the following we shall concentrate in the low and moderate $\tan\beta$ region, which involves interesting phenomenological implications.

We shall now consider that the breakdown of supersymmetry comes through the addition of all possible soft supersymmetry breaking terms. In the framework of minimal supergravity one considers universal soft supersymmetry breaking parameters at the grand unification scale. This includes common soft supersymmetry breaking mass terms m_0 and $M_{1/2}$ for the scalar and gaugino sectors of the theory, respectively, and a common value A_0 (B_0) for all trilinear (bilinear) couplings A_i (B) appearing in the full scalar potential which are proportional to the trilinear (bilinear) terms in the superpotential. In addition, the supersymmetric Higgs mass parameter μ appearing in the superpotential takes a value μ_0 at the grand unification scale M_{GUT}. Knowing the values of the mass parameters at the unification scale, their low energy values may be specified by their renormalization group evolution [4]-[7], which contains also a dependence on the gauge and Yukawa couplings. In the limit of small $\tan\beta$, $\tan\beta < 10$, the following approximate analytical solutions are obtained [8],

$$m_L^2 = m_0^2 + 0.52 M_{1/2}^2 \qquad\qquad m_E^2 = m_0^2 + 0.15 M_{1/2}^2$$

$$m_{Q(1,2)}^2 = m_0^2 + 7.2 M_{1/2}^2 \qquad\qquad m_{U(1,2)}^2 \simeq m_D^2 \simeq m_0^2 + 6.7 M_{1/2}^2$$

$$m_Q^2 = 7.2 M_{1/2}^2 + m_0^2 + \frac{\Delta m^2}{3}$$

$$m_U^2 = 6.7 M_{1/2}^2 + m_0^2 + 2\frac{\Delta m^2}{3} \tag{14}$$

where E, D and U are the right handed leptons, down-squarks and up-squarks, respectively, L and $Q = (\text{T B})^T$ are the lepton and top-bottom left handed doublets and m_η^2, with $\eta = E, D, U, L, Q$ are the corresponding soft supersymmetry breaking mass parameters. The subindices (1,2) are to distinguish the first and second generations from the third one, whose mass parameters receive the top quark Yukawa coupling contribution to their renormalization group evolution, singled out in the Δm^2 term.

$$\Delta m^2 = -\frac{3m_0^2}{2}\frac{Y_t}{Y_f} + 2.3A_0M_{1/2}\frac{Y_t}{Y_f}\left(1 - \frac{Y_t}{Y_f}\right)$$
$$- \frac{A_0^2}{2}\frac{Y_t}{Y_f}\left(1 - \frac{Y_t}{Y_f}\right) + M_{1/2}^2\left[-7\frac{Y_t}{Y_f} + 3\left(\frac{Y_t}{Y_f}\right)^2\right]. \tag{15}$$

For the Higgs sector, the mass parameters involved are

$$m_{H_1}^2 = m_0^2 + 0.52M_{1/2}^2 \qquad m_{H_2}^2 = m_{H_1}^2 + \Delta m^2 , \tag{16}$$

which are the soft supersymmetry breaking parts of the mass parameters m_1^2 and m_2^2 appearing in the Higgs scalar potential (see section 3). Moreover, the renormalization group evolution for the supersymmetric mass parameter μ reads,

$$\mu^2 = 2\mu_0^2\left(1 - \frac{Y_t}{Y_f}\right)^{1/2} , \tag{17}$$

while the running of the soft supersymmetry breaking bilinear and trilinear couplings gives,

$$B = B_0 - \frac{A_0}{2}\frac{Y_t}{Y_f} + M_{1/2}\left(1.2\frac{Y_t}{Y_f} - 0.6\right). \tag{18}$$

$$A = A_0\left(1 - \frac{Y_t}{Y_f}\right) - M_{1/2}\left(4.2 - 2.1\frac{Y_t}{Y_f}\right), \tag{19}$$

respectively. Eq. (17) shows that the RG evolution of μ, which is a supersymmetric preserving parameter, does not involve any dependence on the soft supersymmetry breaking parameters.

The coefficients characterizing the dependence of the mass parameters on the universal gaugino mass $M_{1/2}$ depend on the exact value of the gauge couplings. In the above, we have taken the values of the coefficients that are obtained for $\alpha_3(M_Z) \simeq 0.12$. The above analytical solutions are sufficiently accurate for the purpose of understanding the properties of the mass parameters in the limit $Y_t \to Y_f$. We shall then confront the results of our analytical study with those obtained from the numerical two loop analysis.

3   Mass Parameter Correlations from Radiative Electroweak Symmetry Breaking

The solutions for the mass parameters may be strongly constrained by experimental and theoretical restrictions. The experimental contraints come from the present lower bounds on the supersymmetric particle masses. Concerning the theoretical constraints, many of them impose bounds on the allowed space for the soft supersymmetry breaking parameters in model dependent ways to various degrees. The conditions of stability of the effective potential and a proper breaking of the $\mathrm{SU}(2)_L \times \mathrm{U}(1)_Y$ symmetry are, instead, basic necessary requirements.

The Higgs potential of the Minimal Supersymmetric Standard Model may be written as [3, 28]-[30]

$$V_{eff} = m_1^2 H_1^\dagger H_1 + m_2^2 H_2^\dagger H_2 - m_3^2 (H_1^T i\tau_2 H_2 + h.c.)$$
$$+ \frac{\lambda_1}{2} \left(H_1^\dagger H_1 \right)^2 + \frac{\lambda_2}{2} \left(H_2^\dagger H_2 \right)^2$$
$$+ \lambda_3 \left(H_1^\dagger H_1 \right) \left(H_2^\dagger H_2 \right) + \lambda_4 \left| H_2^\dagger i\tau_2 H_1^* \right|^2 , \tag{20}$$

with $m_i^2 = \mu^2 + m_{H_i}^2$, $i = 1, 2$, and $m_3^2 = B|\mu|$ and where at scales at which the theory is supersymmetric the running quartic couplings λ_j, with $j = 1 - 4$, must satisfy the following conditions:

$$\lambda_1 = \lambda_2 = \frac{g_1^2 + g_2^2}{4} = \frac{M_Z^2}{2\,v^2}, \qquad \lambda_3 = \frac{g_2^2 - g_1^2}{4}, \qquad \lambda_4 = -\frac{g_2^2}{2} = \frac{M_W^2}{v^2}. \tag{21}$$

Hence, in order to obtain the low energy values of the quartic couplings, they must be evolved using the appropriate renormalization group equations, as was explained in Refs. [28]-[31]. The mass parameters m_i^2, with $i = $ 1-3 must also be evolved in a consistent way and their RG equations may be found in the literature [4]-[6, 32, 33]. The minimization conditions $\partial V / \partial H_i|_{<H_i>=v_i} = 0$, which are necessary to impose the proper breakdown of the electroweak symmetry, read

$$\sin(2\beta) = \frac{2m_3^2}{m_A^2} \tag{22}$$

$$\tan^2 \beta = \frac{m_1^2 + \lambda_2 v^2 + (\lambda_1 - \lambda_2)\,v_1^2}{m_2^2 + \lambda_2 v^2}, \tag{23}$$

where m_A is the CP-odd Higgs mass,

$$m_A^2 = m_1^2 + m_2^2 + \lambda_1 v_1^2 + \lambda_2 v_2^2 + (\lambda_3 + \lambda_4)\,v^2. \tag{24}$$

Considering the one loop leading order contribution to the running of the quartic couplings which, in the limit of stop mass degeneracy transforms λ_2 into $\lambda_2 + \Delta\lambda_2$, with $\Delta\lambda_2 = (3/8\pi^2)h_t^4 \ln(m_{\tilde{t}}^2/m_t^2)$, the minimization condition Eq. (23), can be written as:

$$\tan^2 \beta = \frac{m_1^2 + M_Z^2/2}{m_2^2 + M_Z^2/2 + \Delta\lambda_2 v_2^2}. \tag{25}$$

Therefore, using Eq. (25) and considering the approximate analytical expressions for the mass parameters m_i, Eq. (16), the supersymmetric mass parameter μ is determined as a function of five parameters,

$$\mu^2 = \mathcal{F}(m_0, M_{1/2}, A_0, \tan\beta, Y_t/Y_f). \tag{26}$$

Furthermore, the ratio of the top quark Yukawa coupling to its infrared fixed point value may be expressed as a function of the top quark mass and the angle β,

$$\frac{Y_t}{Y_f} = \left(\frac{m_t}{m_t^{IRmax}}\right)^2 \frac{1}{\sin^2 \beta}, \tag{27}$$

where the exact value of m_t^{IRmax}, Eq. (6), depends on the value of the strong guage coupling considered, and for the experimentally allowed range it varies approximately between 190–200 GeV. Depending on the precise value of the running top quark mass m_t and $\tan\beta$, the above equation gives a measure of the proximity to the infrared fixed point solution.

The other minimization condition, Eq. (22), depends on the soft supersymmetry breaking parameter B and, hence, on its boundary condition B_0. Both minimization conditions put restrictions on the soft supersymmetry breaking parameters. However, B (and thus B_0) is not involved in the renormalization group evolution of the (s)particle masses, implying that Eq. (22) is not relevant in defining the range of possible mass values of the Higgs and supersymmetric particle spectra.

Considering the relation between the physical and the running top quark mass, Eq.(7), for a given value of the physical top quark mass, the running top quark mass is fixed and then Eq. (27) fixes the ratio Y_t/Y_f as a function of $\sin\beta$. Then, the correlation implied by the minimization condition, Eqs. (25) and (26), determines the Higgs and supersymmetric spectrum as a function of four parameters. However, if one is at the infrared fixed point, $Y_t \to Y_f$, the model becomes much more predictive. This is partially due to the strong correlation between the top quark mass and the value of $\tan\beta$, Eq.(5), which allows to reduce in one the number of free parameters. Moreover, there is an additional reduction in one in the number of effective free parameters, which follows from the infrared fixed point structure of the theory. Indeed, the expressions for the low energy parameters, Eqs. (14)-(19), show important properties of the solution when $Y_t \to Y_f$ [8]:

a) The term Δm^2 and hence the mass parameters $m_{H_2}^2$, m_Q^2 and m_U^2 become very weakly dependent on the supersymmetry breaking parameter A_0. In fact, the dependence on A_0 vanishes in the formal limit $Y_t \to Y_f$. The only relevant dependence on A_0 enters through the mass parameter m_3^2, that is to say, through B. This leads to property (b).

b) There is an effective reduction in the number of free independent soft supersymmetry breaking parameters. In fact, the dependence on B_0 and A_0 of the low energy solutions is effectively replaced by a dependence on the parameter

$$\delta = B_0 - \frac{A_0}{2}. \tag{28}$$

c) There is also a very interesting dependence of the low energy mass parameters on m_0. For example, the m_0 dependence of the combination $m_Q^2 + m_{H_2}^2$

vanishes in the formal limit $Y_t \to Y_f$. Moreover, the right stop mass m_U^2 becomes itself independent of m_0^2 in this limit, a property which is very important for the analysis of the bounds on the stop sector.

¿From properties a) and b), it follows that, at the infrared fixed point, the dependence of the Higgs and supersymmetric spectrum on the parameter A_0 is negligible. Indeed, considering only the one loop leading order radiative corrections to the quartic couplings, the minimization condition at the infrared fixed point reads,

$$\mu^2 + \frac{M_Z^2}{2} = \left[m_0^2 \left(1 + 0.5 \tan^2 \beta\right) + M_{1/2}^2 \left(0.5 + 3.5 \tan^2 \beta\right) \right.$$
$$\left. - \omega_t \, 0.5 \tan^2 \beta \right] \frac{1}{\tan^2 \beta - 1}, \tag{29}$$

where we define $\omega_t = 2\Delta\lambda_2 v_2^2$, which depends only logarithmically on m_0 and $M_{1/2}$. In the limit $\tan\beta \to 1$, one has $\mu^2 \gg m_0^2, M_{1/2}^2$.

Hence, due to the independence of the spectrum on the parameter A_0 and the strong correlation of the top quark mass with $\tan\beta$, for a given top quark mass the Higgs and supersymmetric particle spectrum is completely determined as a function of only two parameters, m_0 and $M_{1/2}$. It is now possible to perform an scanning of all the possible values for m_0 and $M_{1/2}$, bounding the squark masses to be, for example, below 1 TeV, and the whole allowed parameter space for the Higgs and superparticle masses may be studied.

3.1   Color breaking minima

There are several conditions which need to be fulfilled to ensure the stability of the electroweak symmetry breaking vacuum. In particular, one should check that no charge or color breaking minima are induced at low energies. A well known condition for the absence of color breaking minima is given by the relation [34]

$$A_t^2 \le 3(m_Q^2 + m_U^2 + m_{H_2}^2) + 3\mu^2. \tag{30}$$

At the fixed point, however, since $A_t \simeq -2.1 M_{1/2}$ and $m_Q^2 + m_U^2 + m_{H_2}^2 \simeq 6 M_{1/2}^2$, this relation is trivially fulfilled (see also Ref. [35]).

For values of $\tan\beta$ close to one, large values of μ are induced and a more appropriate relation is obtained by looking for possible color breaking minima in the direction $< H_2 > \simeq < H_1 >$ and $< Q > \simeq < U >$. The requirement of stability of the physically acceptable vacuum implies the following sufficient condition

$$(A_t - \mu)^2 \le 2 \left(m_Q^2 + m_U^2\right) + \tilde{m}_{12}^2 \tag{31}$$

where $\tilde{m}_{12}^2 = \left(m_1^2 + m_2^2\right)(\tan\beta - 1)^2/(\tan\beta^2 + 1)$.

If Eq.(31) is not fulfilled, a second sufficient condition is given by

$$\left[(A_t - \mu)^2 - 2 \left(m_Q^2 + m_U^2\right) - \tilde{m}_{12}^2 \right]^2 \le 8 \left(m_Q^2 + m_U^2\right) \tilde{m}_{12}^2. \tag{32}$$

The above relations, Eqs. (31) and (32) are sufficient conditions since they assure that a color breaking minima lower than the trivial minima does not develop in the theory. If the above conditions are violated, a necessary condition to avoid the existence of a color breaking minima lower than the physically acceptable one is given by

$$V_{col} \geq V_{ph} \tag{33}$$

with

$$V_{col} = \frac{(A_t - \mu)^2 \alpha_{min}^2}{h_t^2 (2\alpha_{min}^2 + 1)^3} \left[(m_Q^2 + m_U^2) - 2\tilde{m}_{12}^2 \alpha_{min}^4 \right], \tag{34}$$

$$V_{ph} = -\frac{M_Z^4}{2(g_1^2 + g_2^2)} \cos^2(2\beta), \tag{35}$$

and

$$\alpha_{min}^2 = \left[(A_t - \mu)^2 - 2(m_Q^2 + m_U^2) - \tilde{m}_{12}^2 \right] / (4\tilde{m}_{12}^2). \tag{36}$$

The above conditions impose strong constraints on the possible fixed point solutions, particularly for low values of $\tan\beta$, for which the value of μ rapidly increases. Therefore, they have implications in determining the Higgs and supersymmetric particle mass predictions of the model.

4 Higgs and Supersymmetric Particle Spectrum

The infrared fixed point solution yields, as we said before, a quite predictive framework for the Higgs and Supersymmetric spectrum of the MSSM. Indeed most of the properties of the masses may be understood analytically through its dependence on the mass parameters m_0 and $M_{1/2}$ which govern their behaviour. Experimental bounds on the sparticle masses as well as theoretical restrictions to avoid inconsistencies in the predicted spectrum are useful in constraining the allowed values of the defining parameters m_0 and $M_{1/2}$. As a matter of fact, the only two sectors of the theory which one should be particularly careful about, at the infrared fixed point, are those related to the Higgs and the stop.

Let us first summarize the results for the relevant low energy mass parameters at the fixed point solution:

$$m_{H_2}^2 \simeq -0.5 m_0^2 - 3.5 M_{1/2}^2 \qquad\qquad m_{H_1}^2 \simeq m_0^2 + 0.5 M_{1/2}^2$$

$$m_Q^2 \simeq 0.5 m_0^2 + 6 M_{1/2}^2 \qquad\qquad m_U^2 \simeq 4 M_{1/2}^2$$

$$A_t \simeq -2.1 M_{1/2}$$

$$\mu^2 \simeq \left[m_0^2 \left(1 + 0.5 \tan^2\beta\right) + M_{1/2}^2 \left(0.5 + 3.5 \tan^2\beta\right) \right] \frac{1}{\tan^2\beta - 1} \tag{37}$$

As we mentioned before, since the whole spectrum may be given as a function of $\tan\beta$, μ and the soft supersymmetry breaking parameters, for low values of $\tan\beta$ and for a given value of the top quark mass, it is completely determined as a function of two free independent parameters, which we take to be m_0 and $M_{1/2}$.

4.1 Stop Sector

We shall first analyse the stop sector, considering the stop mass matrix given by,

$$M_{\tilde{t}}^2 = \begin{bmatrix} m_Q^2 + m_t^2 + D_{t_L} & m_t(A_t - \mu/\tan\beta) \\ m_t(A_t - \mu/\tan\beta) & m_U^2 + m_t^2 + D_{t_R} \end{bmatrix} \tag{38}$$

where $D_{t_L} \simeq -0.35 M_Z^2 |cos2\beta|$ and $D_{t_R} \simeq -0.15 M_Z^2 |cos2\beta|$ are the D-term contributions to the left and right handed stops, respectively. The above mass matrix, after diagonalization, leads to the two stop mass eigenvalues, $m_{\tilde{t}_1}$ and $m_{\tilde{t}_2}$. To avoid a tachyon in the theory, it is necessary to require $\det M_{\tilde{t}}^2 \geq 0$. At the infrared fixed point, the values of the parameters involved in the mass matrix are given in Eq. (37). Already from an analytical study it is possible to conclude that, for values of $\tan\beta$ close to one, the off-diagonal term contribution will be enhanced due to the large values of μ associated with such low values of $\tan\beta$ and, consequently, the mixing may be sufficiently large to yield a tachyonic solution. Thus, depending on the hierarchy between m_0 and $M_{1/2}$ and on the sign of μ, important constraints on the parameter space may be obtained. For example, for $\tan\beta = 1.2$, which implies $M_t \simeq 160$ GeV and for which the value of the supersymmetric mass parameter $\mu^2 \simeq 4m_0^2 + 12M_{1/2}^2$, it is straighforward to show that, if one considers the regime $M_{1/2}^2 \ll m_0^2$, then for both signs of μ a tachyon state will develop unless $M_{1/2} \geq 0.9\, m_t$. If, instead, one considers the regime $M_{1/2}^2 \gg m_0^2$, then for $\mu > 0$ it follows that in order to avoid a tachyon it is necessary to require $M_{1/2} \geq 1.2\, m_t$. For negative values of μ, since there is a partial cancellation of the off-diagonal term which suppresses the mixing, no tachyonic solution may develop and, hence, no constraint is derived. However, as we shall show below, restrictions coming from the Higgs sector will constrain this region of parameter space as well. Observe that for these low values of $\tan\beta$, the necessary and sufficient conditions to avoid a color breaking minima, Eqs. (31), (32) and (33), put strong restrictions on the solutions with large left–right stop mixing.

For slightly larger values of $\tan\beta \simeq 1.8$, which correspond to much larger values of the top quark mass, $M_t \simeq 180 GeV$, the value of $\mu \simeq 1.2m_0^2 + 5.3M_{1/2}^2$ is sufficiently small so that, helped by the factor $1/\tan\beta$ there is no possibility for a tachyon to develop in this case and, hence, no constraints on $M_{1/2}$ are obtained. Of course, this result holds for larger values of $\tan\beta$ as well. Is is interesting to notice that, although there is no necessity to be concerned about tachyons for values of $\tan\beta \simeq 1.8$, it is still possible to have light stops, close to the experimental bound, if the value of $M_{1/2} \leq 100 GeV$. Fig. 4.2 gives the value of the lightest stop quark mass as a function of the gluino mass, $M_{\tilde{g}} \simeq 3M_{1/2}$, while considering various values of $\tan\beta$ and M_t close to the infrared fixed point solution, and it shows the results obtained from the full numerical study [8]. The dots in Fig. 4.2 denote solutions forbidden due to experimental bounds. In particular, for $\tan\beta = 1.2$ the restrictions on $M_{\tilde{g}}$ come through bounds on $M_{1/2}$ derived from the fulfillment of the lower bounds on the lightest CP-even Higgs mass m_h (see below). For larger values of $\tan\beta$ a light stop is not possible any longer. Indeed, for $\tan\beta \geq 5$ the stop mass becomes heavier than the top mass, $M_t \simeq 200$ GeV, for most of the parameter space and the experimental bound on the gluino mass, taken as $M_{\tilde{g}} > 120$ GeV becomes an important constraint for the solutions.

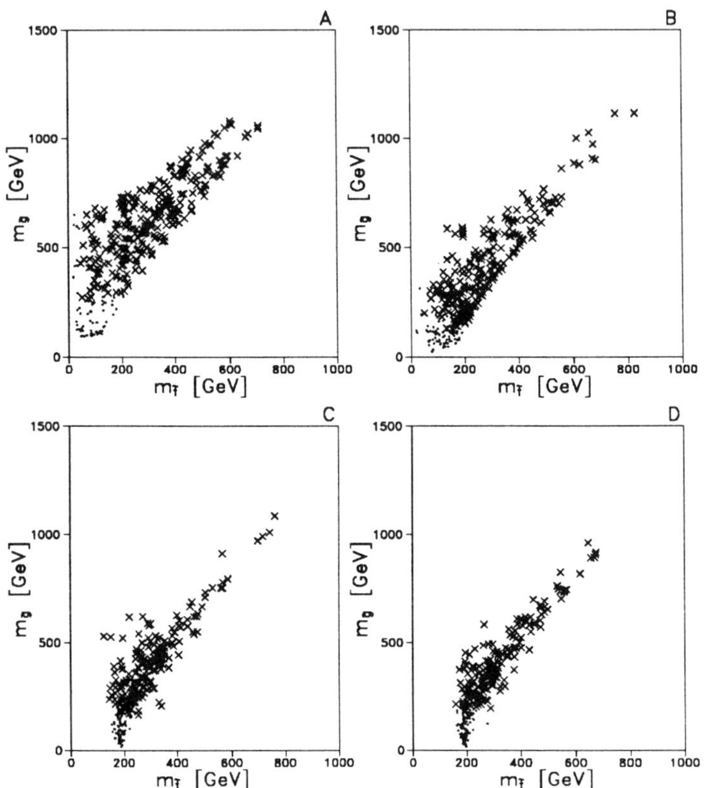

Figure 2. Running gluino mass $M_{\tilde{g}}$ as a function of the lightest stop mass $m_{\tilde{t}}$, for different values of the top quark mass at the infrared fixed point solution: A)$M_t = 160$ GeV, $\tan\beta = 1.2$, B)$M_t = 180$ GeV, $\tan\beta = 1.8$, C) $M_t = 202$ GeV, $\tan\beta = 5$, D) $M_t = 205$ GeV, $\tan\beta = 10$. Crosses (dots) denote solutions allowed (excluded) experimentally.

4.2 Higgs Spectrum

Other important features of the spectrum at the infrared fixed point are associated with the Higgs sector. The Higgs spectrum is composed by three neutral scalar states –two CP-even, h and H, and one CP-odd, A, and two charged scalar states $H^{\pm}$. Considering the one loop leading order corrections to the running of the quartic couplings –those proportional to m_t^4– and neglecting in a first approximation the squark mixing, the masses of the scalar states are given by,

$$m_{h,H}^2 = \frac{1}{2}\left[m_A^2 + M_Z^2 + \omega_t \right.$$

$$\left. \pm\sqrt{\left(m_A^2 + M_Z^2\right)^2 + \omega_t^2 - 4m_A^2 M_Z^2 \cos^2(2\beta) + 2\omega_t \cos(2\beta)\left(m_A^2 - M_Z^2\right)}\right] \tag{39}$$

$$m_A^2 + M_Z^2 = m_1^2 + m_2^2 + M_Z^2 + \frac{\omega_t}{2}$$

$$= \left[\frac{3}{2} m_0^2 + 4 M_{1/2}^2 - \frac{\omega_t}{2} \right] \frac{(1 + \tan^2 \beta)}{(\tan^2 \beta - 1)} \tag{40}$$

$$m_{H^\pm}^2 = m_A^2 + M_W^2 . \tag{41}$$

In the above, we have omitted the one loop contributions proportional to $\omega_t / m_{\tilde{t}}^2$, since for $\tan \beta > 1$ they are negligible with respect to the other contributions. Indeed, the radiative corrections to the Higgs mass become only relevant for large values of the heaviest stop mass, $m_{\tilde{t}}^2 \gg M_Z^2$. To obtain these stop mass values we need moderate values of the soft supersymmetry breaking parameters, which for low values of $\tan \beta \leq 2$ induce large values of the CP-odd mass, $m_A^2 \gg M_Z^2$. In this case, the ω_t contribution is of order M_Z^2 and, hence, it gives a decisive contribution to m_h in Eq. (39) but does not give a relevant contribution to m_A, Eq. (40). If $m_A^2 \gg M_Z^2$, then m_H and $m_{H^\pm}$ are also large, of the order of the CP-odd mass. If, instead, $m_A^2 = \mathcal{O}(M_Z^2)$, then m_0^2 and $M_{1/2}^2$ are also small and, due to the logarithmic dependence of ω_t on these two parameters, its contribution is small, both in Eq. (39) as in Eq. (40).

For the lightest CP-even mass a finite upper bound on its value, m_h^{max}, is reached in the limit of very large values of the CP-odd mass, $m_A^2 \gg M_Z^2$,

$$(m_h^{max})^2 = M_Z^2 cos^2(2\beta) + \frac{3}{4\pi^2} \frac{m_t^4}{v^2} \left[\ln \left(\frac{m_{\tilde{t}_1} m_{\tilde{t}_2}}{m_t^2} \right) + \Delta_{\theta_{\tilde{t}}} \right] \tag{42}$$

In the above, we have now considered the expression in the case of non-negligible squark mixing [35]-[37]. $\Delta_{\theta_{\tilde{t}}}$ is a function which depends on the left-right mixing angle in the stop sector and it vanishes in the limit in which the two mass eigenstates are equal, $m_{\tilde{t}_1} = m_{\tilde{t}_2}$. ¿From Eq. (40) it follows that, for lower values of $\tan \beta$, the value of the CP-odd eigenstate mass is enhanced. This means that in such case the expression for the lightest Higgs mass is given by Eq. (42), and it is independent of the exact value of the CP-odd mass. The fact that values of $\tan \beta$ close to one yield larger values of m_A, implies as well that the charged Higgs and the heaviest CP-even Higgs will become heavier in such regime.

Furthermore, the infrared fixed point solution for the top quark mass has explicit important implications for the lightest Higgs mass. For a given value of the physical top quark mass, the infrared fixed point solution is associated with the minimun value of $\tan \beta$ compatible with the perturbative consistency of the theory. For values of $\tan \beta \geq 1$, lower values of $\tan \beta$ correspond to lower values of the tree level lightest CP-even mass, $m_h^{tree} = M_Z |cos 2\beta|$. Therefore, the infrared fixed point solution minimizes the tree level contribution and after the inclusion of the radiative corrections it still gives the lowest possible value of m_h for a fixed value of M_t [8], [23, 38]. This property is very appealing, in particular, in relation to future Higgs searches at LEP2, as we shall show explicitly below.

Due to the specific dependence of the lightest Higgs mass with $\tan \beta$, it occurs that, for values of $\tan \beta$ close to one, restrictions on the allowed high energy parameter space and, hence, on the spectrum, may be derived by the requirement that m_h is above its experimental bound. Indeed, if $\tan\beta \simeq 1.2$ ($|\cos 2\beta| \simeq 0.2$), the tree level value is very small and, in order to satisfy the experimental constraint on m_h, it is necessary to impose a bound on the radiative correction contribution. One may choose to push m_0 to large values, but this will induce a tachyon in the stop spectrum unless $M_{1/2} \geq m_t$. If, instead, one keeps moderate values of m_0, values of $M_{1/2} > 100 GeV$ are needed to generate the appropriate radiative corrections. Summarizing, for values of $\tan\beta$ close to one, $M_t \leq 160$ GeV ($\tan\beta \leq 1.2$), to avoid conflicts in the Higgs and stop sectors one needs

$$
\begin{aligned}
M_{1/2} &> m_t & if \quad \mu &> 0 \\
M_{1/2} &\geq 100 GeV & if \quad \mu &< 0.
\end{aligned}
\tag{43}
$$

For larger values of $\tan\beta \simeq 1.8$, the Higgs sector constraints are still important, although they do not lead to a lower bound on $M_{1/2}$ independent of the experimental bounds on the gaugino sectors. In this case, one has $m_A^2 \simeq 8M_{1/2}^2 + 3m_0^2$ and for values of the defining parameters consistent with the experimental constraints in the gaugino and slepton sectors, the CP-odd mass is still sufficiently large, so that the lightest CP-even mass is given by its upper bound, Eq. (42). Then, $m_h^{tree} \simeq 50$ GeV and if $m_0 \geq m_t \simeq 170$ GeV, no bounds on $M_{1/2}$ are obtained from the experimental constraint on m_h. In Fig. 4.3 we show the m_A–m_h plane for various values of $\tan\beta$ and M_t extremely close to the infrared fixed point, as derived from the full numerical study. The results from Fig. 4.3 are in perfect agreement with the behaviour described above. Moreover, it follows that for values of $M_t \leq 180$ GeV the lightest Higgs mass is expected to be in the 50 - 100 GeV range, while for larger values of $M_t \simeq 200$ GeV it is mostly larger than 100 GeV with a range $m_h \simeq 125 \pm 25$ GeV. In general, for larger values of $\tan\beta$ the tree level value becomes larger and the experimental bounds on gauginos and gluinos also contribute to push the lightest Higgs mass to larger values.

All the above analysis is done under the assumption of being at the infrared fixed point of the top quark mass. However, it is also interesting to observe how the predictions for the lightest Higgs mass are altered if one considers a departure from the infrared fixed point solution. As we said before, in this case a fixed value of the top quark mass may be considered and still the value of $\tan\beta$ may vary implying in each case a different degree of departure from the infrared fixed point solution. In Fig. 4.4 we show the value of the lightest Higgs mass as a function of $\tan\beta$, performing an scanning over all the possible values of m_0 and $M_{1/2}$ for a top quark mass $M_t = 175$ GeV, so that the squark masses have an upper bound of 1 TeV. (In this plot we have considered the Higgs mass value obtained from the one loop effective potential computation, Eq. (42). The upper bound obtained within this approach differs in approximately 5 GeV with the one obtained through the RG procedure in which the squark mixing is directly considered through the matching conditions for the quartic couplings, as done in Fig. 4.3. These results show the degree of uncertainty in the Higgs mass computation [35].)

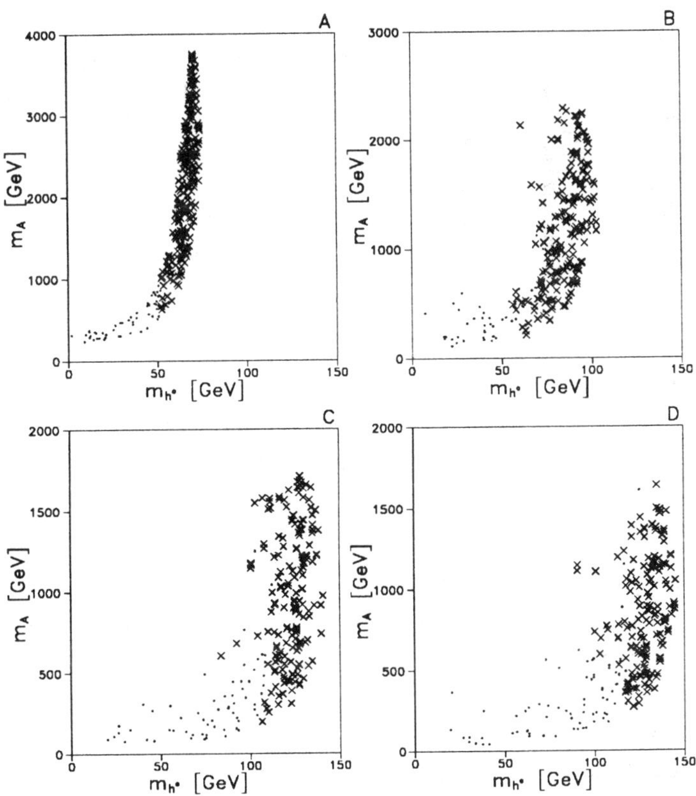

Figure 3. The same as Fig. 4.2 but for the CP-odd Higgs mass m_A vs. the lightest CP-even Higgs mass m_h.

For each value of the top quark mass , the lowest possible value of $\tan \beta$ is associated with the infrared fixed point value. The larger values of $\tan \beta$, for which the solutions are increasingly away from the infrared fixed point, show larger values for the lightest Higgs mass, which, however, become stagnant for values of $\tan \beta$ close to 10. Away from the infrared fixed point solution an scanning over A_0 is also done. The most remarkable feature is that, for solutions which depart from the infrared fixed point, not only the upper bound on m_h becomes larger, but the whole set of solutions lie in a region of the parameter space which renders a lightest Higgs, which is predominantly out of the reach of LEP2. On the contrary, the predictions from the infrared fixed point solution are very appealing in this respect, since for the presently, experimentaly favoured values of $M_t = 174 \pm 16$ GeV *[40]* there are good chances that the lightest Higgs may be at the reach of LEP2.

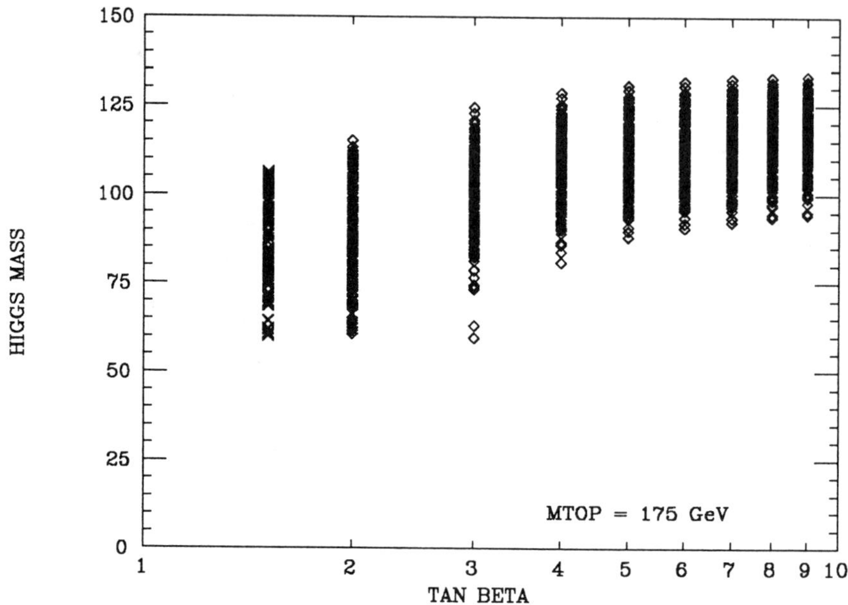

Figure 4. Lightest CP-even Higgs mass for different values of $\tan\beta$. The lowest value of $\tan\beta$ (crosses) corresponds to the infrared fixed point solutions for the considered value of the top quark mass, $M_t = 175$ GeV.

4.3 Chargino and neutralino spectrum

From the restrictions on $M_{1/2}$ which follow from the analysis of the Higgs and stop sectors at the infrared fixed point of the top quark mass, a very interesting result can be obtained. Indeed, due to the large values of the mass parameter μ in this framework, there is small mixing in the chargino and neutralino sectors. Hence, to a good approximation the lightest chargino mass and the lightest and next to lightest neutralino masses are given by $m_{\tilde{\chi}_1^\pm} \simeq m_{\tilde{\chi}_2^0} \simeq 2m_{\tilde{\chi}_1^0} \simeq 0.8 M_{1/2}$. For values of the top quark mass $M_t \leq 160$ GeV, light charginos, with masses very close to its present experimental bounds, are forbidden due to the lower bounds on the gaugino masses, Eq. (43). Quite generally, we obtain $m_{\tilde{\chi}_1^\pm} > 70$ GeV in this case. On the contrary, due to the large mixing in the stop sector, small values of the lightest stop mass, $m_{\tilde{t}_1} \leq 150$ GeV, may be easily achieved (see Fig. 4.2). For values of $M_t \geq 185$ GeV, the situation is basically reversed. As can be observed in Fig.2, light stops are harder to obtain due to the reduced mixing, while light charginos are possible, since there is no constraint on $M_{1/2}$ neither from the stop nor from the Higgs sector. Most interesting, just for the phenomenologically preferred region, 165 GeV $\leq M_t \leq 185$ GeV, both the charginos and the stops may become light.

Fig. 4.5 shows the correlation between the lightest chargino mass and the lightest stop mass, for the infrared fixed point solution, for a value of the top

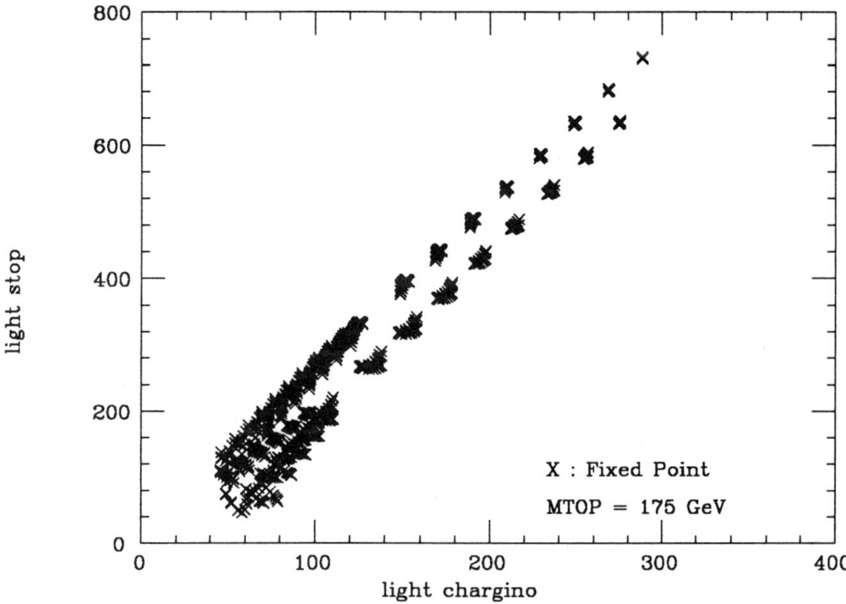

Figure 5. Running lightest stop mass as a function of the lightest chargino mass for a top quark mass $M_t = 175 GeV$ at the infrared fixed point solution.

quark mass $M_t = 175$ GeV. Light stops and charginos are very interesting, both for direct experimental searches as well as for indirect searches through deviations from the Standard Model predictions for the leptonic and hadronic variables measured at LEP.

5 Unification of Couplings and the Infrared Fixed Point

In the above, we have assumed the infrared fixed point solution for the top quark mass, and we have analysed its implications in the MSSM under the general requirement of unification of the gauge couplings at some high energy scale $M_{GUT} = \mathcal{O}(10^{16})$ GeV and considering universal conditions for the soft supersymmetry breaking parameters at the grand unification scale. It is now interesting to investigate which physical scenarios may predict the infrared fixed point solution for the top quark mass. One possibility would be the onset of non-perturbative physics at scales of the order of M_{GUT}, as it occurs for example in the supersymmetric extension of the so-called top condensate models [3]. In this case an analysis of the nonperturbative effects would be necessary before considering the precise unification conditions. Other possibility would be perturbative grand unification with the large values of the top quark Yukawa coupling at the unification scale necessary to induce its infrared fixed point behaviour, followed by the onset of non-perturbative physics for scales just above M_{GUT}. Moreover, the infrared fixed point solution also appears as a prediction in some interesting class of string theories, where the stability of the cosmological constant against corrections of order $M_{SUSY}^2 M_P^2$ is ensured [39].

Another option, and may be the most appealing case to treat, is to have a perturbative theory up to the Planck scale, but with large values of the top quark Yukawa coupling -close to its perturbative limit- and with the onset of new physics above the unification scale. In this context it is possible to consider grand unified models with an SU(5) or S0(10) symmetry, which include also unification of Yukawa couplings. In particular, as we are going to show, the unification of bottom and tau Yukawa couplings at the high energy scale yields a very interesting framework, which naturally renders large values of the top quark Yukawa coupling at M_{GUT}.

The condition of gauge coupling unification in itself gives predictions for the strong gauge coupling as a function of the electroweak gauge couplings. Considering a two loop RG analysis, it is necessary to include the supersymmetric threshold corrections at one loop, to take into account the decoupling of the different supersymmetric particles above M_Z. These supersymmetric threshold corrections may be parametrized in terms of a single effective scale T_{SUSY} [17] which, in the limit of common characteristic values for the masses of electroweak gauginos, $m_{\tilde{w}}$, gluinos, $m_{\tilde{g}}$, sleptons, $m_{\tilde{l}}$, squarks, $m_{\tilde{q}}$, Higgsinos, $m_{\tilde{H}}$, and heavy Higgs doublet, m_H, is given by [18],

$$T_{SUSY} = m_{\tilde{H}} \left(\frac{m_{\tilde{w}}}{m_{\tilde{g}}}\right)^{28/19} \left[\left(\frac{m_{\tilde{l}}}{m_{\tilde{q}}}\right)^{3/19} \left(\frac{m_H}{m_{\tilde{H}}}\right)^{3/19} \left(\frac{m_{\tilde{w}}}{m_{\tilde{H}}}\right)^{4/19}\right] . \tag{44}$$

The above equation shows that the main contribution to the supersymmetric threshold corrections comes from the gaugino and Higgsino sectors. For the models under study, in which a common gaugino mass $M_{1/2}$ at M_{GUT} is assumed and in the case of large values of μ for which the mixing in the gaugino-Higgsino sector is negligible, Eq. (44) reads,

$$T_{SUSY} \simeq |\mu| \left(\frac{\alpha_2(M_Z)}{\alpha_3(M_Z)}\right)^{3/2} \simeq \frac{|\mu|}{6} . \tag{45}$$

The strong gauge coupling at M_Z can then be computed as follows [17, 18],

$$\frac{1}{\alpha_3(M_Z)} = \frac{1}{\alpha_3^{SUSY}(M_Z)} + \frac{19}{28\pi} \ln\left(\frac{T_{SUSY}}{M_Z}\right) , \tag{46}$$

where $1/\alpha_3^{SUSY}(M_Z)$ would be the value of the strong gauge coupling coming from the two loop RG running if the theory were supersymmetric all the way down to M_Z. The effective scale T_{SUSY} is quite useful, since it permits to parametrize the uncertainty about the exact SUSY spectrum in a very general way. Indeed, to vary T_{SUSY} from 15 GeV to 1 TeV is equivalent to consider the supersymmetric threshold corrections due to variations in the sparticle masses within a very conservative wide range.

Performing a complete two loop numerical analysis, we have as inputs the value of $1/\alpha_{em} = 127.9$, which has only a logarithmic dependence on the top quark mass, and the value of $\sin^2\theta_W(M_Z)$, which is given by the electroweak parameters G_F, M_Z and α_{em} as a function of M_t (at the one loop level) by the formula,

$$\sin^2 \theta_W (M_Z) = 0.2324 - 10^{-7} \times GeV^{-2} \times \left(M_t^2 - (138 GeV)^2 \right) \pm 0.0003 . \quad (47)$$

Then, the unification condition implies the following numerical correlation [17, 24],

$$\sin^2 \theta_W (M_Z) = 0.2324 - 0.25 \times (\alpha_3 (M_Z) - 0.123) \pm 0.0025 . \quad (48)$$

The above central value corresponds to $T_{SUSY} = M_Z$ and the error ± 0.0025 is the estimated uncertainty in the prediction arising from possible supersymmetric threshold corrections and including also possible effects from threshold corrections at the unification scale and from higher dimensional operators, but assuming that they are not larger that the supersymmetric threshold corrections. Thus, considering the α_3–$\sin^2 \theta_W$ correlation predicted by the unification of the gauge couplings together with the $\sin^2 \theta_w$–M_t correlation obtained from the fit of the experimental data (both within their uncertainties) a band of correlated values between $\alpha_3 (M_Z)$ and M_t is obtained [24],

$$\alpha_3 (M_Z) = 0.123 + 4 \times 10^{-7} \times GeV^{-2} \times \left(M_t^2 - (138 GeV)^2 \right) \pm 0.01 . \quad (49)$$

As we shall show below, this correlation is crucial in the analysis of the top quark mass predictions coming from bottom-tau Yukawa coupling unification.

For given values of the gauge coupling the requirement of bottom and tau Yukawa coupling unification determines the value of the top quark mass as a function of $\tan \beta$, depending on the input value of the bottom quark mass. Indeed, the additional inputs with respect to the guage coupling unification analysis are the value of the tau mass, $M_\tau = 1.78$ GeV and the value of the bottom mass which involves a large uncertainty. In fact, the range of experimentally allowed values for the physical bottom quark mass is $M_b = 4.6 - 5.2$ GeV [41]. Moreover, a significant difference, of the order of 12 %, between the running bottom quark mass, which is the one directly related to the bottom Yukawa coupling, and the physical bottom mass arises due to QCD corrections. At the two loop level the relation is $M_b = m_b(M_b)[1 + (4/3\pi)\alpha_3(M_b) + 12.4(\alpha_3(M_b)/\pi)^2]$. Assuming bottom-tau Yukawa coupling unification, the exact range of values to be considered for the physical bottom mass as well as the appropriate treatment of the difference between the physical and running bottom quark masses have important consequences on the determination of the top quark Yukawa coupling. This is due to the fact that the bottom mass fixes the overall scale of the bottom quark Yukawa coupling. We shall return to the dependence of our predictions on the exact value of the bottom mass after presenting the numerical study. The other decisive variable in the bottom-tau Yukawa unification scheme is the exact value of the strong gauge coupling. Indeed, for relatively large values of the strong gauge coupling, $\alpha_3(M_Z) \geq 0.115$, large values of the top quark Yukawa coupling at the high energy scale are needed in order to partially contravene the strong renormalization effect of the strong gauge coupling in the running of the bottom quark Yukawa coupling. This is the reason why, for such values of the strong gauge coupling, the condition of bottom-tau unification yields predictions for the top quark mass close to its infrared fixed point values -the exact value of M_b defining the precise degree of closeness.

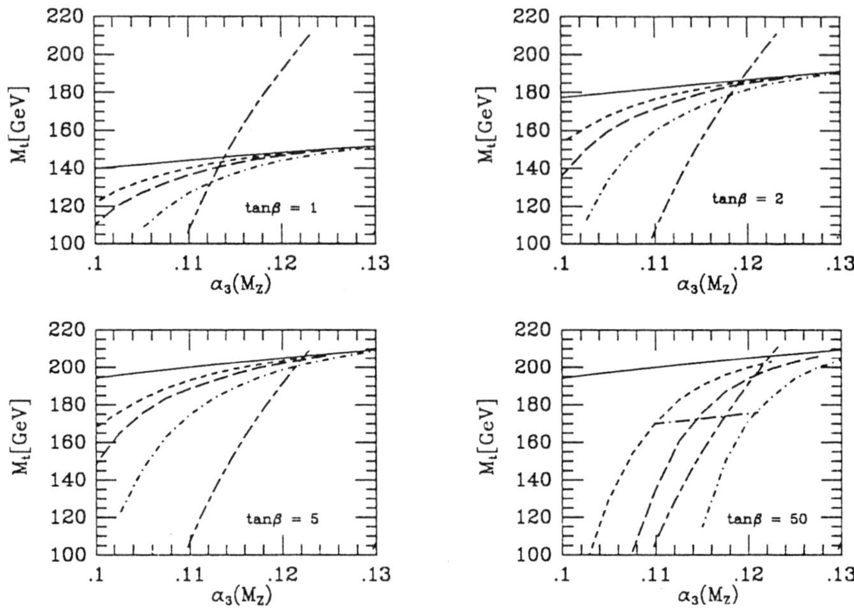

Figure 6. Top quark mass prediction as a function of the strong gauge coupling for the condition of unification of Yukawa couplings $h_b(M_G) = h_\tau(M_G)$, for different values of $\tan\beta$. The solid line shows the infrared fixed point solutions, while the dashed, long dashed and dot-dashed lines show the results for $M_b = 4.7, 4.9$ and 5.2, respectively. Here the unification scale M_G is defined as the scale at which the weak gauge couplings unify and the region to the right of the dash–long dashed line show the regime of $\alpha_3(M_Z)$ preferred by the gauge coupling unification condition.

For smaller values of the strong gauge coupling, $\alpha_3(M_Z) \leq 0.110$, which may still be compatible with its experimental bound, the necessity of a large top quark Yukawa coupling becomes weaker and as a result the infrared fixed point prediction for the top quark mass would not be a necessary outcome of the Yukawa coupling unification condition. However, for those smaller values of $\alpha_3(M_Z)$ the condition of gauge coupling unification is not consistent with the experimentally allowed values for $\sin^2\theta_W$. Therefore, large values of Y_t at M_{GUT} which imply the proximity to the infrared fixed point solution for M_t, are always a necessary outcome in the low and moderate $\tan\beta$ region, if gauge and bottom-tau Yukawa coupling unification are required [17, 24].

In Fig. 5.6 we show a detailed numerical study of the degree of proximity to the infrared fixed point solution implied by the unification conditions. The value of the top quark mass is plotted as a function of the strong gauge coupling for the exact infrared fixed point solution as well as for the case of bottom-tau Yukawa coupling unification for three different values of the bottom quark mass, which define the allowed domain of solutions compatible with the experimental predictions for M_b. Moreover, the condition of gauge coupling unification, Eq.

(49), implies that the region in $\alpha_3(M_Z)$ to the right of the dash-long dashed curve is the allowed one. Indeed, Eq. (49) defines a band whose upper bound is $\alpha_3(M_Z)^u \geq 0.13$ and, thus, it does not appear in the figure. The intersection of this region with the M_t–α_3 curves which follow from $h_b = h_\tau$ at M_{GUT}, for the range $M_b = 4.9 \pm 0.3$ GeV, determines the predicted values for M_t to be within 10% of its infrared fixed point values. The above is fulfilled for small and moderate values of $\tan\beta$. For larger values of $\tan\beta \geq 30$, the behaviour is drastically changed and, as we said, we shall not concentrate in such case here (see Refs. [26, 27] this proceedings). It is interesting to notice that low values of $\alpha_3(M_Z) \simeq 0.113$ are only possible for $M_t \simeq 140$ GeV ($\tan\beta \simeq 1$). For larger values of $\tan\beta$, the lower bound on the strong gauge coupling increases together with the top quark mass, which has then a stronger convergence to its infrared fixed point. For $M_t \simeq 180$ GeV ($\tan\beta \simeq 2$) a value $\alpha_3(M_Z) \geq 0.118$ is already necessary.

Concerning the relevance of the experimental bounds on the physical bottom quark mass, it is worth to mention that, if values of $M_b < 4.6$ GeV were allowed, it would induce a top quark Yukawa coupling which may become too large. For a consistent perturbative treatment of the theory, one requires $Y_t(M_{GUT}) \leq 1$, which implies that the two loop contribution to the renormalization group evolution of h_t is less than 30 % of the one loop one. As a matter of fact, observe that in Fig. 5.6 the curves for $M_b = 4.7$ GeV and $M_b = 4.9$ GeV do not continue up to $\alpha_3(M_Z) = 0.13$, since the top quark Yukawa coupling would then develop a Landau pole before reaching the unification scale. Larger values of the bottom mass, $M_b > 5.2$–5.3 GeV, would destroy the proximity to the infrared fixed point solution. Let us mention, however, that a recent analysis based on QCD sum rules, gives values for the perturbative bottom quark pole mass M_b close to the lower experimental bound considered above ($M_b \simeq 4.6$ GeV) [42].

Concerning possible threshold corrections which may affect the unification of both Yukawa couplings, it follows that a relaxation in the exact unification condition of the order of 10% for $M_b = 4.9$ GeV gives approximately the same behaviour as if one considers exact bottom–tau unification but with $M_b = 5.2$ GeV. Hence, values of $M_b \leq 4.9$ GeV secure the infrared fixed point behaviour even against possible supersymmetric threshold corrections to the Yukawa couplings. It is necessary to say that, if there are large threshold corrections at the grand unification scale, then all the above study can be significantly changed. These high energy threshold corrections depend, however, on the particular physics above the scale M_{GUT} and may not be computed in a general framework. The study of the Higgs and Supersymmetric spectrum performed in section 4 is only based on the infrared fixed point solution or in the proximity to it whithin the MSSM and has general validity. Then, depending on the exact, complete grand unified model under study, one has to compute the degree of proximity to the infrared fixed point solution. In this section we showed that provided the threshold corrections at M_{GUT} are not very large, a grand unified gauged theory with the extra ingredient of bottom-tau Yukawa coupling unification provides a framework in which the infrared fixed point solution of the top quark mass is realized. Most interesting is the fact that this result depends crucially on the values of the bottom mass and the electroweak parameters being exactly within their experimentally allowed range.

6 Conclusions

We have studied the properties of the MSSM with unification of gauge couplings and universal soft supersymmetry breaking parameters, for the case in which the top quark mass is close to its infrared quasi fixed point solution. To study the regime of the infrared quasi fixed point solution for m_t is of interest for various reasons.

i) It appears as a prediction in many interesting theoretical scenarios. In particular, it has been shown that for the presently allowed values of the bottom quark mass and the electroweak parameters, the conditions of gauge and bottom-tau Yukawa coupling unification imply a strong convergence of the top quark mass to its infrared fixed point value.

ii) It gives a very predictive framework in which, given the value of the top quark mass, the properties of the Higgs and supersymmetric spectrum in the minimal supergravity model are determined as a function of two high energy parameters, the common scalar mass m_0 and the common gaugino mass $M_{1/2}$. The implementation of the radiative electroweak symmetry breaking condition is a crucial ingredient for this result.

iii) For the range of top quark mass values suggested by the recent experimental measurements at CDF [40], $M_t = 174 \pm 16$ GeV, the value of $\tan \beta$, within its low and moderate regime, is bounded to be $1 < \tan \beta < 2.5$. For $M_t \leq 175$ GeV, the lightest Higgs mass is bounded to be $m_h \leq 105$ GeV, implying that there are good chances to observe it at the LEP2 experiment. Moreover, light charginos and light stops may appear in the spectrum. If they are present, they will have many interesting phenomenological implications.

iv) The correlations among the free parameters of the theory derived from the conditions of a proper breakdown of the electroweak symmetry may be very useful in probing models of dynamical supersymmetry breakdown, in which the soft supersymmetry breaking parameters are predicted.

Acknowledgements

Part of this work was done in collaboration with S. Pokorski and M. Olechowski, to whom we are grateful. We would like to thank G. Altarelli, C. Kounnas, I. Pavel and N. Polonsky for interesting discussions. This work is partially supported by the Worldlab.

References

[1]   C. T. Hill, Phys. Rev. D24 (1981) 691; C. T. Hill, C. N. Leung and S. Rao, Nucl. Phys. B262 (1985) 517.

[2]   L. Alvarez Gaume, J. Polchinski and M.B. Wise, Nucl. Phys. B221 (1983) 495; Bagger, S. Dimopoulos, E. Masso, Phys. Rev. Lett. 55 (1985) 920.

[3]   M. Carena, T.E. Clark, C.E.M. Wagner, W.A. Bardeen and K. Sasaki, Nucl. Phys. B369 (1992) 33.

[4]   L. Ibañez and C. Lopez, Nucl. Phys. B233 (1984) 511; L. Ibañez, C. Lopez and C. Muñoz, Nucl. Phys. B256 (1985) 218.

[5]   J.P. Derendinger and C.A. Savoy, Nucl. Phys. B237 (1984) 307.

[6]   A. Bouquet, J. Kaplan and C.A. Savoy, Nucl. Phys. B262 (1985) 299.

[7]   R. Barbieri and G. Giudice, Nucl. Phys. B306 (1988) 63.

[8]   M. Carena, M. Olechowski, S. Pokorski and C.E.M. Wagner, CERN preprint, CERN-TH.7060/93, Nucl. Phys. B (in press).

[9]   V. Barger, M.S. Berger and P. Ohmann, Univ. of Wisconsin - Madison preprint, MAD/PH/801, Nov. 1993.

[10]   B. Ananthanarayan, K.S. Babu and Q. Shafi, Bartol Report BA-94-03 (1994).

[11]   G.L. Kane, C. Kolda, L. Roszkowski and J.D. Wells, Michigan Report UM-TH-94-03 (1994).

[12]   G.K. Leontaris and N.D. Tracas, Ioannina Report IOA.303/94 (1994).

[13]   H. Baer et al. Florida Report FSU-HEP-940311 (1994).

[14]   W. de Boer, R. Ehret and D.I. Kazakov, Univ. of Karlsruhe report IEKP-KA/94-07, May 1994.

[15]   J. Gunion and H. Pois, UC Davis Report UCD-94-1 (1994).

[16]   S. Dimopoulos, S. Raby and F. Wilczek, Phys. Rev. D24 (1981) 1681; S. Dimopoulos and H. Georgi, Nucl. Phys. B 193 (1981) 150; L. Ibanez and G.G. Ross, Phys. Lett. 105B (1981) 439.

[17]   P. Langacker and N. Polonsky, Phys. Rev. D47 (1993) 4028; Phys. Rev. D49 (1994) 1454.

[18]   M. Carena, S. Pokorski and C.E.M. Wagner, Nucl. Phys. B406 (1993) 59.

[19]   See, for example, B.D. Wright, Univ. of Wisconsin report MAD/PH/812, March 1994.

[20]   H. Arason, D. J. Castaño, B. Keszthelyi, S. Mikaelian, E. J. Piard, P. Ramond and B. D. Wright, Phys. Rev. Lett. 67 (1991), 2933.

[21]   S. Kelley, J.L. Lopez and D.V. Nanopoulos, Phys. Lett. B278 (1992) 140.

[22]   S. Dimopoulos, L. Hall and S. Raby, Phys. Rev. Lett. 68 (1992) 1984, Phys. Rev. D45 (1992) 4192.

[23]   V. Barger, M.S. Berger and P. Ohmann, Phys. Rev. D 47 (1993) 1093; V. Barger, M.S. Berger, P. Ohmann and R.J.N. Phillips, Phys. Lett. B314 (1993) 351.

[24]   W. A. Bardeen, M. Carena, S. Pokorski and C. E. M. Wagner, Phys. Lett. B320 (1994) 110.

[25]   R. Arnowitt and P. Nath, Phys. Rev. Lett. 69 (1992) 1014; Phys. Lett. B287 (1992) 89; B289 (1992) 368.

[26]   U. Sarid, R. Rattazzi and L.J. Hall, Stanford Univ. report SUITP-94-14, May 1994.

[27]   M. Carena and C. E. M. Wagner, CERN preprint, CERN-TH.7321/94, June 1994.

[28]   M. Carena, K. Sasaki and C. E. M. Wagner, Nucl. Phys. B381 (1992) 66.

[29] P. Chankowski, S. Pokorski and J. Rosiek, Phys. Lett. B281 (1992) 100.

[30] H. Haber and R. Hempfling, Phys. Rev. D48 (1993) 4280.

[31] P. Chankowski, Phys. Rev. D41 (1990) 2877.

[32] K. Inoue, A. Kakuto, H. Komatsu and S. Takeshita, Prog. Theor. Phys. 67 (1982) 590.

[33] M. Olechowski and S. Pokorski, Nucl. Phys. B404 (1993) 509.

[34] J.M. Frere, D.R.T. Jones and S. Raby, Nucl. Phys. B222 (1983) 11; L. Ibañez and C. Lopez, Phys. Lett. 126B (1983) 54; L. Alvarez Gaumé, J. Polchinski and M. Wise, Nucl. Phys. B221 (1983) 495; J. Ellis, D. V. Nanopoulos and K. Tamvakis, Phys. Lett. 121B (1983) 123; C. Kounnas, A. Lahanas, D.V. Nanopoulos and M. Quirós, Nucl. Phys. B236 (1984) 438.

[35] P. Langacker and N. Polonsky, Pennsylvania Report No. UPR-0594 T (1994).

[36] G. Gamberini, G. Ridolfi and F. Zwirner, Nucl. Phys. B331 (1990) 331.

[37] J. Ellis, G. Ridolfi and F. Zwirner, Phys. Lett. 262 (1991) 477; J.L. Lopez and D.V. Nanopoulos, Phys. Lett. B266 (1991) 397.

[38] C.E.M. Wagner, MPI preprint MPI-PH-93-25, April 1993, to appear in Proc. of *Properties of SUSY Particles*, Erice, Italy, Sep. 1992.

[39] S. Ferrara, C. Kounnas and F. Zwirner, CERN preprint CERN-TH.7192/94; C. Kounnas, I. Pavel and F. Zwirner, CERN preprint CERN-TH.7185/94.

[40] F. Abe et al., CDF Collab. FERMILAB-PUB-94/097-E.

[41] K. Hikasa et al., Particle Data Group, Phys. Rev. D45 (1992).

[42] S. Narison, CERN preprint, May 1994, to appear in the *Proceedings of "QCD and High Energy Hadronic Interactions", Méribel, Haute-Savoie, March 1994.*

Soft Supersymmetry Breaking Parameters and Minimal SO(10) Unification*

M. CARENA AND C.E.M. WAGNER[1]
CERN THEORY DIVISION,
1211 GENEVA 23, SWITZERLAND

Abstract: The minimal supersymmetric SO(10) model, in which not only the gauge but also the third generation fermion Yukawa couplings are unified, provides a simple and highly predictive theoretical scenario for the understanding of the origin of the low energy gauge interactions and fermion masses. In the framework of the Minimal Supersymmetric Standard Model with universal soft supersymmetry breaking parameters at the grand unification scale, large values of the universal gaugino mass $M_{1/2} \geq 300$ GeV are needed in order to induce a proper breakdown of the electroweak symmetry. In addition, in order to obtain acceptable experimental values for both the pole bottom mass and the $b \to s\gamma$ decay rate, even larger values of the gaugino masses are required. The model is strongly constrained by theoretical and phenomenological requirements and a heavy top quark, with mass $M_t \geq 170$ GeV, is hard to accomodate within this scheme. We show, however, that it is sufficient to relax the condition of universality of the scalar soft supersymmetry breaking parameters at the grand unification scale to be able to accommodate a top quark mass $M_t \simeq 180$ GeV. Still, the requirement of a heavy top quark demands a very heavy squark spectrum, unless specific relations between the soft supersymmetry breaking parameters are fulfilled.

1   Introduction

The Standard Model provides a very good understanding of the strong and electroweak interactions and, so far, it has withstood all the experimental onslaughts. Yet, it leaves a host of open questions, which require, to be answered, the presence of an underlying structure with a larger symmetry content than the one sufficient to describe the physics at present accelerator energies. In particular, an explanation to the origin of forces and fermion masses, as well as the hierarchy between the Planck and the weak scale is still lacking. Supersymmetric theories have the potential of dealing with these problems and, at the same time, of remaining compatible with the low energy data. This is why, in spite of the lack of experimental evidence for supersymmetry, most theories beyond the Standard Model include this symmetry at some stage. In the simplest supersymmetric theories, gauge invariant, soft supersymmetry breaking terms are present, which yield masses to the unobserved supersymmetric particles, while the standard model fermion and boson fields acquire masses through the usual Higgs mechanism [1]. In addition, supersymmetry ensures the stability of the hierarchy between the weak and the Planck scales. Moreover, in the minimal supersymmetric models the weak scale and the scale of supersymmetry

---

[1]Talk was presented by C.E.M. Wagner.

breaking are interrelated, and a natural explanation of the scale of electroweak symmetry breaking may only be achieved if the supersymmetric particle masses are smaller than, or of the order of, 1 TeV.

It has recently been realized that the weak and strong gauge couplings determined by the most recent measurements at the LEP experiments are consistent with the unification of gauge couplings within the minimal supersymmetric standard model [2]. The presence of gauge coupling unification has strongly revived the interest in softly broken supersymmetric theories, in particular, in the Minimal Supersymmetric extension of the Standard Model. Much work has been done in understanding the influence of threshold corrections at low and high energy scales and the impact of the precise supersymmetric spectrum on the predictions for the strong gauge coupling [3]. In particular, it has been shown that the low energy threshold corrections are strongly dependent on the Higgsino and gaugino masses, and weakly dependent on the squark and sfermion masses [4]. Moreover, the relevance of the experimental correlation between the top quark mass and the weak mixing angle in the obtention of the strong gauge coupling predictions has been emphasized. As we shall discuss below, due to this correlation, a heavy top quark, with mass $M_t > 150$ (170) GeV, is associated with values of the strong gauge coupling $\alpha_3(M_Z) \geq 0.114$ (0.117). The lower bound on $\alpha_3(M_Z)$ increases for larger values of the top quark mass.

Furthermore, the question of fermion masses may also find a natural explanation within minimal supersymmetric Grand Unified Theories (see, for example, Refs. [5],[6]). In fact, minimal supersymmetric GUTs give a natural explanation for the heaviness of the top quark: The condition of bottom–tau Yukawa coupling unification [7],[8] requires large values of the top quark Yukawa coupling, h_t, at the grand unification scale, M_{GUT}, in order to contravene the strong gauge coupling effects on the running of the bottom Yukawa coupling [3]–[5],[9]. Moreover, if the top quark Yukawa coupling acquires large values at the grand unification scale, $Y_t = h_t^2/4\pi \geq 0.1$, its low energy value is completely determined by the infrared fixed point structure of the theory [10]–[12]. Indeed, for small and moderate values of $\tan\beta$ –the ratio of vacuum expectation values of the Higgs fields– for which the effects of the bottom quark Yukawa coupling in the running of h_t may be safely neglected, the infrared quasi-fixed point value of the running top quark mass in the $\overline{MS}$ scheme is approximately given by

$$m_t(M_t)^{IR} \simeq 196 \, \text{GeV} \left[1 + 2\left(\alpha_3(M_Z) - 0.12\right)\right] \sin\beta \qquad (1)$$

where the pole mass is related to the running mass by [13]

$$M_t \simeq m_t(M_t) \left[1 + 4\alpha_3(M_t)/3\pi + 11(\alpha_3(M_t)/\pi)^2\right]. \qquad (2)$$

A careful analysis shows that, for small and moderate values of $\tan\beta$ the conditions of gauge and bottom–tau Yukawa coupling unification lead to a top quark mass which, for the currently acceptable values for the bottom mass and the electroweak parameters, is within 10% of its infrared fixed point value [14],[15].

In general, the condition of bottom–tau Yukawa coupling unification determines the value of the top quark Yukawa coupling and, in the small and moderate $\tan\beta$ regime, implies a strong attraction of the top quark mass to its infrared fixed point. Observe that the infrared fixed point solution does not provide a direct prediction for the top quark mass, but only a strong correlation between M_t and $\tan\beta$. A prediction for both the top quark mass and $\tan\beta$ may only be obtained in theories with a richer symmetry structure than the one provided by the minimal supersymmetric SU(5) theory. In this respect, the minimal SO(10) model, where all Yukawa couplings proceed from a common coupling of the 16 representation of matter fields with a 10 representation of Higgs fields, provides a natural extension of the SU(5) scenario [6],[16]–[20]. Since the top and bottom Yukawa couplings are of the same order, the hierarchy of top and bottom masses is due to a large value of the ratio of vacuum expectation values:

$$\tan\beta \simeq \frac{m_t(m_t)}{m_b(m_t)} \simeq \mathcal{O}(50). \tag{3}$$

The condition of bottom–tau Yukawa unification is implicit within this scheme, therefore, the infrared fixed point attraction is potentially present and the top quark mass tends to get larger values. However, since the bottom and the top Yukawa couplings are of the same order, the bottom Yukawa effect is sufficiently strong by itself to partially contravene the strong gauge coupling effects on its renormalization group running. Then, the top quark Yukawa coupling at the grand unification scale tends to be smaller than for moderate values of $\tan\beta$ and the infrared fixed point attraction becomes weaker. The top quark mass prediction becomes much more sensitive to the actual value of the bottom mass and the strong gauge coupling. Hence, the strong predictivity expected from the combination of large values of $\tan\beta$ and the infrared fixed point attraction is not actually realized within this scheme.

Moreover, for the large values of $\tan\beta$ implied by the above condition, Eq. (3), large corrections to the running bottom mass induced by the supersymmetry breaking sector of the theory are present [6],[21]–[23]. These corrections, which may be as large as 50% of the bottom mass value, make the top quark mass predictions highly dependent on the nature of the supersymmetry breaking sector of the theory. For instance, if the soft supersymmetry breaking parameters are such that these corrections are negligible, larger values of the top quark mass $M_t \geq 170$ GeV are preferred. In the case of universal soft supersymmetry breaking parameters at the grand unification scale, these corrections are, instead, large, and lower values of the top quark mass, $M_t \leq 170$ GeV, are preferred [26]. Hence, no prediction for the top quark mass may be obtained unless a specific framework for the breakdown of supersymmetry is given.

In this talk, we first concentrate on the simplest supersymmetry breaking scenario, with universal soft supersymmetry breaking parameters at the grand unification scale, and we describe the phenomenological and theoretical constraints arising in this model (for a detailed discussion of similar issues at low values of $\tan\beta$, we refer the reader to [24] and [25]). We shall show that strong correlations between the different sparticle masses appear within this scheme. Furthermore, a lower bound on the squark and gaugino masses is obtained from the requirement of a proper $SU(2)_L \times U(1)_Y$ breakdown. We shall also discuss the corrections to the bottom mass, its correlation with the supersymmetric contributions to the $b \to s\gamma$ decay rate and its implication for the top quark mass predictions. Finally, we shall briefly describe the implications of relaxing the condition of universality of the soft supersymmetry breaking parameters, both in the Higgs and the sfermion sector of the theory.

2   Gauge Coupling Unification

In order to analyse the condition of gauge coupling unification, the experimental correlation between the top quark pole mass and the weak mixing angle should be considered. Indeed, taking as input values the Fermi constant, the value of the Z–boson mass M_Z, and the value of $\alpha_{em}(M_Z)$, in the modified $\bar{M}S$ scheme a correlation between the top quark mass and $\sin^2\theta_W(M_Z)$ is induced through the top quark mass dependent radiative corrections to the weak mixing angle [3],

$$\sin^2\theta_W(M_Z) = 0.2324 - 10^{-7}\left(M_t^2 - (138 \text{ GeV})^2\right)\text{GeV}^{-2} \pm 0.003 \ . \qquad (4)$$

In addition, in order to fully understand the implications of gauge coupling unification, a few words about the supersymmetric threshold corrections to the gauge couplings should be said. For a given supersymmetric spectrum, and a fixed value of the weak mixing angle, the value of $\alpha_3(M_Z)$, determined by the gauge coupling unification condition, is given by

$$\frac{1}{\alpha_3(M_Z)} = \frac{(b_1 - b_3)}{(b_1 - b_2)}\left[\frac{1}{\alpha_2(M_Z)} + \gamma_2 + \Delta_2\right] - \frac{(b_2 - b_3)}{(b_1 - b_2)}\left[\frac{1}{\alpha_1(M_Z)} + \gamma_1 + \Delta_1\right]$$
$$- \gamma_3 - \Delta_3 + \Delta^{Sthr}\left(\frac{1}{\alpha_3(M_Z)}\right), \qquad (5)$$

where

$$\Delta^{Sthr}\left(\frac{1}{\alpha_3(M_Z)}\right) = \frac{19}{28\pi}\ln\left(\frac{T_{SUSY}}{M_Z}\right) \qquad (6)$$

is the contribution to $1/\alpha_3(M_Z)$ due to the inclusion of the supersymmetric threshold corrections at the one–loop level, γ_i includes the two–loop corrections to the value of $1/\alpha_i(M_Z)$, Δ_i are correction constants that allow a transformation of the gauge couplings from the minimal $\bar{M}S$ scheme to the dimensional reduction scheme $\bar{D}R$, more appropriate for supersymmetric theories, and b_i are the supersymmetric beta function coefficients associated to the gauge coupling α_i. As becomes clear from Eq. (6), the effective threshold scale T_{SUSY} gives a parametrization of the size of the supersymmetric threshold corrections to the gauge couplings and it would coincide with the overall mass scale M_{SUSY} only if all supersymmetric particles were degenerate in mass.

In order to study the dependence of T_{SUSY} on the different sparticle masses, we define $m_{\tilde{q}}$, $m_{\tilde{g}}$, $m_{\tilde{l}}$, $m_{\tilde{W}}$, $m_{\tilde{H}}$ and m_H as the characteristic masses of the squarks, gluinos, sleptons, electroweak gauginos, Higgsinos and the heavy Higgs doublet, respectively. Assuming different values for all these mass scales, we derive an expression for the effective supersymmetric threshold T_{SUSY}, which is given by [4]

$$T_{SUSY} = m_{\tilde{H}} \left(\frac{m_{\tilde{W}}}{m_{\tilde{g}}}\right)^{28/19} \left[\left(\frac{m_{\tilde{l}}}{m_{\tilde{q}}}\right)^{3/19} \left(\frac{m_H}{m_{\tilde{H}}}\right)^{3/19} \left(\frac{m_{\tilde{W}}}{m_{\tilde{H}}}\right)^{4/19}\right]. \qquad (7)$$

The above relation holds whenever all the particles involved have masses $m_\eta > M_Z$. If, instead, any of the sparticles or the heavy Higgs boson has a mass $m_\eta < M_Z$, it should be replaced by M_Z for the purpose of computing the supersymmetric threshold corrections to $1/\alpha_3(M_Z)$. ¿From Eq. (7), it follows that T_{SUSY} has only a slight dependence on the squark, slepton and heavy Higgs masses and a very strong dependence on the overall Higgsino mass, as well as on the ratio of masses of the gauginos associated with the electroweak and strong interactions. In Table 1, we show the predictions for the strong gauge coupling, for different values of $\sin^2 \theta_W(M_Z)$ and the supersymmetric threshold scale, together with the approximate value of the top quark pole mass associated with each value of $\sin^2 \theta_W(M_Z)$.

M_t [GeV]	$\sin^2 \theta_W(M_Z)$	$\alpha_3(M_Z)$ for $T_{SUSY} = 1$ TeV	$\alpha_3(M_Z)$ for $T_{SUSY} = 100$ GeV
140	0.2324	0.116	0.123
170	0.2315	0.119	0.127
195	0.2305	0.123	0.131

Table 1. Dependence of $\alpha_3(M_Z)$ on $\sin^2 \theta_W(M_Z)$ and T_{SUSY}, in the framework of gauge coupling unification.

The above values of $\alpha_3(M_Z)$ are obtained from our two–loop renormalization group analysis, and they depend slightly on the top quark Yukawa coupling, and hence on $\tan \beta$. In general, for values of the top quark mass $M_t \geq 140$ GeV, the variation induced through the top quark Yukawa coupling contribution is at most 1% of the values given above (with lower values of $\alpha_3(M_Z)$ obtained for larger values of the top quark Yukawa coupling). Observe that in models with universal gaugino masses at the grand unification scale and for large values of the supersymmetric mass parameters, for which the mixing in the neutralino and chargino sectors may be neglected, $T_{SUSY} \simeq \mu/6$, where μ is the supersymmetric mass parameter appearing in the superpotential. Hence, a value of T_{SUSY} of the order of 1 TeV implies that the supersymmetric spectrum contains sparticles with masses far above the TeV scale. If all supersymmetric masses are below or of the order of 1 TeV, the effective supersymmetric threshold scale is below or of the order of the weak scale.

The largest uncertainties associated with the unification scheme come from the threshold corrections at the grand unification scale. We shall not discuss them here (for a detailed discussion, see, for example, Ref. [3]), but we shall assume moderate corrections, of the order of those coming from the supersymmetric spectrum. Thus, for a supersymmetric spectrum with characteristic masses of the order of or below 1 TeV, the condition of gauge coupling unification, together with the experimental correlation between $\sin^2 \theta_W(M_Z)$ and M_t, Eq. (4), imply the following correlation between $\alpha_3(M_Z)$ and the top quark mass,

$$M_t^2 \simeq (138 \text{ GeV})^2 + 0.25 \times 10^7 \text{ GeV}^2 \left(\alpha_3(M_Z) - 0.123 \pm 0.01\right). \qquad (8)$$

It is instructive to compare Eq. (8) with the results presented in table 1. As we discussed in section 1, a lower bound for $\alpha_3(M_Z)$ as a function of the top quark mass is derived. From Eq. (8) it follows that, for a top quark mass $M_t \geq 150$ (170) GeV, the values of the strong gauge coupling are $\alpha_3(M_Z) \geq 0.114$ (0.117).

3 Yukawa Coupling Unification

The general features of Yukawa coupling unification in the Minimal Supersymmetric Standard Model are discussed in section 1. In the following, we shall concentrate on the properties of the minimal supersymmetric SO(10) model, for which not only the bottom and the tau, but also the top quark Yukawa coupling unify at the GUT scale. As we discussed in section 1, large values of $\tan\beta$ ($\simeq \mathcal{O}(50)$), are predicted within this scheme.

The fact that in the minimal supersymmetric SO(10) model, the value of $\tan\beta$ is approximately equal to the ratio of the top and bottom quark masses at the top mass scale comes from the approximate equality of the top and bottom Yukawa couplings at low energies. Such an approximate equality is implied by their unification condition, and the presumption that the bottom and top quarks acquire masses, each of them, only through one of the two Higgs doublet vacuum expectation values: $m_t(m_t) = h_t(m_t)v_2$, $m_b(m_t) = h_b(m_t)v_1$, where v_i is the vacuum expectation value of the Higgs H_i. This property holds in the supersymmetric limit and, within a good approximation, for small and moderate values of $\tan\beta$ when supersymmetry is softly broken. In general, however, a coupling of the bottom (top) quark to the Higgs H_2 (H_1) may be generated at the one–loop level. Although these couplings are small compared to h_b (h_t) (typically lower than 1% of h_b), for large values of $\tan\beta$ –which implies $v_2 \gg v_1$– this may induce important corrections to the bottom mass [21]–[23], [26]:

$$m_b = h_b(v_1 + K_1 v_2) \equiv \tilde{m}_b \left(1 + \frac{\Delta m_b}{\tilde{m}_b}\right), \qquad (9)$$

where K_1 is the coefficient of the one–loop corrections to the bottom mass, $\Delta m_b/\tilde{m}_b = K_1 \tan\beta$, and $\tilde{m}_b$ would be the value of the running bottom mass if the supersymmetric one–loop corrections were negligible. Recalling that, at the two–loop level, the physical and running bottom masses are related by

$$M_b = m_b(M_b) \left(1 + \frac{4}{3\pi}\alpha_3(M_b) + 12.4 \left(\frac{\alpha_3(M_b)}{\pi}\right)^2\right), \qquad (10)$$

due to the low energy renormalization group running of the bottom quark mass the following property is fulfilled,

$$\frac{\Delta M_b}{\tilde{M}_b} \simeq \frac{\Delta m_b(M_b)}{\tilde{m}_b(M_b)} \simeq \frac{\Delta m_b(M_Z)}{\tilde{m}_b(M_Z)}. \qquad (11)$$

In the above, $\tilde{M}_b$ would be the value of the physical bottom mass if no supersymmetric corrections were present, while $\Delta M_b \simeq M_b - \tilde{M}_b$.

The above property, Eq. (11), is relevant for the understanding of the top quark mass predictions coming from the unification of couplings. In Fig. 1, we present the predictions for the top quark mass as a function of $\alpha_3(M_Z)$ for different values of the *uncorrected* bottom quark mass $\tilde{M}_b$, as well as for different values of $\tan\beta$ [26]. From this figure we observe that the top quark mass predictions are strongly dependent on the exact value of the strong gauge coupling and the *uncorrected* bottom mass $\tilde{M}_b$. Notice that, if we restrict ourselves to the region of $\alpha_3(M_Z)$ preferred by gauge coupling unification, Eq. (8) (to the right of the dotted line), for $M_{\tilde{b}} \leq 5$ GeV, the top quark mass is pushed towards large values. In fact, if the bottom mass corrections were small, $\tilde{M}_b \simeq M_b$, for experimentally acceptable values of the bottom quark pole mass, $M_b = 4.9 \pm 0.3$ GeV [27], the top quark mass would be $M_t \geq 165$ GeV. However, as we shall show in section 5, large bottom quark mass corrections may significantly change this prediction.

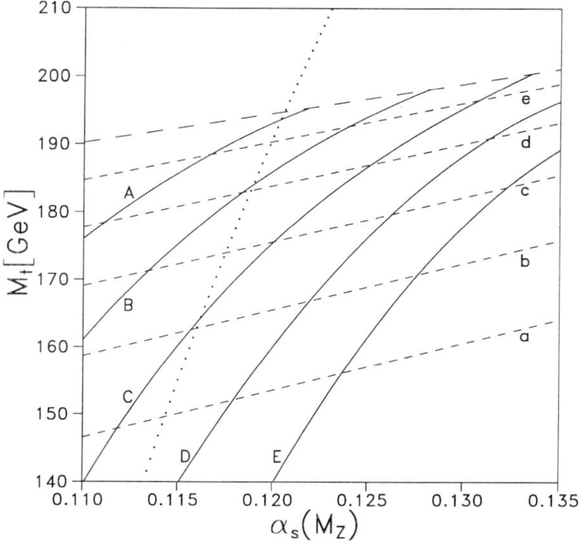

Fig.1. Top quark mass predictions as a function of the strong gauge coupling $\alpha_3(M_Z)$, with unification of the three Yukawa couplings of the third generation. Starting from above, the solid lines represent values of $\tilde{M}_b$ equal to 4.6, 4.9, 5.2, 5.5 and 5.8 GeV. Analogously, the dashed lines represent values of $\tan\beta$ equal to 60, 55, 50, 45 and 40. The long-dashed line represents the top quark mass fixed point value and the dotted line shows the region preferred by the gauge coupling unification condition as explained in the text.

For a clear interpretation of Fig. 1 it is important that, for large values of $\tan\beta$, the fixed point prediction, Eq. (1), is modified due to the non-negligible bottom quark Yukawa coupling effect in the running of the top quark Yukawa coupling *[4],[26]*. Indeed, for $h_b \simeq h_t$ a more appropriate expression than Eq. (1) is

$$m_t(M_t)^{IR} \simeq 182 \text{ GeV} \left[1 + 2 \left(\alpha_3(M_Z) - 0.12\right)\right] \tag{12}$$

which, recalling the relation between the running and the pole masses, Eq. (2), describes within a good approximation the upper bound on the top quark mass (long-dashed line) shown in Fig. 1.

4 Supersymmetry and Electroweak Symmetry Breaking

As we mentioned above, in the large $\tan\beta$ regime, the top quark mass predictions depend strongly on the nature of the soft supersymmetry breaking terms arising at low energies. In principle, the Minimal Supersymmetric Standard Model yields a multiplication of free parameters, which are only constrained by the requirement of avoiding a conflict with present experimental data. One of the strongest requirements is the absence of flavour changing neutral currents, which is fulfilled if the squarks of the first two generations are approximately degenerate in mass. It has been realized long ago that such is naturally the case if all soft supersymmetry breaking squark and gaugino mass parameters are universal at the grand unification scale. This supersymmetry breaking scheme appears naturally in minimal supergravity models, in which not only the squarks but also the Higgs fields acquire common soft supersymmetry breaking terms at high energies, and all the supersymmetry breaking terms at low energies may be given as a function of only four parameters: The universal scalar mass m_0, the universal gaugino mass $M_{1/2}$, and the universal trilinear and bilinear couplings A_0 and B_0, which are associated with contributions to the high energy effective potential proportional to the trilinear and bilinear terms of the superpotential.

We shall first concentrate on the minimal supergravity SO(10) model. In section 6 we shall briefly discuss the situation where the condition of universality of the soft supersymmetry breaking parameters is relaxed. In general, the superpotential reads

$$P = \mu\epsilon_{ab}H_1^a H_2^b + h_t\epsilon_{ab}Q^a U H_2^b + h_\tau\epsilon_{ab}L^a E H_1^b + h_b\epsilon_{ab}Q^a D H_1^b , \tag{13}$$

where $Q^T = (T_L, B_L)$ and $L^T = (\nu_L^\tau, \tau_L)$ are the top–bottom and lepton left–handed doublets, $U = T_L^c$, $D = B_L^c$ and $E = \tau_L^c$ are the right–handed top–squark, bottom–squark and stau, respectively, and ϵ_{ab} is the antisymmetric tensor with $\epsilon_{12} = -1$. The scalar potential is given by

$$V = m_1^2 H_1^\dagger H_1 + m_2^2 H_2^\dagger H_2 + m_Q^2 Q^\dagger Q + m_U^2 U^* U + m_D^2 D^* D + m_L^2 L^\dagger L + m_E^2 E^* E$$
$$+ \epsilon_{ab}(m_3^2 H_1^a H_2^b + A_t h_t Q^a U H_2^b + A_b h_b Q^a D H_1^b + \text{h.c.}) + \text{q.t.} + \dots , \tag{14}$$

where the Higgs field mass terms contain a part coming from the superpotential and another coming from the soft supersymmetry breaking parameters: $m_i^2 = \mu^2 + m_{H_i}^2$, with $i = 1,2$ and $m_3^2 = B\mu$. The dots stand for scalar trilinear F-terms and q.t. characterizes the quartic terms in the scalar fields. The soft supersymmetry breaking parameters $m_{H_i}^2$ have the following renormalization group evolution,

$$4\pi \frac{dm^2_{H_1}}{dt} = 3\alpha_2 M_2^2 + \alpha_1 M_1^2 - 3Y_b M^2_{Deff} - Y_\tau M^2_{Eeff},$$

$$4\pi \frac{dm^2_{H_2}}{dt} = 3\alpha_2 M_2^2 + \alpha_1 M_1^2 - 3Y_t M^2_{Ueff}, \tag{15}$$

where $t = \ln[(M_{GUT}/Q)^2]$, $Y_i = h_i^2/(4\pi)$, $M^2_{Deff} = m_Q^2 + m_D^2 + m^2_{H_1} + A_b^2$, $M^2_{Ueff} = m_Q^2 + m_U^2 + m^2_{H_2} + A_t^2$ and $M^2_{Eeff} = m_L^2 + m_E^2 + m^2_{H_1} + A_\tau^2$.

The rest of the renormalization group equations may be found in the literature *[28]*. Let us just remark that, in the minimal supergravity scheme, M_{Deff} and M_{Ueff} present very similar renormalization group evolutions when $h_t \simeq h_b$. Indeed, for bottom–top Yukawa coupling unification they only differ by the different hypercharge quantum numbers of the right bottom and top quarks, the slightly different running of the bottom and top Yukawa couplings, and the small tau Yukawa coupling effects (recall that the tau Yukawa coupling is renormalized to lower values than the bottom and top ones, due to the absence of strong gauge coupling effects in its one–loop renormalization group evolution).

Considering the renormalization group evolution of the mass parameters $m^2_{H_1}$ and $m^2_{H_2}$, for $M^2_{1/2} \gg m_0^2$, values of $m^2_{H_1} > m^2_{H_2}$ are obtained, mainly due to the difference between the hypercharge quantum numbers of the right bottom and top quarks and their supersymmetric partners. For $m_0^2 \gg M^2_{1/2}$, instead, the inverse hierarchy, $m^2_{H_2} > m^2_{H_1}$, is obtained, mainly due to the τ-Yukawa coupling effects. In general, considering the bottom–top Yukawa coupling unification condition and performing a complete numerical analysis it follows that *[26]*

$$m^2_{H_1} - m^2_{H_2} \simeq \alpha M^2_{1/2} + \beta m_0^2, \tag{16}$$

with $\alpha \simeq -\beta \simeq 0.1 - 0.2$, depending on the proximity of the top quark Yukawa coupling to its infrared fixed point value h_f (the above range is obtained for $Y_t/Y_f = h_t^2/h_f^2 = 0.6 - 0.95$, respectively).

4.1 Radiative Electroweak Symmetry Breaking

In order to induce a proper breakdown of the electroweak symmetry, the following conditions need to be fulfilled,

$$\sin 2\beta = \frac{2m_3^2}{m_A^2} \tag{17}$$

and

$$\tan^2 \beta = \frac{m_1^2 + M_Z^2/2 + \text{r.c.}}{m_2^2 + M_Z^2/2 + \text{r.c.}}, \tag{18}$$

where $m_A^2 \simeq m_1^2 + m_2^2 +$ r.c. is the mass of the CP-odd Higgs eigenstate and r.c. simbolizes one–loop radiative correction contributions, which depend logarithmically on the supersymmetry breaking scale, and become of the order of M_Z^2 for a characteristic supersymmetric scale of the order of 1 TeV. We shall ignore these corrections in the following, since they are unessential for the qualitative understanding of the phenomena under discussion. They are included, however, in the numerical analysis.

In general, independently of the supersymmetry breaking mechanism, since $\tan\beta$ is very large, in order to avoid extremely large values of m_1^2 (or m_A^2), the mass parameters m_2^2 and m_3^2 should fulfil the following properties

$$m_2^2 \simeq -\frac{M_Z^2}{2}, \qquad m_3^2 \ll M_Z^2. \tag{19}$$

As was explained in Ref. [22], the second of these conditions can be obtained by assuming that there is a softly broken symmetry implying the smallness of the parameters B and/or μ. The first condition is just a reflection of the fact that the vacuum expectation value of the Higgs H_2 is the one that determines the electroweak vector boson masses. The fact that the CP-odd Higgs mass squared should be positive, together with Eq. (19), implies that

$$m_1^2 - m_2^2 > M_Z^2 . \tag{20}$$

The above described properties are general in the sense that they do not depend on the supersymmetry breaking mechanism. If we consider the running of the Higgs mass parameters in the minimal supergravity model with bottom–top Yukawa unification, Eq. (16), strong implications follow from Eqs. (19) and (20),

$$M_{1/2} > m_0, \qquad M_{1/2} \geq \frac{M_Z}{\sqrt{\alpha}}. \tag{21}$$

Analysing the RG evolution of the soft supersymmetry breaking parameters, Eq. (15), and taking into account the constraints of Eq. (21), an approximate solution for the Higgs mass parameter m_2^2 is obtained,

$$m_2^2 \simeq \mu^2 + M_{1/2}^2 \left(0.5 - C_1 \frac{Y}{Y_f} + C_2 \left(\frac{Y}{Y_f} \right)^2 \right), \tag{22}$$

where Y/Y_f is the ratio of the top quark Yukawa coupling squared to its fixed point value, and C_1 and C_2 are coefficients that depend on the value of the strong gauge coupling constant and, for $\alpha_3(M_Z) \simeq 0.12$, are approximately given by $C_1 \simeq 6$ and $C_2 \simeq 3$. Hence, the condition $m_2^2 \simeq -M_Z^2/2$, together with the large values of $M_{1/2}$ necessary to achieve unification of couplings, imply a strong correlation between μ and $M_{1/2}$. This correlation, which is shown in Fig. 2, has profound implications for the sparticle spectrum and the determination of the bottom mass corrections.

In the following, we shall summarize the main features of the Higgs and supersymmetric spectrum. For a more detailed discussion, we refer the reader to [26].

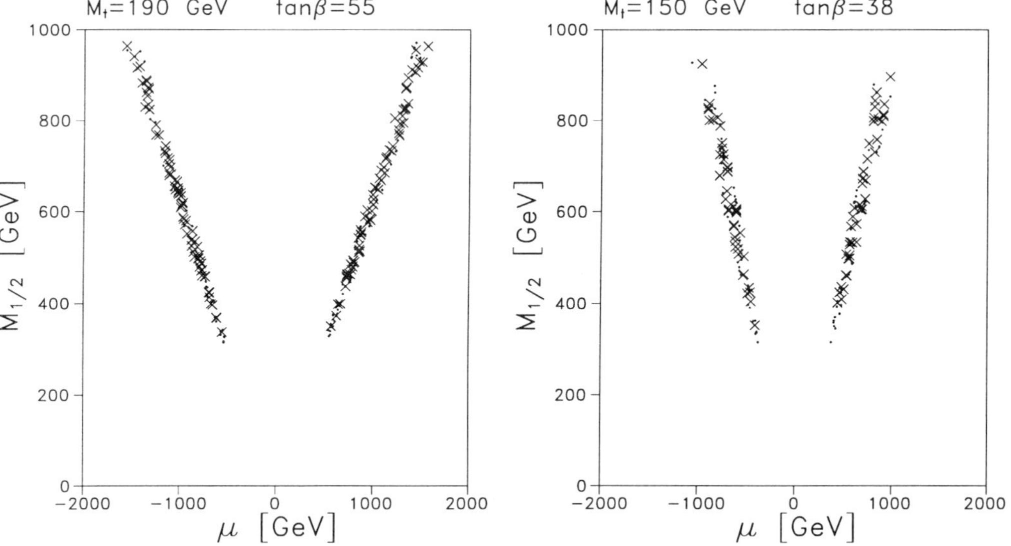

Fig. 2. Supersymmetric mass parameter μ as a function of the gaugino mass $M_{1/2}$ for two different values of the top quark mass in the framework of bottom–top Yukawa unification. Only the lower top quark mass value $M_t \simeq 150$ GeV leads to an acceptable value of the physical bottom mass, once the supersymmetry breaking-induced bottom mass corrections are included.

The main features of the sparticle spectrum are governed by the three properties explained above: a) Strong μ–$M_{1/2}$ correlation, b) Large values of $M_{1/2} \geq 300$ GeV, and c) $M_{1/2} \geq m_0$. This yields:

1) Small mixing in the chargino and neutralino sectors, which naturally follows from properties a) and b). The lightest supersymmetric particle is mainly a bino, with mass $M_{\tilde{B}} \simeq 0.4 M_{1/2}$. The second–lightest neutralino and the lightest chargino are winos and hence almost degenerate in mass $M_{\tilde{W}^+} \simeq M_{\tilde{W}^0} \simeq 2 M_{\tilde{B}}$. The heaviest neutralino and chargino are Dirac (pseudo–Dirac in the neutralino case) particles, with masses approximately equal to the parameter $|\mu|$.

2) Strong stop (sbottom)–gluino running mass correlations, with $M_{\tilde{t}} \simeq M_{\tilde{b}} \simeq 0.75$–$0.8 M_{\tilde{g}}$ (0.85–$0.9 M_{\tilde{g}}$) for the lightest (heaviest) squark mass eigenstate. The running gluino mass is related to the common gaugino mass by $M_{\tilde{g}} \simeq 2.6$–$2.8 \, M_{1/2}$.

3) Large mixing in the stau sector, due to the off–diagonal left–right stau matrix element $m^2_{\tilde{\tau}_{LR}} \simeq -h_\tau \mu v_2 \simeq -m_t \mu / \sqrt{3}$. The large stau mixing leads to a lower

bound on m_0,

$$m_0^2 \geq -0.15 M_{1/2}^2 + \left((0.15 M_{1/2}^2)^2 + \mu^2 m_t^2/3 \right)^{1/2} \tag{23}$$

in order to avoid a stau being the lightest supersymmetric particle.

4) The Higgs spectrum is characterized by relatively low values for the CP-odd Higgs mass, $m_A \leq M_{\tilde{t}} \, (\alpha/5)^{1/2}$, where α is the coefficient characterizing the dependence of $m_{H_1}^2 - m_{H_2}^2$ on $M_{1/2}^2$, Eq. (16). Observe that, due to the characteristic values of $\alpha \simeq 0.2$ (0.1) for $M_t \simeq 190$ (160) GeV, if the squark masses are lower than a few TeV, the CP-odd and charged Higgs masses will be lower than a few hundred GeV.

5) The lightest CP-even scalar would become lighter than the Z-boson if the CP-odd Higgs scalar were light, $m_A \leq 150$ GeV. However, in the minimal supergravity SO(10) model, such low values of m_A induce large values of the $b \to s\gamma$ decay rate, since the chargino contribution enhances the rate with respect to that of the Standard Model with an extra Higgs doublet, if the correct bottom mass predictions are required. Therefore, in the minimal supergravity SO(10) model the lightest CP-even Higgs mass is larger than the Z boson mass. We shall come back to this issue in section 5.

5   Bottom Mass Predictions and the $b \to s\gamma$ Decay Rate

In section 3 we have mentioned the possibility of inducing non–negligible one–loop corrections to the running bottom mass, Eq. (9). In this section we want to explicitly show which are the relevant one–loop contributions to the running bottom mass and how they are correlated with the chargino contributions to the $b \to s\gamma$ decay rate. As we show in section 3, in order to estimate the size of the bottom mass corrections, we need to compute the supersymmetry breaking-induced effective coupling of the bottom quark to the H_2 field, which amounts to computing the coefficient K_1, with $\Delta m_b/\tilde{m}_b = K_1 \tan\beta$. The contributions to the coefficient K_1 come from the sbottom–gluino and stop–chargino graph contributions to the bottom quark self energies. It is easy to show that the dominant contributions are given by

$$K_1 = \frac{2\alpha_3}{3\pi} \, M_{\tilde{g}}\mu \, I(m_{\tilde{b}_1}^2, m_{\tilde{b}_2}^2, M_{\tilde{g}}^2) + \frac{Y_t}{4\pi} \, A_t\mu \, I(m_{\tilde{t}_1}^2, m_{\tilde{t}_2}^2, \mu^2) , \tag{24}$$

where $m_{\tilde{q}_i}^2$, with $i = 1, 2$, are the squark mass eigenstates and the integral factor $I(a, b, c)$ is given by

$$I(a, b, c) = \frac{ab\log(a/b) + bc\log(b/c) + ac\log(c/a)}{(a - b)(b - c)(a - c)}. \tag{25}$$

The renormalization group equation of the A_t parameter shows that its low energy values are strongly correlated with the universal gaugino mass. In fact, using the relation $M_{\tilde{g}} \simeq 2.6$–2.8 $M_{1/2}$ it follows [26] that

$$A_t \simeq -\frac{2}{3} M_{\tilde{g}}, \qquad \text{for} \ \ Y/Y_t \simeq 1$$

$$A_t \simeq -M_{\tilde{g}} \qquad \text{for} \ \ Y/Y_f \simeq 0.6. \tag{26}$$

Observe that, due to the minus sign appearing in Eq. (26), there is a partial cancellation between the two different contributions to K_1, which yields a significant reduction in the bottom mass corrections (typically of the order of 25%). As will be shown below, this partial cancellation, although important, is by far not sufficient to render the bottom mass corrections small.

5.1 Conditions for a small K_1 and Minimal Supergravity

To get an estimate of the size of the bottom mass corrections, it is important to observe that the integral factors $I(a, b, c)$ are always of the order of the inverse of the largest mass squared appearing in the integral. Since the coupling constant dependent factors of both contributions to K_1 are of order 0.01, and $\tan \beta$ is of order 50, to get a small bottom mass correction the following properties should be fulfilled [22]:

$$M_{\tilde{g}} \ll m_{\tilde{q}}, \qquad \text{and/or} \qquad \mu \ll m_{\tilde{q}}, \tag{27}$$

where $m_{\tilde{q}}$ represents the heaviest third generation squark mass eigenstates. In addition, the requirement of electroweak symmetry breaking implies that the mass parameter B should be small in comparison to m_A, unless μ itself is much smaller than m_A. All these requirements may be satisfied by imposing a softly broken Peccei-Quinn symmetry, which implies the smallness of the mass parameter μ, together with an approximate continuous R symmetry, present in the limit $B \to 0$, $M_{\tilde{g}} \to 0$, $A_t \to 0$, whose breaking is characterized by the (assumed) small parameter

$$\epsilon_R \simeq \frac{B}{m_{\tilde{q}}} \simeq \frac{A_t}{m_{\tilde{q}}} \simeq \frac{M_{\tilde{g}}}{m_{\tilde{q}}} . \tag{28}$$

As we shall discuss in section 6, the above conditions may only be reached by relaxing the condition of universality of the soft supersymmetry breaking parameters at the grand unification scale.

In the framework of minimal supergravity, with exact unification of the third generation quark and lepton Yukawa couplings, the strong correlations between the parameters μ, A_t, $M_{\tilde{g}}$ and the third generation squark masses –derived from their renormalization group equations and the condition of a proper breakdown of the electroweak symmetry– imply that the above symmetries are not present in the low energy spectrum. Hence, the bottom mass corrections are large within this framework [26]:

$$\frac{\Delta m_b}{\tilde{m}_b} \simeq 0.0045 \tan \beta \left(\frac{\mu}{M_{1/2}} \right) . \tag{29}$$

Taking into account Eq. (22) and the numerical results from Fig. 1, we see that the corrections are of order 45 % for $M_t \simeq 190$ GeV and of order 20 % for $M_t \simeq 150$ GeV.

The corrections are sufficiently large to rule out any solution with $\tilde{M}_b < M_b$. This is simply due to the impossibility to accommodate a physical bottom mass in the experimentally allowed range, while keeping the top Yukawa coupling in the perturbative domain at energies of the order of the grand unification scale. Hence, the coefficient K_1 should be negative, implying that the only acceptable branch is that with negative (positive) values of $M_{\tilde{g}} \times \mu$ $(A_t \times \mu)$. Consequently, an upper bound on the top quark mass M_t may be obtained. This corresponds, for a given value of the strong gauge coupling, to the maximum value of the top quark Yukawa coupling (and $\tan \beta$) consistent with a physical bottom mass M_b equal to its lower experimental bound $M_b^L \simeq 4.6$ GeV. There are small uncertainties in the computation of this upper bound, associated with the size of the low energy threshold corrections to the top quark mass, small tau mass corrections analogous to the bottom ones and QCD scale uncertainties. Conservative upper bounds for the top quark mass are given by [26]

$$M_t \leq 165 \ (175) \ (185) \ \text{GeV}, \qquad \text{for} \quad \alpha_3(M_Z) = 0.12 \ (0.125) \ (0.13). \quad (30)$$

These bounds go rapidly down if the bottom mass is larger than 4.6 GeV. Let us remark again that these bounds do not apply in general in the supersymmetric SO(10) model, but are only a consequence of the particular supersymmetry breaking scheme under study. Relaxing the high energy boundary conditions for the soft supersymmetry breaking parameters, these bounds on M_t may be diluted.

5.2 $b \to s\gamma$ Decay Rate

The dominant supersymmetric contributions to the $b \to s\gamma$ decay rate have been recently analysed by several authors [29]–[33]. In the Minimal Supersymmetric Standard Model, there is a contribution coming from the charged Higgs, which, for low values of the charged Higgs mass, enhances the Standard Model decay rate. The dominant effect from supersymmetric particles comes from the one–loop chargino–stop contributions to the $b_R \to s_L\gamma$ transition. In the supersymmetric limit, $\tan \beta = 1$ and $\mu = 0$, the chargino contributions exactly cancel the $W^\pm$ and charged Higgs ones, and the $b \to s\gamma$ transition element vanishes. This behaviour is not preserved once supersymmetry is broken.

In particular, the large $\tan \beta$ scenario is far from being close to the supersymmetric limit and the dominant chargino contribution to the $b \to s\gamma$ decay rate may have both signs. In general, in the large $\tan \beta$ regime it is proportional to

$$A_{\gamma,g} \simeq \frac{m_t^2}{m_{\tilde{t}}^2} \frac{A_t\mu}{m_{\tilde{t}}^2} \tan \beta \ g_{\gamma,g}\left(\frac{m_{\tilde{t}}^2}{\mu^2}\right), \quad (31)$$

where A_γ and A_g are the coefficients of the effective operators for bs–photon and bs–gluon interactions, as defined in Ref. [30], $g_{\gamma,g}(x)$ is a function of x proportional to the derivative of the function $f_{\gamma,g}^{(3)}(x)$ defined in Ref. [30], and we have assumed a small mixing in the stop sector, with mass eigenstates $m_{\tilde{t}_{1(2)}} = m_{\tilde{t}}^2 + (-)A_t m_t$. ¿From Eq. (31) it follows that the sign of the chargino contribution to the $b \to s\gamma$ decay amplitude depends on the sign of $A_t \times \mu$, and hence is correlated in sign with the bottom mass corrections discussed above. Observe that the chargino (charged Higgs) contribution to the decay amplitude is always small if the supersymmetric (charged Higgs) spectrum is sufficiently heavy.

One can show that for positive (negative) values of $A_t \times \mu$ the supersymmetric rate becomes larger (smaller) than that of the Standard Model plus one extra Higgs doublet. Since in the minimal supergravity SO(10) model, once the supersymmetry breaking-induced bottom mass corrections are included, positive values of $A_t \times \mu$ are required to obtain acceptable values for the physical bottom mass, then, the $b \to s\gamma$ decay rate may be significantly enhanced with respect to the SM one. Consequently, one can obtain sparticle mass bounds by requiring the decay rate to be smaller than the present experimental bounds, $BR(b \to s\gamma) < 5.4 \times 10^{-4}$ [34]. One should recall, however, that there are theoretical uncertainties associated with the decay rate computations, which may be as large as 20–30% [35]. Taking them into account at the 2–σ level [26], for a top quark mass M_t of the order of 150 GeV, the following lower bound on the universal mass parameter $M_{1/2}$ may be obtained:

$$M_{1/2} \geq 550 \text{ GeV}. \tag{32}$$

This implies a lower bound for the masses of the third generation squarks, charginos and charged Higgses larger than the ones already demanded by the condition of electroweak symmetry breaking,

$$m_{\tilde{q}} \geq 1.2 \text{ TeV}, \quad m_{\tilde{\chi}^+} \geq 450 \text{ GeV}, \quad m_{H^+} \geq 180 \text{ GeV}. \tag{33}$$

Observe that, in this case, the lightest CP-even Higgs mass is constrained to be in the range $m_h \simeq 110$–130 GeV.

6 Non-Universal Soft Supersymmetry Breaking Parameters

In the above, we have analysed in detail the implications of having universal soft supersymmetry breaking parameters at the grand unification scale. In this case, the bottom mass corrections and the chargino contributions to the $b \to s\gamma$ decay rate are correlated in sign and they are sufficiently large to put constraints on the particle spectrum as well as an upper bound on the top quark mass as a function of the strong gauge coupling. The phenomenological implications of relaxing the condition of universality depend on the nature of the soft supersymmetry breaking parameters. Instead of performing a detailed study of the general case, we shall concentrate on understanding which should be the pattern of supersymmetry breaking parameters at the grand unification scale in order to induce only small bottom mass corrections (lower than, say, 10% for a heavy top quark) and small chargino contributions to the $b \to s\gamma$ decay rate.

We assume that the universality of the soft supersymmetry breaking gaugino masses is kept and we investigate possible violations of the universality condition in the scalar sector. As was shown in section 5, to obtain an effective cancellation of the bottom quark mass corrections and of the chargino contributions to the $b \to s\gamma$ decay rate, we need either small values of A_t and $M_{\tilde{g}}$ and/or small values of μ in comparison with the squark masses. Hence, the soft supersymmetry breaking parameters should fulfil the following property

$$A_{t,b}(0) \simeq B(0) \simeq M_{1/2} \ll m_S(0), \tag{34}$$

where $m_S(0)$ represents the characteristic scalar soft supersymmetry breaking mass parameters. Assuming bottom–top–tau Yukawa coupling unification and the fulfilment of Eq. (34), and neglecting the small tau Yukawa coupling effects in the bottom quark Yukawa coupling renormalization group running as well as the small right top–bottom hypercharge difference, the following approximate analytical solutions for the Yukawa couplings and mass parameters are obtained,

$$m^2_{H_2} = m^2_{H_2}(0) - 3I_U, \qquad m^2_{H_1} = m^2_{H_1}(0) - 3I_D - I_L,$$
$$m^2_Q = m^2_Q(0) - I_U - I_D, \qquad m^2_U = m^2_U(0) - 2I_U,$$
$$m^2_D = m^2_D(0) - 2I_D, \qquad Y_b \simeq Y_t = \frac{4\pi Y(0)E}{4\pi + 7Y(0)F}, \tag{35}$$

where

$$I_U = \frac{Y}{7\,Y_f}\left(m^2_{H_2}(0) + m^2_Q(0) + m^2_U(0)\right)$$
$$+ \frac{1}{10}\left(m^2_{H_1}(0) - m^2_{H_2}(0) + m^2_D(0) - m^2_U(0)\right)\left[\frac{5Y}{7Y_f} + \left(1 - \frac{Y}{Y_f}\right)^{5/7} - 1\right],$$
$$I_D = \frac{Y}{7\,Y_f}\left(m^2_{H_1}(0) + m^2_Q(0) + m^2_D(0)\right)$$
$$- \frac{1}{10}\left(m^2_{H_1}(0) - m^2_{H_2}(0) + m^2_D(0) - m^2_U(0)\right)\left[\frac{5Y}{7Y_f} + \left(1 - \frac{Y}{Y_f}\right)^{5/7} - 1\right],$$
$$I_L \simeq \frac{1}{6}(m^2_{H_1}(0) + m^2_L(0) + m^2_E(0))\left[1 - \left(1 - \frac{Y}{Y_f}\right)^{1/4}\right]. \tag{36}$$

In the above, $Y = Y_t \simeq Y_b$, with $Y_i = h^2_i/4\pi$, E and F are functions of the gauge couplings –which at the weak scale take values $E \simeq 14$ and $F \simeq 290$– and Y_f is the fixed point value for the top quark Yukawa coupling, $Y_f \simeq 4\pi E/7F$, obtained whenever $Y(0) \geq 0.1$ and whose associated mass expression at the two–loop level is given in Eq. (12). The functions I_U and I_D are obtained from an exact solution to the coupled renormalization group equations in the limit of negligible tau Yukawa coupling effects. The function I_L provides a good approximation to the Yukawa coupling effect in the range $Y/Y_f \simeq 0.6$–0.95, and has been obtained assuming that Y_τ rapidly acquires its low energy hierarchy with respect to Y_b and Y_t.

The interrelation among the different Higgs, squark and slepton masses is quite relevant. For instance, even if we start with values of $m^2_{H_2}(0)$ lower than $m^2_{H_1}(0)$ at the grand unification scale, depending on the relation between the squark mass parameters $m^2_U(0)$ and $m^2_D(0)$, one may or may not find an acceptable solution. Close to the infrared fixed point of the top quark mass, Eq. (12), the following relations

$$m^2_Q + m^2_U + m^2_{H_2} \simeq 0$$
$$m^2_Q + m^2_D + m^2_{H_1} \simeq 0 \tag{37}$$

are fulfilled, implying a strong correlation between the low energy values of the Higgs and squark mass parameters. In particular, for large values of the squark masses $m_{\tilde{q}}^2 \gg M_Z^2$, since $m_2^2 \simeq -M_Z^2/2$, the above yields:

$$\mu^2 \simeq -m_{H_2}^2 \simeq m_Q^2 + m_U^2. \tag{38}$$

Hence, for values of the top quark mass close to its fixed point value, the bottom mass corrections and the chargino contributions to the $b \to s\gamma$ decay rate will only be suppressed if $M_{\tilde{g}}$ is much smaller than the top and bottom squark masses. Since the gluino mass should be larger than 140 GeV, then, to have bottom mass corrections smaller than 10%, the heavy squark mass values should be larger than 1–2 TeV (observe that, since we are assuming universality of the gaugino masses, the gluino mass constraint may be directly inferred from the chargino mass constraints, without relying on the CDF bounds). Therefore, the top quark mass may be close to its fixed point value, but a very heavy squark spectrum is needed. Observe that this is a general statement based only on the assumption of dominance of the scalar soft supersymmetry breaking terms and has no dependence on the supersymmetry breaking scheme. Larger gaugino masses do not help, since they not only increase the bottom mass corrections, but they also make $|m_{H_2}^2|$ larger.

Away from the infrared fixed point, the situation becomes more relaxed. In general, the closer one gets to the fixed point value, the heavier should the spectrum be. In order to minimize the bottom mass corrections with low values of the squark masses, the following condition needs to be fulfilled:

$$m_{H_2}^2(0)\left(\frac{7}{3}\frac{Y_f}{Y} - 1\right) \simeq m_Q^2(0)+m_U^2(0)-\frac{7}{10}\frac{Y_f}{Y}\left(1 - \frac{5Y}{7Y_f} - \left(1 - \frac{Y}{Y_f}\right)^{5/7}\right)\Delta_{DU}, \tag{39}$$

where $\Delta_{DU} = m_D^2(0) - m_U^2(0) + m_{H_1}^2(0) - m_{H_2}^2(0)$. The above condition ensures that $\mu^2 \simeq m_{H_2}^2 \ll m_{\tilde{q}}^2$. In addition, the condition $m_{H_1}^2 > m_{H_2}^2$ implies

$$m_{H_1}^2(0)-m_{H_2}^2(0) > \frac{1}{6}M_{E_{eff}}^2(0)\left[1 - \left(1 - \frac{Y}{Y_f}\right)^{1/4}\right]+\frac{3}{5}\left[1 - \left(1 - \frac{Y}{Y_f}\right)^{5/7}\right]\Delta_{DU}. \tag{40}$$

Finally, in order to avoid tachyonic solutions in the squark sector, we need,

$$m_U^2(0) > \frac{2}{3}m_{H_2}^2(0),$$

$$m_Q^2(0) > \frac{2}{3}m_{H_2}^2(0) + \frac{1}{5}\Delta_{DU}\left(1 - \left(1 - \frac{Y}{Y_f}\right)^{5/7}\right)$$

$$m_D^2(0) > \frac{2}{3}m_{H_2}^2(0) + \frac{2}{5}\Delta_{DU}\left(1 - \left(1 - \frac{Y}{Y_f}\right)^{5/7}\right). \tag{41}$$

The requirement that μ^2 is not correlated with the squark masses, Eq. (39), implies, for a given top quark mass, a very precise relationship between the different supersymmetry breaking parameters at the grand unification scale, which may only be fulfilled in particular supersymmetry breaking scenarios. The necessity of fulfilling the additional constraints given above reduces even more the degree of arbitrariness of the supersymmetry breaking parameters at high energies. The further away from the top quark mass fixed point we are, the easier the conditions are to fulfil. For example, for a top quark mass $M_t \simeq 175$ GeV, which corresponds to $Y/Y_f \simeq 0.8$, the above conditions may be fulfilled by a particularly simple set of supersymmetry breaking parameters: All the scalars acquire a universal soft supersymmetry breaking term m_0^2, apart from $m_{H_1}^2(0) \simeq 2m_0^2$.

One may wonder about the source of the non–universality of the soft super-symmetry breaking scalar masses. In principle, the SO(10) symmetry assures that all squark and sleptons belonging to a single representation of SO(10) ac-quire a common soft supersymmetry breaking term m_0^2, while the two Higgs doublets, belonging to a 10 of SO(10), have a common supersymmetry breaking mass $m_H^2(0)$ at the grand unification scale. There may be, however, additional sources of supersymmetry breaking at the SO(10) breaking scale. A particular natural one is the presence of a D–term associated with the necessary $U(1)_X$ gauge symmetry breakdown to reduce the rank of the SO(10) group to that of SU(5). If this term were present, it would break supersymmetry in a very specific way. The following boundary conditions would be obtained in this case [36],

$$m_{H_1}^2(0) = m_H^2(0) + 2m_X^2 \qquad\qquad m_{H_2}^2(0) = m_H^2(0) - 2m_X^2$$
$$m_U^2(0) = m_Q^2(0) = m_E^2(0) = m_0^2 + m_X^2 \qquad m_D^2(0) = m_L^2(0) = m_0^2 - 3m_X^2 \tag{42}$$

where m_X is the extra supersymmetry breaking term associated with the SO(10) gauge symmetry breaking. Observe that scalar fields transforming with the same quantum numbers under $SU(5)$ have degenerate masses. In this specific case, $\Delta_{DU} = 0$ and the above conditions take a particularly simple form. In particular, Eq. (39) leads to

$$2m_0^2 \simeq m_H^2(0) \left(\frac{7Y_f}{3Y} - 1 \right) - \frac{14Y_f}{3Y} m_X^2. \tag{43}$$

In addition, the requirement $m_{H_1}^2 > m_{H_2}^2$ puts a lower bound on m_X^2, while the requirement of a positive m_D^2 puts an upper bound on m_X^2; these are given by

$$\frac{72Y/Y_f}{7 \left[1 - (1 - Y/Y_f)^{1/4} \right]} + 2 > \frac{m_H^2(0)}{m_X^2} > \frac{2 \left(1 + 5Y/7Y_f\right)}{(1 - Y/Y_f)}. \tag{44}$$

The above conditions cannot be fulfilled for $Y/Y_f > 0.88$ ($M_t \geq 185$–190 GeV for $\alpha_3(M_Z) \simeq 0.12$–0.13), while for $Y/Y_f \simeq 0.8$ ($M_t \simeq 175$–180 GeV) they are only fulfilled for $m_H^2(0) \simeq 20\, m_X^2$ ($m_0^2 \simeq 0.8 m_H^2(0)$).

7   Conclusions

In this talk, we have analysed the radiative breakdown of the electroweak sym-metry within the Minimal Supersymmetric Standard Model, in the case in which the three Yukawa couplings of the third generation unify at the grand unification scale. We have shown that, due to large bottom quark mass corrections induced through the supersymmetry breaking sector of the theory at the one–loop level, the top quark mass predictions depend strongly on the nature of the soft supersymmetry breaking parameters at low energies. Moreover, the chargino contribution to the bottom mass corrections is strongly correlated with the contribution of these sparticles to the $b \to s\gamma$ decay rate.

In the interesting case of universal soft supersymmetry breaking parameters at the grand unification scale, the requirement of electroweak symmetry breaking leads to strong correlations between the mass parameter μ and the gaugino masses. The universal scalar mass is constrained to be smaller than the universal gaugino mass, with $M_{1/2}$ larger than 300 GeV. The result is a heavy sparticle spectrum with strong correlations between the different mass parameters. These correlations allow a precise computation of the bottom mass corrections, which turn out to be very significant. In fact, an upper bound on the top quark mass is derived, $M_t \leq 165$ (175) (185) GeV for $\alpha_3(M_Z) \simeq 0.12$ (0.125) (0.13). Moreover, in order to have the $b \to s\gamma$ decay rate within its experimental bounds, the common gaugino mass needs to be further constrained, $M_{1/2} > 550$ GeV. This implies a very heavy squark and gaugino spectrum, with the lightest squark mass larger than 1 TeV. A relatively light Higgs spectrum is obtained, with the lightest CP-even Higgs mass in the range 110–130 GeV and a charged Higgs mass bounded to be $C_{\tilde{q}} m_{\tilde{q}} \geq m_{H^\pm} \geq 180$ GeV, where $C_{\tilde{q}} = 0.15$–0.2 and $m_{\tilde{q}}$ is the characteristic third generation squark masses. For squark masses lower than 2 TeV, for example, the charged Higgs mass cannot be larger than 400 GeV.

The above conditions may be modified if non-universal soft supersymmetry breaking parameters at the grand unification scale are considered. We have shown that it is possible to minimize the supersymmetric bottom mass corrections and the contributions to the $b \to s\gamma$ decay rate, which allows to accommodate a heavier top quark mass. This demands the presence of light gauginos in the spectrum. The squarks and sleptons are in general heavy, with masses larger than or of the order of 1 TeV. A very heavy squark and slepton spectrum may only be avoided if very specific relations between the soft supersymmetry breaking parameters are fulfilled. Nevertheless, these relations may only be satisfied if there is a departure of the top quark mass from its infrared fixed point value. For example, if the only source of non–universality is associated with the breakdown of the SO(10) symmetry down to SU(5) via the corresponding U(1) D–term, the necessary requirements to avoid a heavy squark spectrum may be fulfilled only if $M_t \leq 185$–190 GeV for $\alpha_3(M_Z) \simeq 0.12$– 0.13, and for a very specific relation among the high energy mass parameters, $m_0^2 \simeq 0.8 m_H^2(0) \simeq 16 m_X^2$, where m_0^2 and $m_H^2(0)$ are the universal squark and Higgs mass parameters before the SO(10) breakdown, while m_X^2 is the D–term contribution. In the general case the condition of bottom–top Yukawa coupling unification requires specific relations among the soft supersymmetry breaking parameters of the theory, which demand a departure from the infrared fixed point solution of the top quark mass in order to allow moderate values for the squark mass spectrum.

Acknowledgements. The results presented here were partially obtained in collaboration with M. Olechowski and S. Pokorski, to whom we are grateful. We would also like to thank S. Dimopoulos, S. Raby, U. Sarid and R. Rattazzi for useful discussions. This work is partially supported by the Worldlab.

References

[1]   S. Dimopoulos, S. Raby and F. Wilczek, Phys. Rev. **D24** (1981) 1681;
S. Dimopoulos and H. Georgi, Nucl. Phys. **B193** (1981) 150;
L. Ibañez and G.G. Ross, Phys. Lett. **105B** (1981) 439; N. Sakai, Z. Phys. **C11** (1981) 153;
E. Witten, Nucl. Phys. **B188** (1981) 513.

[2]   J. Ellis, S. Kelley and D.V. Nanopoulos, Phys. Lett. **B260** (1991) 131;
P. Langacker and M.X. Luo, Phys. Rev. **D44** (1991) 817;
U. Amaldi, W. de Boer and H. Fúrstenaru, Phys. Lett. **B260** (1991) 817;
F. Anselmo, L. Cifarelli, A. Peterman and A. Zichichi, Nuovo Cimento **104A** (1991) 817.

[3]   P. Langacker and N. Polonsky, Phys. Rev. **D47** (1993) 4028.

[4]   M. Carena, S. Pokorski and C.E.M. Wagner, Nucl. Phys. **B406** (1993) 59.

[5]   S. Dimopoulos, L.J. Hall and S. Raby, Phys. Rev. Lett. **68** (1992) 1984; Phys. Rev. **D45** (1992) 4192;
G.W. Anderson, S. Raby, S. Dimopoulos and L. Hall, Phys. Rev. **D47** (1993) 3702.

[6]   G. Anderson, S. Raby, S. Dimopoulos, L.J. Hall and G.D. Starkman, LBL report LBL-33531, August 1993, to be published in Phys. Rev. **D**.

[7]   H. Arason, D. J. Castaño, B. Keszthelyi, S. Mikaelian, E. J. Piard, P. Ramond and B. D. Wright, Phys. Rev. Lett. **67** (1991) 2933.

[8]   S. Kelley, J.L. Lopez and D.V. Nanopoulos, Phys. Lett. **B278** (1992) 140.

[9]   V. Barger, M.S. Berger and P. Ohmann, Phys. Rev. **D 47** (1993) 1093;
V. Barger, M.S. Berger, P. Ohmann and R.J.N. Phillips, Univ. Wisconsin preprint MAD/PH/755, April 1993.

[10]   C. T. Hill, Phys. Rev. **D24** (1981) 691;
C. T. Hill, C. N. Leung and S. Rao, Nucl. Phys. **B262** (1985) 517.

[11]   J. Bagger, S. Dimopoulos and E. Masso, Phys. Rev. Lett. **55** (1985) 920.

[12]   M. Carena, T.E. Clark, C.E.M. Wagner, W.A. Bardeen and K. Sasaki, Nucl. Phys. **B369** (1992) 33.

[13]   See, for example, H. Arason, D.J. Castaño, B. Keszthelyi, S. Mikaelian, E.J. Piard, P. Ramond and B.D. Wright, Phys. Rev. **D46** (1992) 3945.

[14]   P. Langacker and N. Polonsky, Phys. Rev. **D49** (1994) 1454.

[15]   W.A. Bardeen, M. Carena, S. Pokorski and C.E.M. Wagner, Phys. Lett. **B320** (1994) 110.

[16] M. Olechowski and S. Pokorski, Phys. Lett. **B214** (1988) 393;
S. Pokorski, "How is the isotopic symmetry of quark masses broken?", in Proc. XIIth Int. Workshop on Weak Interactions and Neutrinos, Ginosar, Israel, (Nucl. Phys. **B13** Proc. Supp. 1990 p. 606).

[17] B. Anantharayan, G. Lazarides and Q. Shafi, Phys. Rev. **D44** (1991) 1613.

[18] H.P. Nilles, "Beyond the Standard Model", in Proc. of the 1990 Theoretical Advanced Study Institute in Elementary Particle Physics, eds. M. Cvetic and P. Langacker, (World Scientific, Singapore, 1990, p. 633);
P.H. Chankowski, Diploma Thesis, University of Warsaw (1990);
A. Seidl, Diploma Thesis, Technical University, Munich (1990);
W. Majerotto and B. Mösslacher, Z. Phys. **C48** (1990) 273;
S. Bertolini, F. Borzumati, A. Masiero and G. Ridolfi, Nucl. Phys. **B353** (1991) 591;
M. Drees and M.M. Nojiri, Nucl. Phys. **B369** (1992) 54;
B. Anantharayan, G. Lazarides and Q. Shafi, Phys. Lett. **300B** (1993) 245.

[19] B. Anantharayan and Q. Shafi, Bartol Research Institute preprint BA-93-25-REV, Nov. 1993.

[20] M. Olechowski and S. Pokorski, Nucl. Phys. **B404** (1993) 590.

[21] T. Banks, Nucl. Phys. **B303** (1988) 172.

[22] L.J. Hall, R. Rattazzi and U. Sarid, preprint LBL - 33997, June 1993.

[23] R. Hempfling, preprint DESY-93-092, July 1993.

[24] M. Carena, M. Olechowski, S. Pokorski and C.E.M. Wagner, Nucl. Phys. **B419** (1994) 213.

[25] M. Carena and C. Wagner, preprint CERN.TH.7320/94, these proceedings.

[26] M. Carena, M. Olechowski, S. Pokorski and C.E.M. Wagner, preprint CERN-TH.7163/94, to appear in Nucl. Phys. **B**.

[27] K. Hikasa et al., Particle Data Group, Phys. Rev. **D45** (1992).

[28] K. Inoue, A. Kakuto, H. Komatsu and S. Takeshita, Prog. Theor. Phys. **67** (1982) 889 and **68** (1983) 927;
J.P. Derendinger and C.A. Savoy, Nucl. Phys. **B253** (1985) 285;
B. Gato, J. Leon, J. Perez–Mercader and M. Quirós, Nucl. Phys. **B253** (1985) 285;
N.K. Falck, Z. Phys. **C30** (1986) 247.

[29] S. Bertolini, F. Borzumati, A. Masiero and G. Ridolfi, Nucl. Phys. **B353** (1991) 591.

[30] R. Barbieri and G. R. Giudice, Phys. Lett. **B309** (1993) 86.

[31] F.M. Borzumati, preprint DESY 93-090, August 1993.

[32] M. A. Diaz, Phys. Lett. **B322** (1994) 207.

[33] S. Bertolini and F. Vissani, preprint SISSA 40/94/EP, March 1994.

[34] R. Ammar et al., Phys. Rev. Lett. **71** (1993) 674.

[35] S. Pokorski, Munich preprint MPI-PH-93-76, Oct. 1993.
A.J. Buras, M. Misiak, M. Munz and S. Pokorski, Munich preprint MPI-Ph/93-77, Nov. 1993.

[36] R. Rattazzi, U. Sarid and L.J. Hall, Stanford Univ. report SU-ITP-94-15, May 1994, these proceedings.

Aspects of Radiative Electroweak Breaking in the Minimal Supersymmetric Extension of the Standard Model

DIEGO J. CASTAÑO
CENTER FOR THEORETICAL PHYSICS
M.I.T., CAMBRIDGE, MA 02139

Abstract: The dependence of the top quark mass predictions of the MSSM with soft supersymmetry breaking on various parameters is discussed. It is found that the top mass is very sensitive to the precise value of the ratio of bottom to τ Yukawa couplings at the unification scale. Furthermore, it is shown that the model may appear unnatural when fine tuning is assessed by the conventional criterion.

1 Introduction

In this talk, aspects of the Minimal Supersymmetric Extension of the Standard Model (MSSM) are discussed. This model has grown in popularity since it was shown using accurate measurements of the strong coupling that gauge coupling unification was possible for supersymmetry (SUSY) breaking masses in the TeV range.[1, 2, 3] In the Standard Model (SM), the gauge couplings fail to meet at a common scale, thus forming a "GUT triangle." Furthermore, constraints inspired by predictions of Grand Unified Models (GUTs) [4, 5, 6, 7, 8] have been used to predict the top quark mass in the MSSM.[9, 10, 11, 12, 13, 14]

Supersymmetry is broken explicitly in the MSSM through the addition of soft breaking terms motivated by the low energy limit of supergravity (SUGRA) models. The complete sparticle spectrum is then calculable using the renormalization group (RG). Furthermore, electroweak breaking can occur radiatively in the model, if the top quark Yukawa is large enough. This model introduces several new unknown input parameters associated with the SUSY breaking soft terms. Many papers have studied this model and its phenomenological consequences.[15, 16, 17, 18, 19, 20]

After a brief introduction to the MSSM and radiative electroweak breaking, some of the factors that affect the value of the top quark mass are discussed. It is shown that the top mass is extremely sensitive to some parameters. A discussion on the issue of fine tuning and the naturalness constraint follows. Finally, there are some concluding remarks.

This talk is based on work done in Ref. [19] and some preliminary, ongoing work [21]. A full description of the methods employed can be found in Refs. [22, 19].

2 The MSSM

The MSSM is based on the same gauge group as the Standard Model, $SU(3)_c \times SU(2)_W \times U(1)_Y$. It contains a second Higgs field to be consistent with supersymmetry (SUSY) and to avoid both triangle and $SU(2)$ global anomalies. All the chiral interactions of the MSSM are described by the superpotential

$$W = \hat{\bar{u}} Y_u \hat{\Phi}_u \hat{Q} + \hat{\bar{d}} Y_d \hat{\Phi}_d \hat{Q} + \hat{\bar{e}} Y_e \hat{\Phi}_d \hat{L} + \mu \hat{\Phi}_u \hat{\Phi}_d{}' \tag{1}$$

where the hat indicates a chiral superfield and all fermions are left-handed Weyl fields. The μ mixing term is the standard and simplest way to break the Peccei-Quinn symmetry and to thereby avoid a phenomenologically disastrous axion. The model contains all the particles of the SM plus thirty-one new ones. Twenty-eight of these are supersymmetric partners of SM particles and three are new Higgs bosons.

Supersymmetry must be broken to reconcile the model with the lack of experimental evidence for superparticles. Since no adequate model of spontaneously broken global SUSY exists, supersymmetry is customarily broken through the introduction of explicit soft terms which do not reintroduce quadratic divergences into the theory. Low energy supergravity provides the motivation for the introduction of these soft breaking terms. The spontaneous breaking of the local $N = 1$ supersymmetry in some "hidden" sector is communicated to the "visible" sector by weak, gravitational interactions and manifests itself as explicit soft breaking terms in the scalar potential of the low energy effective theory. The most general form of the soft SUSY breaking potential is (including gaugino mass terms)

$$V_{soft} = m^2_{\Phi_u}\Phi_u^\dagger\Phi_u + m^2_{\Phi_d}\Phi_d^\dagger\Phi_d + B\mu(\Phi_u\Phi_d + h.c.)$$

$$+ \sum_i \left(m^2_{\tilde{Q}_i}\tilde{Q}_i^\dagger\tilde{Q}_i + m^2_{\tilde{L}_i}\tilde{L}_i^\dagger\tilde{L}_i + m^2_{\tilde{u}_i}\tilde{\bar{u}}_i^\dagger\tilde{\bar{u}}_i + m^2_{\tilde{d}_i}\tilde{\bar{d}}_i^\dagger\tilde{\bar{d}}_i + m^2_{\tilde{e}_i}\tilde{\bar{e}}_i^\dagger\tilde{\bar{e}}_i \right)$$

$$+ \sum_{i,j} \left(A_u^{ij}Y_u^{ij}\tilde{\bar{u}}_i\Phi_u\tilde{Q}_j + A_d^{ij}Y_d^{ij}\tilde{\bar{d}}_i\Phi_d\tilde{Q}_j + A_e^{ij}Y_e^{ij}\tilde{\bar{e}}_i\Phi_d\tilde{L}_j + h.c. \right) ,$$

$$+ \frac{1}{2}\sum_{l=1}^{3} M_l\lambda_l\lambda_l + h.c. .$$

A generic feature of these SUGRA inspired models is universality in the the soft terms. One assumes the following boundary conditions for the masses and trilinears

$$m_{\tilde{Q}_i} = m_{\tilde{\bar{u}}_i} = m_{\tilde{\bar{d}}_i} = m_{\tilde{L}_i} = m_{\tilde{\bar{e}}_i} = m_{\Phi_u} = m_{\Phi_d} = m_0 , \tag{2}$$

$$A_u^{ij} = A_d^{ij} = A_e^{ij} = A_0 . \tag{3}$$

This universality is important in avoiding unwanted flavor changing neutral current effects.

One of the great successes of the MSSM is gauge coupling unification at scales consistent with proton decay bounds.[1, 2, 3] In this presentation, the unification scale is fixed to be $M_X = 10^{16}$ GeV. Furthermore, universality of the soft terms will be assumed at this scale. Since the gauge couplings unify, it is natural to take the gaugino masses equal here as well

$$M_1 = M_2 = M_3 = m_{1/2} . \tag{4}$$

The model introduces five new parameters, m_0, A_0, $m_{1/2}$, B_0, and μ_0. However, it is very predictive since these account for the masses of thirty-one new particles.[23]

3 Radiative Electroweak Breaking

A feature of these models is electroweak breaking through radiative effects.
[24, 25, 26, 27] The starting point is the 1-loop effective Higgs potential

$$V_{1-loop} = V_0 + \Delta V_1 \ . \tag{5}$$

The tree level potential is given by

$$
\begin{aligned}
V_0 = {}& (m_{\Phi_d}^2 + \mu^2)|\Phi_d|^2 + (m_{\Phi_u}^2 + \mu^2)|\Phi_u|^2 \\
& + B\mu(\Phi_u\Phi_d + h.c.) \\
& + \frac{g'^2}{8}(|\Phi_u|^2 - |\Phi_d|^2)^2 \\
& + \frac{g_2^2}{8}(\Phi_u^\dagger \vec{\tau}\Phi_u + \Phi_d^\dagger\vec{\tau}\Phi_d)^2 \ ,
\end{aligned}
\tag{6}
$$

and the expression for the 1-loop correction is

$$\Delta V_1 = \frac{1}{64\pi^2}\sum_i (-1)^{2s_i}(2s_i + 1)m_i^4\left(\ln\frac{m_i^2}{Q^2} - \frac{3}{2}\right) \tag{7}$$

where the m_i represent the field dependent masses of the particles of the model
and the s_i the associated spins. Using the renormalization group, the parameters are evolved to low energies where the potential attains validity. This
RG improvement uncovers electroweak symmetry breaking. The exact low energy scale at which to minimize is unimportant as long as the 1-loop effective
potential is used and the scale is in the expected electroweak range. The minimization scale will arbitrarily be taken to be M_Z. If the electroweak symmetry
is broken, minimization yields non-zero values for the vacuum expectation values (VEVs) of the the two Higgs fields, v_u and v_d, or equivalently $v = \sqrt{v_u^2 + v_d^2}$
and $\tan\beta = v_u/v_d$. For the purposes of this presentation, the two minimization
conditions are expressed as follows

$$\mu^2(M_Z) = \frac{\overline{m}_{\Phi_d}^2 - \overline{m}_{\Phi_u}^2\tan^2\beta}{\tan^2\beta - 1} - \frac{1}{2}m_Z^2 \ , \tag{8}$$

$$B(M_Z) = \frac{(\overline{m}_{\Phi_u}^2 + \overline{m}_{\Phi_d}^2 + 2\mu^2)\sin 2\beta}{2\mu(M_Z)} \ , \tag{9}$$

where $\overline{m}_{\Phi_{u,d}}^2 = m_{\Phi_{u,d}}^2 + \partial\Delta V_1/\partial v_{u,d}^2$.

Clearly in these models, demanding electroweak symmetry breaking puts
constraints on the parameters. It was recognized from the models' inception
that the top quark Yukawa coupling was one parameter that had to be large
enough in order to achieve the desired radiative breaking. Imposing the minimization conditions at M_Z effectively removes two of the five new parameters
at M_X discussed above. The new unknowns are then taken to be A_0, m_0, $m_{1/2}$,
and the $sign(\mu)$ since it is undetermined in Eq. (8). Furthermore, there is also
$\tan\beta$ which is absent in the SM. The exact method used to solve for B_0 and
μ_0 is discussed in [19]. Basically, a solution for B_0 and μ_0 is found consistent
with the RG and such that Eqs. (8) and (9) are satisfied given inputs for A_0,
m_0, $m_{1/2}$, $\tan\beta_0$, and the $sign(\mu)$.

4 GUT Constraints

One of the best motivated predictions of GUTs is the mass relation

$$m_b = m_\tau \ . \tag{10}$$

In the SM, this equality occurs at approximately 10^9 GeV. However, in the MSSM this relation holds at the gauge unification scale for appropriate values of the top quark mass. Consequently, as a constraint, it provides a means of predicting the top mass.[11] In this presentation, the ratio $R_0 = m_b(M_X)/m_\tau(M_X)$ will taken equal to 1, but can be a free input of the model.

The results being reported on are based on an analysis using $M_b = 4.9$ GeV and $M_\tau = 1.78$ GeV. Also, the value of the strong coupling at M_Z is taken as $\alpha_3(M_Z) = .113$. The corresponding value of the strong coupling at M_X is determined based on this low energy constraint. The values of $\alpha_1(M_X)$ and $\alpha_2(M_X)$ are set equal and fixed. This constant value for $\alpha_{1,2}$ at M_X never leads to more than about 5% error in α_{em} and $\sin^2\theta_W$. The difference in $\alpha_3(M_X)$ and $\alpha_{1,2}(M_X)$ can be accounted for by GUT thresholds.

5 Implications for the Top Quark Mass

The model being considered has four free input parameters A_0, m_0, $m_{1/2}$, $\tan\beta_0$, and $sign(\mu)$. Taken as constraints on the system are: Electroweak breaking in the form of two minimization conditions at M_Z, the masses of the bottom quark and τ lepton, the mass of the Z boson, and the value of the strong coupling at M_Z. Solutions for B_0, μ_0, $y_\tau(M_X)$, $v(M_X)$, $\alpha_3(M_X)$, and M_t are then found consistent with the RG, the constraints, and values for the free input parameters. Furthermore, a non-degenerate spectrum of sparticle masses also results.

Not all input values will yield solutions, and the four dimensional parameter space must be explored and restricted using other constraints such as experimental bounds on sparticle masses. A very exhaustive search of the whole parameter space has recently been preformed.[20] For this presentation, it will suffice to look at the following representative section: $0 < A_0/m_{1/2} < 3$, $0 < m_0/m_{1/2} < 3$, $90\,GeV < m_{1/2} < 500\,GeV$, $.65 < \tan\beta_0 < 15$, $sign(\mu) = +$. All solutions in this section yield top quark masses in the range $150\ GeV < M_t < 200\ GeV$, and $1.4 < \tan\beta < 30$. Because of the GUT constraint, the the top quark mass results are sensitive to the mass of the bottom quark used. In Fig. 1, the variation in M_t with $m_b(m_b)$ is displayed in a particular case with $A_0 = 0$, $m_0 = 0$, $m_{1/2} = 120$ GeV, $\tan\beta = 4.3$. Given that $\alpha_3(M_Z) = .113$, the left end of the curve corresponds to a pole mass $M_b = 4.6$ GeV, and the right end to a mass $M_b = 5.1$ GeV. The top mass varies by ~ 15 GeV over this range.

Another parameter that affects the value of the top mass is $\alpha_3(M_Z)$. From the one loop β-functions for the Yukawa couplings in the MSSM, it follows that

$$\frac{d\ln(y_b/y_\tau)}{d\ln Q} = \frac{1}{(4\pi)^2}\left(y_t^2 - \frac{16}{3}g_3^2\right)\ , \tag{11}$$

where only the top Yukawa and strong gauge couplings are kept. Increasing the value of the strong coupling drives the scale of bottom-τ unification towards lower energy and requires larger top quark masses to restore. Figure 2 displays this behavior. Again the top quark mass varies by ~ 15 GeV in this case as the strong coupling varies from .110 to .120.

The next figure displays the very significant variation in M_t with the ratio, R_0, of bottom to τ masses at M_X. The top mass is highly sensitive to the exact value of R_0 as is evident from the figure. The top mass varies by ~ 100 GeV as the ratio varies from .8 to unity.

6   Fine Tuning

Since for certain values of the input parameters, some parameters, such as y_t, have to be unnaturally tuned to achieve the correct electroweak breaking, a criterion was put forward to assess fine tuning.[28] Fine tuning is avoided provided the quantities

$$c_i = \left| \frac{a_i}{M_Z^2} \frac{\partial M_Z^2}{\partial a_i} \right| , \tag{12}$$

where the a_i are the most general parameters of the theory, are less than some upper limit, Δ. Reasonable values for Δ have varied in the literature from 10 to as high as 100. The fine tuning constraint has been used to restrict the available parameter space and often amounts to setting an upper limit on the masses of the superparticles. It would be interesting to know the range of validity of this criterion, and how well it differentiates between harmless sensitivity and unwanted fine tuning.[21]

We find that over the section of parameter space explored here, the c-parameter associated with g_3 is much larger than any other. Since all the c's are ostensibly on equal footing, imposing $c_{g_3} < \Delta$ can be overly restrictive. Figure 4 is the typical plot of M_t versus $\tan \beta$ for the case $A_0 = m_0 = 0$, $m_{1/2} = 120$ GeV. Each point on the curve represents a viable scenario in the MSSM.

In Fig. 5, the c_{y_t}'s for these points are plotted against $\tan \beta$. There is a range of points with $c_{y_t} < 50$, but there are no points with $c_{y_t} < 25$.

Figure 6 displays c_{g_3} versus $\tan \beta$. The values of c_{g_3} are dramatically larger than the corresponding c_{y_t}. Even using $\Delta = 100$ would rule out all points in this case using the fine tuning criterion as a constraint.

7   Conclusions

The Minimal Supersymmetric Extension of the Standard Model was reviewed. Supersymmetry breaking was implemented using explicit soft breaking terms motivated by the low energy limit of supergravity models. This model has many attractive features. The gauge couplings unify at a scale $\sim 10^{16}$ GeV. Imposing the GUT-inspired relation $m_b = m_\tau$ leads to predictions for the top quark mass. In the present analysis, it was found that $150\,GeV < M_t < 200\,GeV$. The exact value of the top mass is sensitive, however, to several input parameters such as M_b and $\alpha_3(M_Z)$. It is extremely sensitive to the value of the ratio m_b/m_τ at M_X. A percent variation in this ratio can shift the top mass by as much as 10 GeV.

Another feature of this model is radiative electroweak symmetry breaking. Renormalization group improvement of the 1-loop effective potential will uncover electroweak breaking, although the breaking appears absent from the theory at tree level at M_X. One important aspect of this model is fine tuning. Some ranges of the parameters require fine tuning to achieve the desired electroweak symmetry breaking. The standard criterion used to determine this kind of unnaturalness was reviewed. Over the region of soft parameter space explored, the model appeared unreasonably fine tuned with respect to the strong coupling. It is unclear whether this criterion is reflecting fine tuning and can always be trusted.[21]

8 Acknowledgements

I thank Greg Anderson, Jonathan Bagger, Steve Martin, Eric Piard, and Pierre Ramond for discussions that have contributed to this talk.

References

[1] U. Amaldi, W. de Boer, and H. Fürstenau, Phys. Lett. **260B** (1991) 447 .

[2] J. Ellis, S. Kelley, and D. Nanopoulos, Phys. Lett. **260B** (1991) 131.

[3] P. Langacker and M. Luo, Phys. Rev. D **44** (1991) 817.

[4] J. C. Pati and A. Salam, Phys. Rev. D **10** (1974) 275.

[5] H. Georgi and S. Glashow, Phys. Rev. Lett. **32** (1974) 438.

[6] H. Georgi, in *Particles and Fields-1974*, edited by C. E. Carlson, AIP Conference Proceedings No. 23 (American Institute of Physics, New York, 1975) p. 575.

[7] H. Fritzsch and P. Minkowski, Ann. Phys. NY **93** (1975) 193.

[8] F. Gürsey, P. Ramond, and P. Sikivie, Phys. Lett. **60B** (1975) 177.

[9] M. B. Einhorn and D. R. T. Jones, Nuc. Phys. **B196** (1982) 475.

[10] J. E. Björkman and D. R. T. Jones, Nuc. Phys. **B259** (1985) 533.

[11] H. Arason, D. J. Castaño, B. Keszthelyi, S. Mikaelian, E. J. Piard, P. Ramond, and B. D. Wright, Phys. Rev. Lett. **67** (1991) 2933.

[12] B. Ananthanarayan, G. Lazarides, and Q. Shafi, Phys. Rev. D **44** (1991) 1613.

[13] A. Giveon, L.J. Hall, and U. Sarid, Phys. Lett. **271B** (1991) 138.

[14] S. Kelley, J. L. Lopez, and D. V. Nanopoulos, Phys. Lett. **274B** (1992) 387.

[15] G. G. Ross and R. G. Roberts, Nuc. Phys. **B377** (1992) 571.

[16] S. Kelly, J. Lopez, H. Pois, D. V. Nanopoulos, and K. Yuan, Phys. Lett. **273B** (1991) 423.

[17] R. Arnowitt and P. Nath, Phys. Lett. **287B** (1992) 89; Phys. Rev. Lett. **69** (1992) 725.

[18] P. Nath and R. Arnowitt, Phys. Lett. **289B** (1992) 368.

[19] D. J. Castaño, E. J. Piard, and P. Ramond, Report No. UFIFT-HEP-93-13, to be published in Phys. Rev. D.

[20] G. L. Kane, C. Kolda, L. Roszkowski, and J. D. Wells, Report No. UM-TH-93-24.

[21] G. W. Anderson and D. J. Castaño, work in progress.

[22] H. Arason, D. J. Castaño, B. Keszthelyi, S. Mikaelian, E. J. Piard, P. Ramond, and B. D. Wright, Phys. Rev. D **46** (1992) 3945.

[23] S. Martin and P. Ramond, Phys. Rev. D **48** (1993) 5365.

[24] L. E. Ibáñez and G. G. Ross, Phys. Lett. **110B** (1982) 215.

[25] K. Inoue, A. Kakuto, H. Komatsu, and S. Takeshita, Prog. Theor. Phys. **68** (1982) 927.

[26] L. Alvarez-Gaumé, J. Polchinski, M. B. Wise, Nuc. Phys. **B221** (1983) 495.

[27] J. Ellis, J. S. Hagelin, D. V. Nanopoulos, and K. Tamvakis, Phys. Lett. **125B** (1983) 275.

[28] R. Barbieri and G. F. Giudice, Nuc. Phys. **B306** (1988) 63.

Fermion Masses from Superstring Models with Adjoint Scalars.

SHYAMOLI CHAUDHURI
DEPT. OF PHYSICS AND INSTITUTE FOR THEORETICAL PHYSICS
UNIVERSITY OF CALIFORNIA,
SANTA BARBARA, CA 93106

STEPHEN-WEI CHUNG AND JOSEPH LYKKEN
FERMI NATIONAL ACCELERATOR LABORATORY
P.O. BOX 500, BATAVIA, IL 60510

Abstract: We explore the possibility of embedding supersymmetric GUT texture ideas into superstring models. We discuss the construction of GUT models in the free fermionic string construction. We find $SO(10)$ models with adjoint scalars, three generations of chiral fermions, and a fundamental Higgs with a Yukawa coupling to only the third generation.

1   Introduction

Supersymmetric grand unified (GUT) texture models have had considerable success reproducing the existing data for fermion masses and mixings, starting from a restricted set of effective operators at the GUT scale. Two generic features of such schemes suggest that it may be desirable, or even necessary, to eventually embed these structures into full-fledged superstring models.

The first feature has to do with how SUSY GUT texture models generate small numbers. Fermion masses and mixings exhibit a number of distinct hierarchies, characterized by ratios in the range roughly 1/10 - 1/100 or even smaller, depending on how one (arbitrarily) chooses to parametrize them. A popular idea in texture models[1, 2] is that most of these small numbers arise, not from small Yukawa couplings, but rather from replacing Yukawas with higher dimension operators suppressed by powers of

$$\frac{\langle \Phi_{adjoint} \rangle}{M_X} \, , \tag{1}$$

where the numerator is the vev of an adjoint scalar, assumed to be the GUT scale of about 10^{16} GeV, while M_X is an even higher mass scale, assumed to be roughly 10^{17} GeV. Thus, for example, in one of the models of Anderson et al[3], the following $SO(10)$ invariant dimension 6 operator resides in the 2-3 component of the charged fermion mass matrices:

$$O_{23} = 16_2 \, 10 \, \frac{45_{B-L}^2}{45_1^2} \, 16_3 \quad . \tag{2}$$

The superheavy scale M_X is verging on the string scale, which is estimated as $M_{string} \sim 5 \times 10^{17} \text{GeV} \times g_{string}$. At any rate, the use of nonrenormalizable operators sensitive to such high scales entails the risk of being overwhelmed by

"Planck slop". Since superstring theory is the only known method of controlling Planck slop, it may be necessary to invoke strings in order to make valid statements about SUSY GUT textures.

The second feature of texture models which strongly suggests a string interpretation are the textures themselves. The usefulness of these schemes requires that the number of effective GUT scale operators which make the dominant contributions to the fermion mass matrices is rather few. In particular, the number of effective operators up to, say, dimension 6, should be considerably less than the full set of effective operators allowed by the unbroken gauge symmetries at the GUT scale. This requires that there are nontrivial flavor-sensitive selection rules, which provide the extra contraints and thus the desired texture. Superstrings are an obvious source for such selection rules.

In string models, couplings correspond to correlators of vertex operators on the worldsheet, and their vanishing depends on worldsheet symmetries. Since many of these symmetries do not correspond to unbroken gauge symmetries in spacetime, the effective spacetime field theory below the string scale is generically subject to a number of selection rules. Of course, it is possible to avoid strings and obtain selection rules from extra broken $U(1)$'s, discrete symmetries, R symmetries, and the like. However superstring theory not only provides similar symmetries and selection rules, but also provides a more fundamental understanding of them. This is true even though we at present have no clue how nonperturbative string dynamics selects among the vast number of perturbatively degenerate string vacua.

To see why, observe that concrete phenomenological inputs can help narrow down the allowed realistic string vacua to some reasonable number. Given the SUSY GUT texture framework, as well as improved low energy data, it is plausible that this can be done. What is required is that one translate the phenomenological constraints on the GUT scale effective Lagrangian into constraints on the world sheet symmetries of the string models. Then, for each particular string model, which fixes a choice of the string vacuum, one can hope to extract relationships between, say, the worldsheet structure that guarantees three light generations of chiral fermions, and the worldsheet structure that guarantees one of the observed hierarchies in the fermion mass matrices. Any such relationships (if valid) would be a profound new insight into particle physics. It is also important to note that such relationships depend only on order-of-magnitude computation, and thus do not (necessarily) require the ability to make precise determinations of string moduli.

2 Superstring GUT's

Historically, superstring model builders have made very little contact with conventional GUT's. One of the reasons for this is that, in string theory, gauge coupling unification occurs at the string scale independently of whether or not matter fields assemble into GUT multiplets[4]. Thus the agreement between LEP data and minimal SUSY gauge coupling unification does not signify the existence of GUT's in a string context[5, 6, 7]. Many quasi-realistic string models thus attempt to identify the gauge coupling unification scale with the string scale, either by raising the former or by effectively lowering the latter, and do not obviously present two independent superheavy scales.

A technically important reason why so little effort has been put into superstring GUT's is the fact that most string model constructions simply do not allow the appearance of adjoint GUT scalars in the massless spectrum at the string scale. Thus conventional GUT symmetry-breaking vevs are not available. This is because in most string constructions the GUT gauge group is realized at Kac-Moody level one. The Kac-Moody level is a positive integer label needed to specify unitary irreducible representations (irreps) when Lie algebras are combined with worldsheet conformal symmetry to produce Kac-Moody algebras. For fixed level there is a constraint on the allowed irreps for massless matter fields. Thus at level one the allowed irreps of $SU(5)$ and $SO(10)$ are given by:

$$SU(5) : 1, 5, \bar{5}, 10, \overline{10}$$
$$SO(10) : 1, 10, 16, \overline{16}$$

A more precise statement is that massless adjoint scalars are incompatible with the presence of massless chiral fermions in level one string models[8].

It is possible to build level one string models with unconventional structures which are similar to GUT's. The flipped $SU(5)$ model[9] is the most developed example of this; there the breaking of $SU(5)$ is accomplished by vevs of a Higgs 10, rather than the adjoint. Since the $SU(3) \times SU(2)$ singlet in the 10 has nonzero hypercharge, flipped $SU(5)$ requires an extra $U(1)$ and a nonstandard treatment of hypercharge. While it is not a pure GUT, flipped $SU(5)$ is a useful prototype for quasi-realistic models in the free fermionic superstring construction. Indeed we have borrowed some of its worldsheet structure in building the superstring GUT's described below.

Yet another reason why superstring GUT's have been neglected is the fear of exotics. A conventional superstring GUT requires a string construction with Kac-Moody level at least two. Higher levels are required if scalars in other non-fundamental irreps are desired. For example, to obtain a massless 126 of $SO(10)$ would require[10] a string model with $SO(10)$ at level $\geq$ five. Higher levels introduce the possibility that modular invariance of the string model will require the presence of other exotic scalars whenever, say, the adjoint or the 126 are present. Indeed the simplest Kac-Moody modular invariants, the left-right symmetric diagonal invariants, require that all *allowed* irreps do in fact appear in the spectrum. Thus one might worry that for $SO(10)$ at level two the 54 would appear in addition to the 45, while at level five the 54, 120, 144, etc would appear in addition to the 45 and 126. We will show below that this expectation is false for fermionic string models, and thus that there is no problem with unwanted exotics.

There are two basic examples of superstring GUT constructions in the literature. The first, due to Lewellen[8], is a free fermionic string model with $SO(10)$ at Kac-Moody level two, adjoint scalars, and chiral fermions. The second, due to Font, Ibanez, and Quevedo[10], is an orbifold construction that realizes the GUT group G at level n starting with n copies of G at level one. We do not know of any particular reason to prefer one of these constructions over the other. However, we have chosen to base our exploration of superstring

GUT's on Lewellen's free fermionic string model.

3   Model Building

Four dimensional closed free fermionic string models are heterotic superstring vacua described by a worldsheet lagrangian for 64 real (Majorana-Weyl) free fermions, together with the bosons that embed the 4-d spacetime, and ghosts. This construction is described in detail in refs *[11, 12, 13, 14]*; we will, for the most part, follow the notation and conventions of refs *[14, 8]*. Models are conveniently specified by their one-loop partition functions; these involve a sum over spin structures:

$$Z_{fermion} = \sum_{\alpha,\beta} C^\alpha_\beta \, Z^\alpha_\beta \quad , \tag{3}$$

where the C^α_β's are numerical coefficients, while α and β are 64-dimensional vectors labelling different choices of boundary conditions for the fermions around the two independent cycles of the worldsheet torus. For each real fermion there are two possible choices of boundary conditions around a given cycle: either periodic (Ramond) or antiperiodic (Neveu-Schwarz). However for fixed α and β the real fermions always pair up into either Majorana or Weyl fermions; if a particular Weyl pairing occurs consistently across *all* α and β, then this pair can be regarded as a single *complex* fermion. For such complex fermions more general boundary conditions -any rational "twists"- are then allowed*[11, 13, 15]*. A useful notation denotes a complex Ramond fermion as a $-1/2$ twist, while a general m/n twist indicates the complex fermion boundary condition

$$\Psi \to exp\left[2\pi i\,\frac{m}{n}\right]\Psi \quad . \tag{4}$$

It is convenient to regard the partition function as a sum over physical "sectors" labelled by the α's. The contribution of any sector α to the partition function contains a generalized GSO projection operator. Up to an overall constant, this is given by:

$$\sum_\beta C^\alpha_\beta \, exp\left[-2\pi i\beta \cdot \hat{N}(\alpha)\right] \quad , \tag{5}$$

where $\hat{N}(\alpha)$ is the fermion number operator defined in the sector α. There are subtleties in the proper definition of $\hat{N}(\alpha)$ for real Ramond fermions; these are discussed in ref *[14]*.

Thus building a fermionic string model amounts to choosing an appropriate set of α's, β's, and C^α_β's, then performing the GSO projections to find the physical spectrum. These choices are greatly constrained by the requirement of modular invariance of the one-loop partition function; in addition, higher loop modular invariance imposes a factorization condition on the C^α_β's. Together these requirements imply that the $\{\beta\}$ are the same set of vectors as the $\{\alpha\}$, and that, if two sectors α_1 and α_2 appear in the partition function, then the sector $\alpha_1 + \alpha_2$ must also appear. These facts allow one to specify the full set of α's and β's by a list of "basis vectors", denoted V_i.

Of the 64 real fermions, 20 are right-moving and 44 are left-moving. The first pair of right-movers are spacetime fermions (corresponding to the two transverse directions in 4-d), while the other 18 right-movers are "internal". The requirement of a worldsheet supercurrent contructed out of the right-movers and the spacetime bosons is a consistency contraint on model building. As a result there is always a sector -denoted V_1- that contains massless gravitinos, corresponding to an $N=4$ spacetime supersymmetry before the GSO projection. After the projection one may have $N=4$, $N=2$, $N=1$, or no spacetime SUSY at all. We will only consider models with $N=1$ spacetime SUSY.

Fermionic string models always contain an "untwisted sector", with 32 complex Neveu-Schwarz fermions. The untwisted sector contains the graviton, dilaton, and antisymmetric tensor field. It generally also contains gauge bosons, and massless scalars including gauge-singlet moduli.

The first step in building a fermionic string model for a SUSY GUT is to find an embedding of the GUT root lattice in the left-moving fermions. As discussed in ref *[8]*, this means identifying the root vectors with vectors of "fermionic charges" for n complex left-movers, where n is $\geq$ the rank of the gauge group. The fermionic charge of the Neveu-Schwarz vacuum is 0; it is ± 1 for an excited Neveu-Schwarz fermion/antifermion. For the Ramond vacuum the charge is $\pm 1/2$, corresponding to the two degenerate vacuum states. Thus using complex Neveu-Schwarz and Ramond fermions, root vectors have components taking only the values 0, $\pm 1/2$, and ± 1. The length-squared of each root vector is equal to $2/k$, where k is the Kac-Moody level.

Consider a particular example. A level two embedding of $SO(10)$ using, let's say, only Neveu-Schwarz and Ramond fermions, requires that we find 5 simple roots, each with length-squared one, whose inner products reproduce the Cartan matrix of $SO(10)$, and whose components take only the values 0, $\pm 1/2$, and ± 1. Such embeddings exist for any number of complex fermions ≥ 6 (although $SO(10)$ has rank five, there is no solution with only 5 complex fermions).

The minimal embedding, using 6 complex fermions, has simple roots given by *[8]*:

$$(0,0,0,1,0,0), \quad (\tfrac{1}{2},-\tfrac{1}{2},-\tfrac{1}{2},-\tfrac{1}{2},0,0),$$
$$(0,0,1,0,0,0), \quad (0,\tfrac{1}{2},-\tfrac{1}{2},0,-\tfrac{1}{2},\tfrac{1}{2}),$$
$$(0,\tfrac{1}{2},-\tfrac{1}{2},0,\tfrac{1}{2},-\tfrac{1}{2}).$$

Since there is an additional vector orthogonal to the space spanned by these roots, These 6 fermions actually embed $SO(10) \times U(1)$.

The maximal nontrivial embedding, using 10 complex fermions, has simple roots given by:

$$(\tfrac{1}{2},\tfrac{1}{2},-\tfrac{1}{2},-\tfrac{1}{2},0,0,0,0,0,0),$$
$$(0,0,\tfrac{1}{2},\tfrac{1}{2},-\tfrac{1}{2},-\tfrac{1}{2},0,0,0,0),$$
$$(0,0,0,0,\tfrac{1}{2},\tfrac{1}{2},-\tfrac{1}{2},-\tfrac{1}{2},0,0),$$
$$(0,0,0,0,0,0,\tfrac{1}{2},\tfrac{1}{2},\tfrac{1}{2},\tfrac{1}{2}),$$

$$(\ 0,0,0,0,0,0,\tfrac{1}{2},\tfrac{1}{2},-\tfrac{1}{2},-\tfrac{1}{2}),$$

We adopt the convention that the first $2n$ real left-movers of our string models will correspond to the n complex fermions that define an embedding. Having chosen a particular embedding, the next challenge is to find a set of basis vectors, consistent with modular invariance and the GSO projections, such that we generate sectors with massless vector states whose fermionic charges reproduce the root vectors of the embedding. Assuming that we can in fact produce all the gauge bosons of some higher level GUT group, we are then guaranteed that all physical states will assemble into irreps of the GUT group. Note that it is a simple matter to translate from the Dynkin basis to the basis defined by the simple roots written as fermionic charge vectors. Thus we can determine which irreps physical states belong to by simply mapping their weights back to the Dynkin basis.

4 Three Generations

We have begun a systematic search for level two $SO(10)$ and $SU(5)$ fermionic string GUT models which employ Lewellen's minimal embedding of $SO(10)$. So far this search has been limited to models whose fermions are Neveu-Schwarz, Ramond, or complex with $\pm 1/4$ twists. Since the introduction of other rational twists *loosens* the modular invariance constraints, we expect that we have merely scratched the surface of possible models with this embedding.

We find that the requirement of three light chiral generations (three 16_L's in $SO(10)$) is very restrictive in our models, much more so than, e.g., the requirement of adjoint scalars. This is not surprising, since other fermionic string model builders have encountered the same difficulty[9, 16, 17]. ¿From an exhaustive search, we find that there exist *no* examples of three generation level two $SO(10)$ models, with the minimal embedding, using *only* Neveu-Schwarz and Ramond fermions. Thus three generations requires fermions with other twists.

To date, we have only succeeded in obtaining three generations by implementing the $Z_2 \times Z_2$ symmetric orbifold structure described by Faraggi in ref *[18]*. As pointed out in *[18, 19]*, this structure is present in all known level one fermionic string models that have three generations. It is quite possible that there are many other ways of obtaining three generation fermionic string GUT's, but we have not yet found any.

The orbifold structure consists of two Z_2 twists θ_1, θ_2, acting symmetrically on left and right-moving $[SO(4)]^3$ lattices. ¿From now on it will be very convenient to write 0, 1 for Neveu-Schwarz and Ramond fermions, respectively. In this notation:

$$\theta_1 = (1100)(1100)(0000)$$
$$\theta_2 = (0000)(0011)(1100)$$

When this structure is realized in a full fermionic string model, the partition function can obviously be decomposed into a sum of four pieces with respect to this orbifold: an untwisted sector, and the three twisted sectors θ_1, θ_2, and

$\theta_1\theta_2$. For level one models, there is a straightforward way (the "NAHE" set) of constructing basis vectors such that one chiral generation resides in each of the three twisted sectors.

At first sight, it is not at all obvious whether this orbifold trick for obtaining three generations is compatible with realizing $SO(10)$ at Kac-Moody level two. The left-moving $[SO(4)]^3$ lattice cannot, of course, overlap with the 12 left-mover slots reserved for the $SO(10)$ embedding. However for a sector containing a massless chiral 16_L fermion *all* of the left-moving structure is rather severely constrained by the requirements of $SO(10)$ and modular invariance. Furthermore, even if one realizes the orbifold structure in three twisted sectors (which one might as well take to be three basis vectors), the additional basis vectors which produce $SO(10)$ gauge bosons will not respect the left-right symmetry of the $Z_2 \times Z_2$ orbifold.

In spite of these difficulties, we have found a way to simultaneously realize both $SO(10)$ level two and the three generation orbifold structure. This is demonstrated in the next section with a specific model.

5 Features of Superstring GUT's

Since we have only a sampling of models, and have made particular choices of embedding, level, gauge group, and string construction, it would be imprudent to try to make any general statements about the properties of superstring GUT's. Instead, we will be content making a few observations about features of our models.

As an example, we list below the basis vectors of a particular level two $SO(10)$ fermionic string model. Here 0, 1 denote real Neveu-Schwarz or Ramond fermions, and $\pm$ denotes a real fermion which pairs with another $\pm$ real fermion to make a complex fermion with $\pm 1/4$ twist. The 20 right-movers are separated from the 44 left-movers by a double vertical line. A vertical line separates out the 12 left-movers that embed the $SO(10)$ weights. The first two right-movers are the spacetime fermions.

V_0 is required in all fermionic string models by modular invariance. The V_1 sector contains the gravitino, as already discussed. Superpartners of states in some sector α will be found in the sector $V_1 + \alpha$. The 45 massless gauge bosons of $SO(10)$ level two are contained in the untwisted sector, V_2, V_3, V_4, V_2+V_3, V_2+V_4, V_3+V_4, and $V_2+V_3+V_4$. This type of model can always be transformed into a similar level two $SU(5)$ model, either by adding a basis vector or altering existing ones. This is because the 24 roots of $SU(5)$ contained in $SO(10)$ appear precisely in the gauge boson sectors listed above that do not contain V_3.

$$
\begin{aligned}
V_0 &= (11111111111111111111 \mid \mid 111111111111 \mid 11111111111111111111111111111111) \\
V_1 &= (11100100100100100100 \mid \mid 000000000000 \mid 00000000000000000000000000000000) \\
V_2 &= (00000000000000000000 \mid \mid 111111110000 \mid 11111111000000000000000000000000) \\
V_3 &= (00000000000000000000 \mid \mid 000000000000 \mid 00001111111100000000000000000000) \\
V_4 &= (00000000000000000000 \mid \mid 110000111111 \mid 11001100110011000000000000000000) \\
V_5 &= (11100100010010010010 \mid \mid 111100001100 \mid 10101010101010111000000000000000) \\
V_6 &= (11010010100100001001 \mid \mid 111100001100 \mid 101001011010010000010000{+}{+}{-}{-}{+}{+}{+}{+}) \\
V_7 &= (11001001001001100100 \mid \mid 111100001100 \mid 11110000111100000000110000000000)
\end{aligned}
$$

$$V_8 = (00110110110110000000 \mid |000000000000|0101010101010101000001000000000000)$$
$$V_9 = (00{+}{-}0{+}{-}0{+}{+}1{+}{+}11{+}{+}1{+}{+} \mid |00000000{+}{+}{+}{+}|00001111000011001001{+}{+}{+}{+}{+}{+}{+}{+}0000)$$

Three generations of 16_L fermions are contained in V_5, V_6, V_7, plus 3×6 additional sectors obtained by adding V_2, V_4, V_2+V_3, V_2+V_4, V_3+V_4, or $V_2+V_3+V_4$ to these (the massless states in $V_3+V_{5,6,7}$ are GSO projected out). To describe the $[SO(4)]^3$ lattice of the $Z_2 \times Z_2$ orbifold structure, let $\{r_1 \ldots r_{20}\}$ denote the right-movers, and $\{l_1 \ldots l_{44}\}$ the left-movers. Then the right and left-moving $[SO(4)]^3$ lattices consists of the fermions

$$\{r_4, r_5, r_7, r_8, r_{10}, r_{11}, r_{13}, r_{14}, r_{16}, r_{17}, r_{19}, r_{20}\}$$
$$\{l_{14}, l_{16}, l_{17}, l_{18}, l_{19}, l_{20}, l_{22}, l_{24}, l_{25}, l_{26}, l_{27}, l_{32}\}$$

¿From this it is apparent that the three generations of 16_L fermions reside in the three twisted sectors of the $Z_2 \times Z_2$ orbifold. However this structure *does not* guarantee three generations. In fact it is difficult to avoid generating extra chiral fermions in other twisted sectors. This has proven so far to be the most severe constraint on model building.

The observable gauge group of this model is $SO(10) \times [U(1)]^6$. The hidden sector gauge group is $[SU(2)]^3$. The appearance of extra $U(1)$'s seems rather general, and also occurs in level one models. They are interesting as they can be broken by vevs at the string scale and are generation sensitive. The hidden sector gauge group and matter content are very model dependent; however it appears that the rank of the hidden sector gauge group is always ≤ 5.

This model contains a 45 of adjoint scalars contained in sector V_8 and the seven other sectors obtained by adding the gauge boson sectors to V_8. Although this model has 246 different sectors that contain massless particles (before the GSO projections), there are no additional 45's and no 54's. This seems to be a robust feature: singlets, 10's, and 16's proliferate in these models, but 45's and 54's appear once, twice, or not at all. This is easily explained by writing the weights of various irreps in our fermionic charge basis. One finds that the 45's and 54's have weights which require Neveu-Schwarz excited fermions and antifermions. These make large contributions to the mass formula, and it then becomes difficult to satisfy all the constraints while keeping these irreps massless.

The above argument should generalize for (at least) most higher level fermionic string models. Thus we conclude that these models do not have a generic problem with exotics -i.e., large massless irreps tend *not* to occur. This fact also poses a serious problem for fermionic string GUT model building, since realistic $SO(10)$ GUT models, which e.g. solve the doublet-triplet splitting problem, seem to require[20] several 45's and 54's ($SU(5)$ models are similar).

The most important feature of the fermionic string GUT's which we have looked at so far, is that they naturally possess one of the key properties of most SUSY GUT texture schemes. Unlike level one models, our models quite often have only a single set of Higgs in the fundamental which is allowed by its $U(1)$ charge assignments to have a Yukawa coupling to chiral fermions. One can then use conservation of fermionic charge[11, 21] to see that in fact there is only one Yukawa coupling, that of the (by definition) third generation:

$$\phi_{10}\,\Psi_{16}^{(3)}\,\Psi_{16}^{(3)} \tag{6}$$

For example, in the model above, there is a single 10 with weights distributed among the untwisted sector, V_3, V_4, V_2+V_4, V_3+V_4, and $V_2+V_3+V_4$. All of these states have a Neveu-Schwarz fermion/antifermion excitation in r_9/r_{12}. Conservation of fermionic charge then tells us that the only allowed Yukawa coupling to this 10 is the diagonal coupling to the fermion generation in V_6.

It is very gratifying to see the key feature of most texture schemes appear so naturally in fermionic string GUT's. Furthermore we see that this feature is intimately related to the structure of the chiral fermion sectors which was needed to produce three generations!

Of course, level one models with three generations have similar properties, although they typically contain several Higgs, each with only one allowed Yukawa. One must also be extremely cautious about attaching too much significance to Yukawas. String models generate operators which are higher dimension but contain scalars that can get Planck scale vevs. Such terms are then unsuppressed and contribute to the GUT scale effective action just like an ordinary Yukawa. Clearly we need a detailed analysis of higher dimension operators in fermionic string GUT models before we can draw any definite conclusions. Further work is also needed to understand fermion mixings and the sources of masses for the first and second generations.

References

   [1]   C. Frogatt and H. Nielson, *Nucl. Phys.* **B147** (1979) 277.

   [2]   S. Dimopoulos and F. Wilczek, ITP Santa Barbara preprint, 1981.

   [3]   G. Anderson, S. Dimopoulos, L. Hall, S. Raby, and G.D. Starkman, *Phys. Rev.* **D49** (1994) 3660.

   [4]   P. Ginsparg, *Phys. Lett.* **197B** (1987) 139.

   [5]   L. Ibanez, *Phys. Lett.* **318B** (1993) 73.

   [6]   A. Faraggi, *Phys. Lett.* **302B** (1993) 202.

   [7]   J. Lykken and S. Willenbrock, *Phys. Rev.* **D49** (1994) 4902.

   [8]   D. Lewellen, *Nucl. Phys.* **B337** (1990) 61.

   [9]   I. Antoniadis, J. Ellis, J. Hagelin, and D.V. Nanopoulos, *Phys. Lett.* **231B** (1989) 65.

 [10]   A. Font, L. Ibanez, and F. Quevedo, *Nucl. Phys.* **B345** (1990) 389.

 [11]   H. Kawai, D. C. Lewellen, and S.-H. H. Tye, *Nucl. Phys.* **B288** (1987) 1.

 [12]   I. Antoniadis, C. P. Bachas, and C. Kounnas, *Nucl. Phys.* **B289** (1987) 87.

 [13]   I. Antoniadis and C. P. Bachas, *Nucl. Phys.* **B298** (1988) 586.

 [14]   H. Kawai, D. C. Lewellen, J. A. Schwartz, and S.-H. H. Tye, *Nucl. Phys.* **B299** (1988) 431.

 [15]   S. Chaudhuri, H. Kawai and S.-H. H. Tye, *Nucl. Phys.* **B322** (1989) 373.

[16]   A. Faraggi, *Phys. Lett.* **278B** (1992) 131; *Phys. Lett.* **274B** (1992) 47; *Phys. Rev.* **D47** (1993) 5021; *Nucl. Phys.* **B387** (1992) 239.

[17]   I. Antoniadis, G.K. Leontaris, and J. Rizos, *Phys. Lett.* **245B** (1990) 161.

[18]   A. Faraggi, *Phys. Lett.* **326B** (1994) 62.

[19]   A. Faraggi and D.V. Nanopoulos, *Phys. Rev.* **D48** (1993) 3288.

[20]   See, for example, K. S. Babu and S. M. Barr, Delaware preprint, BA-94-04 (1994).

[21]   J. Schwartz, *Phys. Rev.* **D42** (1990) 1777.

Running of the Top Yukawa with and without Light Gluinos

L. CLAVELLI
DEPARTMENT OF PHYSICS AND ASTRONOMY
UNIVERSITY OF ALABAMA
TUSCALOOSA, ALABAMA 35487

Abstract: We investigate correlations among various parameters in the solution space of minimal supersymmetric grand unification. In particular the extent to which the top quark Yukawa coupling exhibits fixed point behavior is discussed and we compare various analytic approximations to its value at the top mass with its exact value in numerical solutions.

One of the successes of the idea of supersymmetric grand unification is the prediction of the b quark to τ lepton mass ratio and the accompanying prediction of a top quark mass significantly above that of the Z. These predictions are based on the solution of a set of coupled renormalization group differential equations involving the gauge and Yukawa couplings. At least in the case of small $\tan(\beta), (< 5)$, the solution space of minimal SUSY unification is a ten dimensional space defined by the values of the following ten parameters.

1) A unification scale M_X
2) A unified gauge coupling $\alpha_0(M_X)$
3) A top Yukawa at M_X, $\alpha_t(M_X)$
4) A Susy scale M_S
5) A ratio $\tan(\beta)$ of the Higgs vacuum expectation values
6) The weak angle $\sin^2(\theta_W)$
7) The fine structure constant $\alpha(M_Z)$
8) The strong coupling constant $\alpha_3(M_Z)$
9) The value of the top quark mass M_t
10) The value of the b/τ mass ratio at the b quark scale

A "solution" is defined as a set of values for these ten parameters which is consistent with the renormalization group running and with the experimental constraints:

$$\alpha^{-1}(M_Z) = 127.9 \pm 0.2 \tag{1}$$

$$\sin^2(\theta_W(M_Z)) = .2328 \pm .0007 \tag{2}$$

$$m_b/m_\tau = 2.39 \pm 0.10 \tag{3}$$

The latter corresponds to a physical b quark mass of $4.95 \pm .15 GeV$. There are some who feel that the uncertainty in this quantity is much smaller than taken here but out of respect for the complications of confinement we content ourselves with this 4.5% uncertainty. The uncertainties on the other two quantities are below 1% and once the top quark mass is known the current data will specify them to about 0.15% due to the correlation

$$\sin^2(\theta_W) = .2324 - .002(M_t^2/(138 GeV)^2 - 1) \pm .0003 \tag{4}$$

The hope is that the increasing precision with which these numbers are known will shed light on the physics at the scales M_S and M_X and will enable predictions to be made for m_t and $\alpha_3(M_Z)$.

In addition to the experimental constraints eqs. 1,3,4 we assume that $\alpha_t(Q)$ < 1 for all Q ("perturbativity") and that

$$100 GeV < M_S < M_{S,max} \tag{5}$$

If the Susy scale were below $100 GeV$, with the expected degeneracy splittings among the Susy particles, we would have expected unacceptably large contributions to the Z width due to Susy decay modes. If M_S is too large the theoretical benefits of Susy in explaining the stability of a low scale for electroweak symmetry breaking is lost. $M_{S,max}$ is variously taken to be $1 TeV$ or $10 TeV$. Ideally this argument would prefer M_S below one TeV since the electroweak scale is in the hundred GeV range. In addition, a Susy solution of the dark matter problem would require the mass of the LSP to be no higher than $200 GeV$ again suggesting an average Susy mass below one TeV. Nevertheless we will, for the sake of conservatism, take $M_{S,max} = 10 TeV$.

In the light gluino scenario, $\tan(\beta)$ is restricted to be between 1 and 2.3[1] and, for simplicity, we limit our investigation in the heavy case, to the range $1 < \tan(\beta) < 5$ so that we can neglect the effect of the b Yukawa on the running of the couplings. This range is also preferred by proton decay. Radiative electroweak breaking would predict a value of $\tan(\beta)$ very close to 1.8 in the light gluino case.

In the simplest version of SUSY unification one assumes that all the GUT scale particles are degenerate at M_S and all the SUSY partners of the standard model particles are degenerate at M_S. In the light gluino variant one assumes that the partners of the squarks and sleptons are at M_S together with the heavy Higgs, while the photino and gluino are in the low energy region below M_b and the other neutralinos and charginos are at the scale of M_Z as suggested by the $M_{1/2} = 0$ model. This is consistent with the current experimental gluino searches which leave open (at least) the three windows shown in $fig.1$. $Fig.1$ updates the chart published by the $UA1$ [2] group in 1987 to include the LEP results which are probably the most model independent constraints together with the results of the $HELIOS$ [3] collaboration which searched for weakly interacting neutral particles.

In higher level variations, the GUT scale spectrum is assumed to be non-degenerate and possibly richer than in the minimal model and/or the Susy scale M_S is split into different masses for the various particles. In the latter schemes for each non-degenerate spectrum of Susy particles there is an effective degenerate scale M_S which leads to the same unification solution apart from small two loop effects [4]. Each solution in the degenerate case corresponds to a family of solutions with different splittings among squarks and sleptons and,

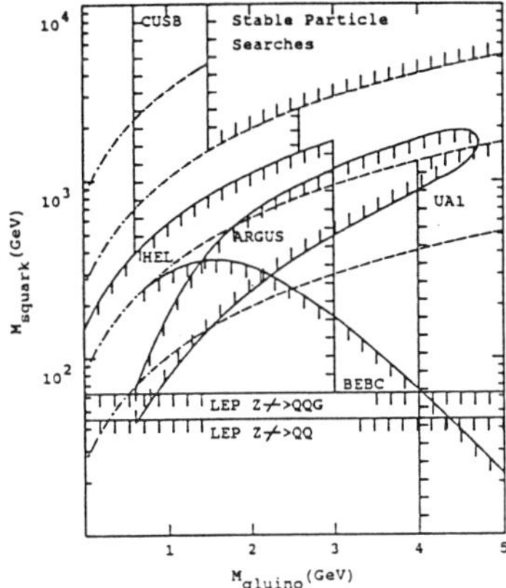

Figure 1. Low mass windows for the gluino mass as a function of the squark masses. The hatched areas are disfavored by the indicated experiments. The dot-dashed curves represent the loci of expected gluino lifetimes $10^{-6}s, 10^{-8}s, 10^{-10}s$, and $10^{-12}s$ respectively from the highest to the lowest curve.

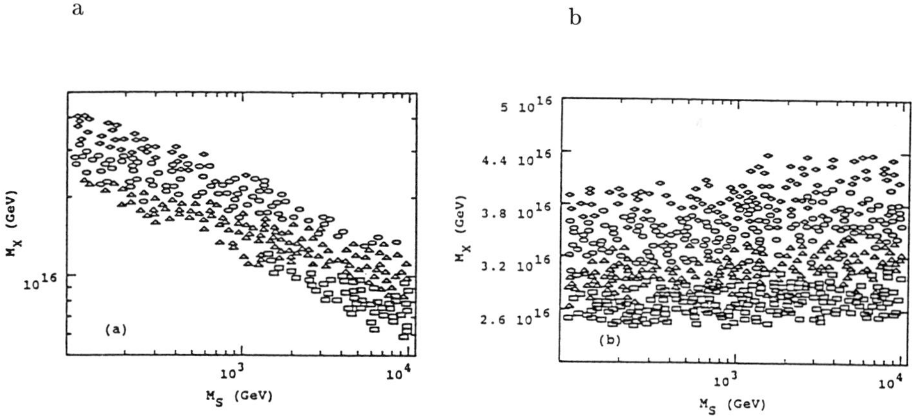

Figure 2. (a) The correlation between the SUSY scale, M_S, and the GUT scale, M_X, in the heavy gluino case, $m_{\tilde{g}} = M_S$. Each solution corresponds to an allowed point in the ten dimensional space discussed in the introduction. For given M_S, no solutions exist outside of the broad band shown. The width of the band corresponds to summing over all other eight parameters. The $\alpha_3(M_Z)$ value for each solution is indicated by the shape coding (see text). (b) The same correlation in the case of the light gluino ($m_{\tilde{g}} < M_b$).

as long as M_S is above M_Z , no solutions for the ten parameters above are lost by assuming degeneracy.

The two-loop differential equations to be solved are summarized in the papers *[5, 6]*. We follow a "top-down" approach where one begins by choosing random values for the first 5 quantities in the list above, extrapolating to low energies and discarding the choices which are inconsistent with the constraints of eqs. 1,3,4,5. The surviving solutions are stored in a data set that can be queried for correlations among the various parameters. The solution space forms a small connected region in the ten dimensional space. Careful checking is required around the borders of this region to insure that these are in fact the borders and that no solutions exist outside this region. In $fig.2a$ for example we show the solution space of the minimal Susy model projected onto the $M_S - M_X$ plane in a random non-overlapping sample from 813 solutions while $fig.2b$ shows the same projection for a non-overlapping subset of 1200 solutions in the light gluino scenario. The shape coding labels the $\alpha_3(M_Z)$ value of each solution. In the heavy gluino case solutions are found only for $0.111 < \alpha_3(M_Z) < .134$ while in the light gluino case the range is $0.122 < \alpha_3(M_Z) < .133$. These ranges are divided into quadrants indicated in the solutions of $fig.2$ by rectangles, triangles, ovals, and diamonds for $\alpha_3(M_Z)$ in the lowest to the highest quadrant respectively. One sees that with the light gluino option, The GUT scale is restricted to values comfortably above $10^{16}GeV$ while this is not the case in the standard Susy picture. In Susy unification, proton decay via lepto-quark gauge bosons of mass as low as $10^{15}GeV$ is not in contradiction with current limits. However, proton decay via the super-heavy scalars requires these particles to have masses in excess of $10^{16}GeV$. Since the GUT scale, above which the theory is grand-unified, is the maximum mass of the GUT scale particles, solutions with M_X below $10^{16}GeV$ are probably not acceptable. One could therefore add a sixth experimental constraint

$$M_X > 10^{16}GeV \tag{6}$$

which would cut the solution spaces of $figs.2a, b$ at the corresponding limit. The solution sets of $fig.2a, b$ were generated in the course of work reported in *[7]*. One sees in $figs.2a, b$ the effect reported there that the α_3 values in minimal Susy unification are significantly more constrained in the light gluino scenario than in the usual picture. This result, however, is critically dependent on the super-gravity inspired prediction that the charginos and neutralinos are relatively light (below the Z) when the gluino is light. In these data sets, the constraint of eq. 2 was not enforced, although that of eq.4 was, so that a Susy prediction for $\sin^2(\theta_W)$ could be made. This prediction can be read from $figs.3a, b$ which show the solutions in the $\sin^2(\theta_W) - \alpha_3(M_Z)$ plane. One sees that in both heavy and light gluino cases, the predicted values of $\sin^2(\theta_W)$ lie between 0.230 and 0.2325. Thus the minimal model predicts the weak angle with a 1% accuracy and agrees with experiment. The shape coding indicates the quadrant values of M_S in the range from $100GeV$ to $10TeV$ (rectangle, triangle, oval, diamond from lowest to highest quadrant). In figures $4a, b$ we show the heavy and light gluino solutions respectively in the $\tan(\beta) - M_t$ plane with the values of M_S indicated by the shape coding. One sees that in both cases the top quark mass is bounded below by about $143GeV$ and that in the

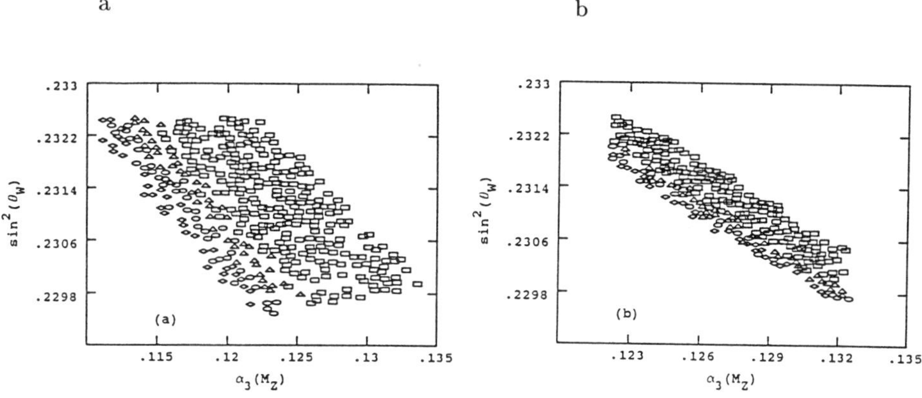

Figure 3. (a) The correlation between $\alpha_t(M_t)$ and $\sin^2(\theta_W)$ in the heavy gluino case. Solutions in the 1st, 2nd, 3rd, and 4th quadrant of the M_S range, $100GeV < M_S < 10TeV$, are printed as rectangles, triangles, ovals, and diamonds respectively. (b) The same as (a) but for the light gluino scenario.

light gluino case the band of solutions is appreciably broader. It is interesting to note that the preliminary evidence from Fermilab for $M_t \simeq 174GeV$ suggests a $\tan(\beta) \simeq 1.8$ as required in the light gluino scenario with radiative electroweak breaking *[8]*. In the current work, however, and that of *[7]*, radiative breaking is not assumed.

Much has been written *[6, 9, 10]* about the quasi-fixed point behavior of

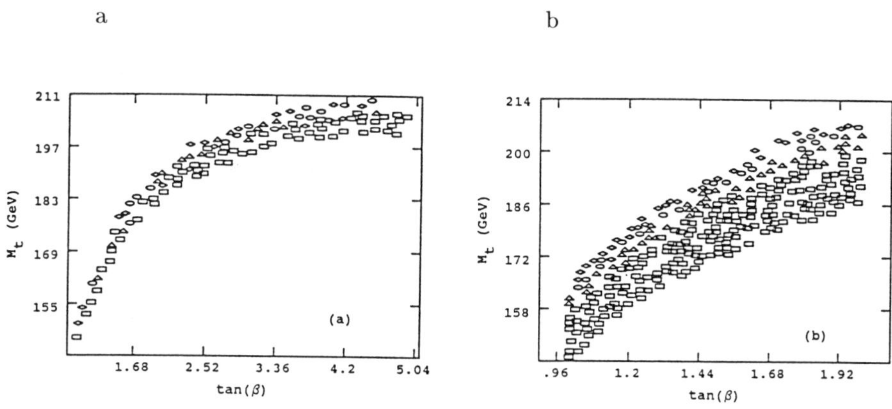

Figure 4. (a) The correlation between $\tan(\beta)$ and M_t in the heavy gluino case. The M_S quadrants are indicated as in *Fig*.3. (b) The same as (a) but for light gluinos.

the top Yukawa. This can be roughly defined by the statement that the value of the top Yukawa at the top mass given by

$$\alpha_t(M_t) = M_t^2 \cdot (173 GeV \sin(\beta)(1 + 4\alpha_3(M_t)/(3\pi) + 11(\alpha_3(M_Z)/\pi)^2))^{-2} \quad (7)$$

is dependent only on the gauge couplings at the top scale and is independent of the GUT scale parameters at least for some range of those parameters that is consistent with the Susy unification solutions. Our present purpose is to clarify and quantify this statement.

Neglecting the effect of the b and τ Yukawas on the running of α_t and two loop contributions, the top Yukawa in the Susy region satisfies

$$2\pi d\alpha_t/dt = \alpha_t(6\alpha_t - c_{t,i}\alpha_i) \quad (8)$$

where the α_i are the three gauge couplings, $t = ln(Q)$, and

$$c_{t,i} = (13/15, 3, 16/3) \quad (9)$$

for $i = 1, 2, 3$ in that order. If α_t is initially (at the GUT scale) higher than $c_{t,i}\alpha_i/6$ the naive prediction is that it will fall with decreasing Q until the right hand side of eq. 8 vanishes. If α_t is below $c_{t,i}\alpha_i/6$ initially, it will rise toward that value as Q decreases. At the most naive level, one might expect from eq. 8 that $\alpha_t(Q)$ will approach the fixed point expression

$$\alpha_{t;f}^{(1)}(Q) = c_{t,i}\alpha_i/6 \quad (10)$$

From our unification solution set it is a simple matter to calculate for each solution the top Yukawa at M_t from eq. 7. The gauge couplings at M_t can be related to the $\alpha_i(M_Z)$ from the renormalization group expressions. We can therefore check how well α_t approaches eq. 10 at $Q = M_t$. In figures $5a, b$, for the heavy and light gluino scenarios respectively, we show the correlation between $\alpha_t(M_t)$ and $\alpha_{t;f}^{(1)}(M_t)$. The correlation is far from the close equality one might have expected. The shape coding, rectangles, triangles, ovals, diamonds, indicates an M_S value in the lowest to the highest quadrant respectively with the total range being $100 GeV < M_S < 10 TeV$. We will return later to discuss the tail of solutions out to high $\alpha_{t;f}^{(1)}(M_t)$ evident in figures $5a, b$.

Ibanez and Lopez[9] have shown that $eq.8$ is satisfied by a top Yukawa given by

$$\alpha_t(Q) = \frac{-\alpha_t(M_X)\dot{F}(Q)}{1 + 6\alpha_t(M_X)F(Q)/4\pi} \quad (11)$$

where

$$F(Q) = -\int_{M_X}^{Q} \frac{dQ'}{Q'} \exp\left(-\int_{M_X}^{Q'} \frac{dQ''}{4\pi Q''} c_{t,i}\alpha_i(Q'')\right) \quad (12)$$

and

$$\dot{F}(Q) \equiv Q\frac{d}{dQ}F(Q) \quad (13)$$

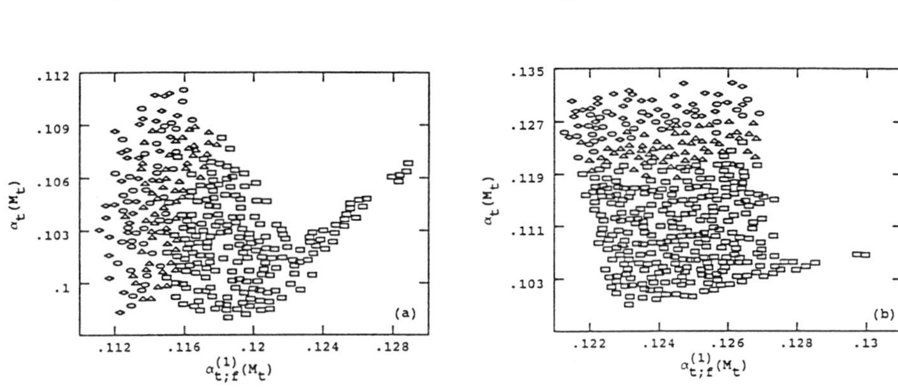

Figure 5. The correlation between the "naive" top Yukawa fixed point $\alpha_{t;f}^{(1)}(M_t)$ and the actual top Yukawa $\alpha_t(M_t)$ found in the numerical solutions in the heavy gluino case (a) and in the light gluino case (b). 10% to 20% departures are observed. The quadrant values of M_S are indicated by the shape coding as in $fig.3$.

$Eq.11$ is of course not an analytic solution of eq. 8 since the Q' integral in $eq.12$ is not analytically soluble. However it does illustrate the quasi-fixed-point behavior in that $\alpha_t(Q)$ becomes independent of $\alpha_t(M_X)$ if $F(Q)$ is sufficiently large as would happen, for example, for large enough M_X.

From eq. 11 one can write a second expression for the fixed point by dropping the 1 in the denominator of eq. 11.

$$\alpha_{t;f}^{(2)} = \frac{-2\pi}{3} Q \frac{d}{dQ} \ln F(Q) \qquad (14)$$

Because we don't have here an analytic solution for $F(Q)$, it is difficult to predict the extent of the independence of $\alpha_t(M_t)$ on M_X and $\alpha_t(M_X)$. Eq. 14 does indicate dependence of $\alpha_t(M_t)$ on M_X and indirectly on $\alpha_t(M_X)$ through the two loop effects on the running of the gauge couplings. The empirical dependence of $\alpha_t(M_t)$ on M_X is substantial as can be shown by projecting the solution space onto the $\alpha_t(M_t) - M_X$ plane. The Ibanez-Lopez expression for the quasi-fixed-point does not yield analytically the dependence of $\alpha_t(M_t)$ on the ten parameters of the unification solution. We seek preferably a quasi-fixed-point expression analogous to that of eq. 10.

One of the problems with both of the above treatments is that the running of the top Yukawa changes dramatically below the Susy scale. Then instead of eqs. 8 and 9 we have

$$2\pi \frac{d\alpha_t^{sm}}{dt} = \alpha_t^{sm} \left(\frac{9}{2}\alpha_t^{sm} - c_{t,i}^{sm}\alpha_i \right) \qquad (15)$$

where below M_S one defines the effective top Yukawa

$$\alpha_t^{sm} = \alpha_t \sin^2 \beta \qquad (16)$$

and

$$c_{t,i}^{sm} = (17/20, 9/4, 8) \tag{17}$$

If M_S is above M_t one extrapolates from M_S to M_t using eqs. 15 and 17 and then redefines $\alpha_t(M_t)$ through eq. 16 before substituting in eq. 7. This rescaling has only a higher order effect on the final answer so we use a rescaling by the fixed value $\sin\beta = .7641$ in both the numerical running and the analytic expressions below.

Thus, if the top Yukawa is following a quasi-fixed-point given approximately by eq. 10 down to M_S, below the Susy scale we would expect it to be drawn toward a naive fixed point corresponding to

$$\alpha_{t;f}^{(1)sm}(Q) = \frac{2}{9} c_{t,i}^{sm} \alpha_i(Q) \tag{18}$$

The corresponding $\alpha_{t;f}$ is more than twice the fixed point of the Susy regime suggested by eq. 10. This effect could partially explain the large deviations of $\alpha_t(M_t)$ from $\alpha_{t;f}^{(1)}(M_t)$ evident in figures $5a, b$. However, it is clear that even with M_S as large as $10TeV$, the top Yukawa does not have time to reach a standard model quasi-fixed-point. From figures $5a, b$ one sees a tendency for the top Yukawa to rise with increasing M_S as would be expected from this effect but it never achieves the doubling expected naively from eq. 18. Thus, if M_S is in the 1 to $10TeV$ region, the top quark Yukawa is unlikely to have reached a limiting behavior. The tail of events at large $\alpha_{t;f}^{(1)}$ in figure 5 represents the events in which $M_S < M_t$ so the Susy quasi-fixed-point should be most accurately attained. Paradoxically, it is here that the discrepancy between $\alpha_t(M_t)$ and $\alpha_{t;f}^{(1)}$ is largest. Clearly a more accurate representation of the top Yukawa behavior is required.

In the SUSY regime the gauge couplings change according to the law

$$2\pi \frac{d\alpha_i}{dt} = \alpha_i^2 b_i \tag{19}$$

with

$$b_i = (33/5, 1, -3) \tag{20}$$

We can combine eqs. 8 and 19 with arbitrary parameter to write

$$2\pi \frac{d(\alpha_t - \lambda_i \alpha_i)}{dt} = 6\alpha_t^2 - c_{t,i}\alpha_i\alpha_t - \lambda_i\alpha_i^2 b_i \tag{21}$$

where summation over repeated indices is intended and, for the purpose of an analytic approximation, we have again neglected two loop contributions to the running. Ideally, one would seek λ_i such that when α_t reaches

$$\alpha_{t;f} = \lambda_i \alpha_i \tag{22}$$

the right hand side of eq. 21 would vanish. Thus we would seek solutions to the equation

$$6(\lambda_i\alpha_i)^2 - c_{t,i}\alpha_i\lambda_j\alpha_j - \lambda_i\alpha_i^2 b_i = 0 \tag{23}$$

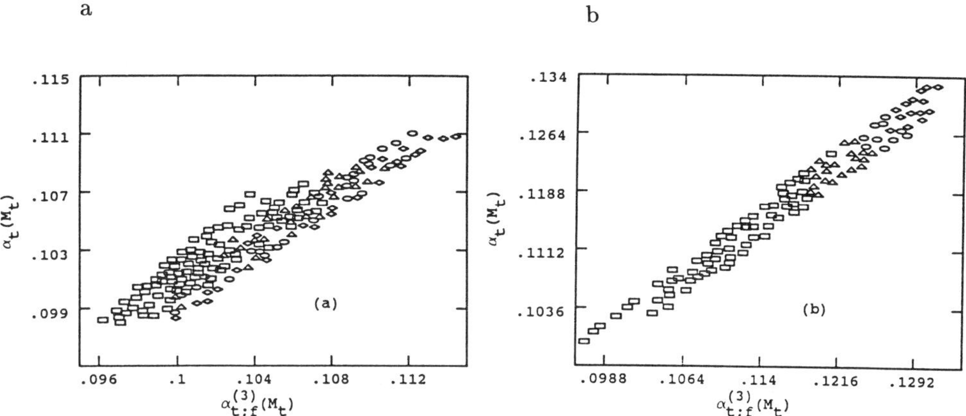

Figure 6. The correlation between the approximate analytic value, $\alpha_{t;f}^{(3)}(M_t)$, of the top Yukawa and the exact numerical value of $\alpha_t(M_t)$ in the heavy gluino solution space (a) and the light gluino solution space (b). Agreement within 2% is found. The quadrant values of M_S are indicated by the shape coding as in $fig.3$.

However, no set of λ_i exists that will satisfy eq. 23 for arbitrary values of the α_i. We propose therefore to write eq. 21 in the form

$$\frac{\pi}{3}\frac{d(\alpha_t - \lambda_i\alpha_i)}{dt} = (\alpha_t - c_{t,i}\alpha_i/12)^2 - \delta^2 \tag{24}$$

where

$$\delta^2 = (c_{t,i}\alpha_i/12)^2 + \lambda_i b_i \alpha_i^2/6 \tag{25}$$

We will now choose

$$\lambda_i = c_{t,i}/12 \tag{26}$$

and make the approximation of ignoring the Q dependence of δ. Then eq. 24 can be integrated to write

$$\alpha_t(Q) = \lambda_i\alpha_i(Q) + \delta\frac{y_0 + \delta + (y_0 - \delta)e^{6\delta \ln(Q/M_X)/\pi}}{y_0 + \delta - (y_0 - \delta)e^{6\delta \ln(Q/M_X)/\pi}} \tag{27}$$

where $y_0 \equiv \alpha_t(M_X) - \lambda_i\alpha_i(M_X)$ and δ is defined at Q. This is an identity at $Q = M_X$ and gives an excellent approximation to $\alpha_t(Q)$ for all $Q > M_S$. In the limit $M_X/Q \to \infty$, the exponentials in eq. 27 become negligible and $\alpha_t(Q)$ is determined solely by the $\alpha_i(Q)$ becoming independent of the GUT scale values. In this case one could talk about a quasi-fixed-point behavior. However in practice M_X is never large enough to justify dropping the exponentials. Furthermore, when $M_t < M_S$ we must face the complication of integrating α_t in the standard model region. In this case we use eq. 27 for $\alpha_t(M_S)$ and write the approximate form using eq. 15

$$\alpha_t(M_t) = \alpha_t(M_S)\left(1 + \frac{\ln M_t/M_S}{2\pi}\left(0.7641 \cdot \frac{9}{2}\alpha_t(M_S) - c_{t,i}^{sm}\alpha_i(M_t)\right)\right) \tag{28}$$

In $fig.6a, b$ we show the correlation between the right hand side of this equation which we may call $\alpha_{t;f}^{(3)}$ and the exact numerically calculated left hand side. The prediction is seen to hold within a few percent over the entire solution space. Given M_t and M_S , the gauge couplings at those scales can be found to sufficient accuracy by the first order extrapolation from their values at M_Z . If the exponentials in eq. 27 were negligible, eqs. 7 and 28, together with knowledge of M_t , M_S and the gauge couplings at the Z would yield an accurate measure of $\tan(\beta)$ independent of the GUT scale parameters. However, only to a crude approximation $\simeq 20\%$ can the exponentials in eq. 27 be neglected.

The author is indebted to Philip Coulter for discussions in the course of this work. The research reported here was supported in part by the Department of Energy under Grant No. $DE - FG05 - 84ER40141$.

References

[1]   L. Clavelli, P. Coulter, and K. Yuan, Phys. Rev. D47, 1973 (1993)

[2]   C. Albajar et al., Phys. Lett. B262, 109 (1987)

[3]   T. Akesson et al., Zeit. f. Phys. C52, 219 (1991)

[4]   P. Langacker and N. Polonsky, Phys. Rev. D47, 4028 (1993)

[5]   H. Arason et al., Phys. Rev. Lett 67, 2933 (1991); Phys. Rev. D46, 3945 (1992)

[6]   V. Barger, M.S. Berger, and P. Ohnmann, Phys. Rev. D47, 1093 (1993)

[7]   M. Carena, L. Clavelli, D. Metalliotakis, H.P. Nilles, and C. Wagner, Phys. Lett. B317, 346 (1993)

[8]   J. Lopez, D.V. Nanopoulos, and X. Wang, Phys. Lett. B313, 241 (19

[9]   L.E. Ibanez and C. Lopez, Phys. Lett. B126, 54 (1983); Nucl. Phys. B233, 511 (1984)

[10]   M. Carena, S. Pokorski, and C. Wagner, Nucl. Phys. B406, 59 (1993); M. Carena, M. Olechowski, S. Pokorski, and C. Wagner, Max Planck Institute preprint MPI-Ph/93-103

Spontaneous CP Violation and Dedicated Studies of B Mesons

PAUL H. FRAMPTON
INSTITUTE OF FIELD PHYSICS,
DEPARTMENT OF PHYSICS AND ASTRONOMY,
UNIVERSITY OF NORTH CAROLINA,
CHAPEL HILL, NC 27599-3255, USA

Abstract: A model in which CP is spontaneously broken is described; it solves the strong CP problem and provides an additional mechanism for weak CP violation. This is the aspon model. Application to B decays leads to predictions for CP asymmetries much smaller than expected from the KM mechanism of explicit CP breaking.

In this presentation [as in one I gave last month in Madras at WHEPP3, to be published] I shall cover the three topics all of which pertain to the aspon model:

1)The motivation, arising from the strong CP problem.

2)The phenomenology of CP violation, giving upper bounds on the masses.

3)Estimation of CP asymmetries in B meson decays.

Topic (3), done in collaboration with Ackley, Kayser and Leung, is the most recent work in the model so I shall emphasize it.

1 Solution of Strong CP Problem

In solving the strong CP problem there are three alternatives:

1) Axion

2) $m_u = 0$

3) Spontaneous (or soft) CP violation

(1) is looking less and less plausible experimentally since the allowed window is $10^{-5}eV \leq m_a \leq 10^{-3}eV$.

(2) is at odds with pseudoscalar meson masses and spontaneous chiral symmetry breaking.

The *aspon model* is an economical (minimal) version of (3).

The strong CP problem is due to the term $\Theta_{QCD}G_{\mu\nu}\tilde{G}^{\mu\nu}$ in the QCD lagrangian. The physical quantity $\bar{\Theta} = (\Theta_{QCD} + \Theta_{QFD}) \leq 2 \times 10^{-10}$ for $d_n \leq 10^{-25}e.cm.$ where $\Theta_{QFD} = -argdetM$, M being the quark mass matrix.

With a CP invariant lagrangian $\Theta_{QCD} = 0$ so we must arrange that $\Theta_{QFD} \leq 2 \times 10^{-10}$.

To do this we introduce $U(1)_{new}$ which will be local (gauged). Thus the model has gauge group $SU(3)_C \times SU(2)_L \times U(1)_Y \times U(1)_{new}$. All the particles in the three family standard model are neutral under this $U(1)_{new}$. In the simplest version of the aspon model we add one doublet of vector-like quarks which carry the new charge $Q_{new} = 1$. There are also two complex scalars χ_α with $\alpha = 1, 2$ and $Q_{new} = 1$. The reason for having at least two is so that the phase of the VEVs cannot be removed by a gauge transformation - assuming there is a nonzero relative phase between them.

The new quarks have gauge-invariant (heavy) Dirac masses and Yukawa couplings to the light quarks:

$$h_i^\alpha U_L^C u_L^i \chi_{alpha}; \; h_i^\alpha D_L^C d_L^i \chi_{alpha}. \tag{1}$$

Here $\alpha = 1, 2$ and i $= 1,2,3$. The superscript C implies charge conjugate. There are *no* Yukawa couplings to U_L or to D_L because of the $U(1)_{new}$ gauge invariance.

The gauge invariance under $U(1)_{new}$ forces the 4×4 quark mass matrices (both *up* and *down* sectors) to have a texture such that the determinant is real. The diagonal blocks are real because of CP conservation; let the heavy quarks have mass M. The off-diagonal terms are of the form:

$$F_i = \sum_\alpha h_i^\alpha < \chi_\alpha > \tag{2}$$

and arise from the off-diagonal Yukawa couplings listed earlier. Only one of the two off-diagonal blocks has this form, the other being zero; this is how the determinant is real.

2   The Phenomenology of CP Violation, Giving Upper Bounds on the Masses

The Yukawa couplings h_α^i are restricted to be $<< 1$ by the limit on $\bar{\Theta}$ which is generated by one-loop corrections. In diagonalization of the quark mass matrix the relevant expansion parameter is $x_i = F_i/M$ ($i = 1, 2, 3$ for the three families). In the aspon model these satisfy $x_i << 1$.

Since $U(1)_{new}$ is clearly anomaly free, it can be gauged and its gauge boson, the *aspon* become massive by the Higgs mechanism. Combining the upper limit on $\bar{\Theta}$ and the values of ϵ and ϵ' from K^0 decay it is possible to show that both the aspon and the vector-like quarks have masses $\leq 600 GeV$. The analysis also bounds the values of $|x_i|$ to be $\leq 10^{-2}, 10^{-3}, 10^{-4}$ for i $= 1,2,3$ respectively.

I shall not go through the details here. In the treatment of CP violation, there is a parallel to the recent paper of Deshpande and He (OITS-529 12/93) which uses a $S_3 \times Z_3$ family symmetry introduced in 1991 by Ernest Ma. Their case (a) has hard explicit CP violation, but their case (b) has spontaneous violation and has too small a KM phase to account for the effect in the K system which is instead explained by charged Higgs exchange [the analog of aspon exchange].

In the MSSM new CP phases occur and must be suppressed to avoid too large a value of d_n. Babu and Barr (1993) suggest spontaneous CP violation at a high scale to solve this by providing phases only in gaugino mass terms.

These are two examples of other recent work suggesting spontaneous CP violation. They both differ from the aspon model as can be seen by the fact that they both require an axion to solve strong CP.

3 Estimation of CP Asymmetries in B Decays

In B decays, using the KM mechanism it has been estimated that with 10^8 samples one could see CP asymmetries and begin to check the unitarity relations required for the (complex) CKM matrix elements.

As shown in *[1]*, the details are different in the aspon model because aspon exchange can compete with the usual box diagrams for the decay amplitudes. In the K system, the mixing is dominated by aspon exchange and that led to the upper limits on the masses mentioned above. The large phase there compensated for the small phase in the $K \to \pi\pi$ amplitude from W exchange.

For the B system, the mixing acquires a sizeable contribution from the W^+W^- box graph and the compensation is incomplete for the small phase occurring in the decay amplitude. The reason that the aspon exchange is relatively less important in the mixing can be traced to two large suppression factors: (1) $(m_c/m_t)^2 \simeq 10^{-4}$ and (2) $|x_3/x_2|^2 \simeq 10^{-2}$.

As a result, in a sample of 10^8 B decays one would *not* see the CP asymmetry if the aspon model were correct. This is good in that it will refute the standard model if one defines the KM mechanism as a part thereof (this may be unfair to the standard model). It is bad because there would still be no example of CP violation seen outside of the neutral K system.

To confirm the aspon model, the most convincing would be the production of the aspon particle in a collider.

In conclusion, if CP asymmetries are undetectably tiny in B decay, spontaneous CP violation is an alternative to the KM mechanism.

This work was supported in part by the US Department of Energy under Grant No. DE-FG05-85ER-40219.

References

[1] A. W. Ackley, P. H. Frampton, B. Kayser and C. N. Leung, Phys. Rev. D50 (1994) 3560, and references cited therein.

The Yukawa Coupling Hierarchy Problem

C. D. FROGGATT[1]
DEPARTMENT OF PHYSICS AND ASTRONOMY
UNIVERSITY OF GLASGOW
GLASGOW G12 8QQ, U.K.

Abstract: We briefly review the ideas of mass-protection and approximate chiral flavour symmetries as a framework for resolving the Yukawa coupling hierarchy problem. A classification is presented of extensions of the Standard Model gauge group having no new fermions. It is argued that the Yukawa coupling hierarchy points to a particular class, characterised by having the same irreducible representations as exist in the Standard Model. We consider how well the observed hierarchy can be predicted, order of magnitude wise, by anti-grand unified Standard Model extensions based on the gauge group $SMG^3 \times U(1)$.

1 Introduction

The most striking qualitative features of the mass spectrum of the quarks and charged leptons are:

1. The large mass ratios between fermions of a given electric charge, i.e. of the same family; the fermion mass hierarchy.

2. The similarity between the mass spectra of the three quark and charged lepton families; the fermion generation structure.

3. The smallness of the off-diagonal elements of the quark weak coupling matrix V_{CKM}; the fermion mixing hierarchy.

The strong hierarchical structure is a reflection of the hierarchy of quark and lepton Yukawa coupling constants y_i, which are related to the fermion masses by:

$$m_i = y_i \langle \phi \rangle_{ws} \tag{1}$$

where the Weinberg-Salam Higgs VEV is $\langle \phi \rangle_{ws} = 174$ GeV. These Yukawa couplings are arbitrary parameters in the Standard Model (SM) and would naively be expected to be of order unity but in fact range over 5 orders of magnitude:

$$y_t \sim 10^0, \ y_c \sim y_b \sim y_\tau \sim 10^{-2}, \ y_s \sim y_\mu \sim 10^{-3}, \tag{2}$$

$$y_d \sim y_u \sim 10^{-4}, \ y_e \sim 10^{-5} \tag{3}$$

Only the top quark mass is unsuppressed relative to its natural mass scale $\langle \phi \rangle_{ws}$; so it is possible that we may obtain a dynamical understanding of y_t, e.g. as a fixed point of the renormalisation group equations, before understanding the electron mass and the rest of the spectrum. More precisely, of course, we consider the three fermion mass matrices:

$$M_{U,D,l}(i,j) = y_{ij}^{U,D,l} \langle \phi \rangle_{ws} \tag{4}$$

[1]Research partially funded by the UK Science and Engineering Research Council

and the corresponding Yukawa coupling matrices y_{ij}.

Neutrino masses, if non-zero, would appear to have a different origin to those of the charged fermions. It is usually assumed that there are no right-handed weak isosinglet neutrino states in the SM and the Higgs mechanism cannot then generate a neutrino mass term. In extensions of the SM it is possible to generate Majorana mass terms, via a super heavy lepton L^0 intermediate state, by the so-called see-saw mechanism[1]. There is not time to discuss neutrino masses further here.

There are three main approaches to the fermion mass problem:

1. Attempts to derive a fermion mass or mass relation exactly from some dynamical or theoretical principle; e.g. Veltman's condition for the cancellation of one loop quadratic divergences in the SM [2].

2. Searches for relationships between mass and mixing angle parameters using symmetries and/or ansätze to make detailed fits to the data; e.g. the phenomenologically successful relation for the Cabibbo angle $\theta_c \simeq \sqrt{\frac{m_d}{m_s}}$ derived from mass matrix ansätze with texture zeros [3].

3. Attempts to naturally explain all the qualitative features of the fermion spectrum, fitting all the data within factors of order unity.

This talk will concentrate on the third approach, in which the aim is to explain all the order of magnitude features of the observed fermion mass spectrum.

2 Chiral Flavour Symmetries and Mass Protection

It is natural to try to explain the occurrence of large mass ratios in terms of selection rules due to approximate conservation laws. A Dirac mass term:

$$- m\overline{\psi_R}\psi_L + h.c. \tag{5}$$

connects a left-handed fermion component ψ_L to its right-handed partner ψ_R. If ψ_L and ψ_R have different quantum numbers, i.e. belong to inequivalent irreducible representations (IRs) of a symmetry group G (G is then called a *chiral* symmetry), then the mass term is forbidden in the limit of exact G symmetry and they represent two massless Weyl particles. G thus "protects" the fermion from gaining a mass. Note that this is exactly the situation for all the SM fermions, which are mass-protected by $SU(2)_L \times U(1)_Y$ (but not by $SU(3)_c$). The $SU(2)_L \times U(1)_Y$ symmetry is spontaneously broken and the SM fermions gain masses suppressed relative to the presumed fundamental (GUT or Planck) mass scale M by the symmetry breaking parameter:

$$\epsilon = \langle\phi\rangle_{ws}/M \tag{6}$$

The extreme smallness of this parameter ϵ constitutes, of course, the gauge hierarchy problem.

Here we are interested in the further suppression of the quark and lepton mass matrix elements relative to $\langle\phi\rangle_{ws}$: the SM Yukakawa coupling hierarchy problem. We take the view that this hierarchy is due to the existence of further approximately conserved chiral quantum numbers beyond those of the Standard Model Group ($SMG \equiv SU(3) \times SU(2) \times U(1)$) [4]. The SMG is then a low energy remnant of some larger group G and the fermion mass and mixing hierarchies are consequences of the spontaneous breaking of G to the SMG. The component chiral flavour symmetries of G which mass-protect the SM fermions, in addition to the mass protection they receive from the SMG itself, are called partially conserved chiral symmetries (PCCSs) below. The mass matrix element suppression factors depend on how the fermions behave *w.r.t.* these PCCSs and on the symmetry breaking mechanism itself.

Consider, for example, an $SMG \times U(1)_f$ model, whose fundamental mass scale is M, broken to the SMG by the VEV of a scalar field ϕ_S where $\langle \phi_S \rangle < M$ and ϕ_S carries $U(1)_f$ charge $Q_f(\phi_S) = 1$. Suppose further that $Q_f(\phi_{ws}) = 0$, $Q_f(b_L) = 0$ and $Q_f(b_R) = 2$. Then it is natural to expect the generation of a b mass of order:

$$\left(\frac{\langle \phi_S \rangle}{M} \right)^2 \langle \phi \rangle_{ws} \tag{7}$$

via the exchange of two $\langle \phi_S \rangle$ tadpoles, in addition to the usual $\langle \phi \rangle_{ws}$ tadpole, through two appropriately charged vector-like superheavy (i.e. of mass M) fermion intermediate states *[4]*. We identify

$$\epsilon_f = \frac{\langle \phi_S \rangle}{M} \tag{8}$$

as the $U(1)_f$ flavour symmetry breaking parameter. In general we expect mass matrix elements of order

$$M(i,j) \simeq \epsilon_f^{n_{ij}} \langle \phi \rangle_{ws} \tag{9}$$

where

$$n_{ij} = \mid Q_f(\psi_{L_i}) - Q_f(\psi_{R_j}) \mid \tag{10}$$

is the degree of forbiddenness due to the $U(1)_f$ quantum number difference between the left- and right-handed fermion components. So the *effective* SM Yukawa couplings of the quarks and leptons to the Weinberg-Salam Higgs

$$y_{ij} \simeq \epsilon_f^{n_{ij}} \tag{11}$$

can consequently be small even though all *fundamental* Yukawa couplings of the "true" underlying theory are of O(1). We are implicitly assuming here that there exists a superheavy spectrum of states which can mediate all of the symmetry breaking transitions; in particular we do not postulate the *absence* of appropriate superheavy states in order to obtain exact texture zeroes in the mass matrices as was done in *[5]*.

The use of such an abelian chiral flavour symmetry group $U(1)_f$ to generate the fermion mass hierarchy in supersymmetric extensions of the SM has been discussed in other contributions to this workshop *[6]*. Here we will consider non-supersymmetric extensions of the SM and assume that only the 45 Weyl fermions of the SM exist in the low energy regime.

3   Extensions of the Standard Model Group

Our basic assumptions in extending the SMG are:
1.   There exist no new mass-protected fermions beyond those of the three generation SM.

2. All gauge and mixed anomalies cancel among the 45 Weyl fermion SM states.

3. The SM Yukawa coupling hierarchy is due to the suppression of mass matrix elements by PCCSs of the extended group G.

We are only interested in that part of any extension which acts non-trivially on at least one of the SM Weyl states and thus G is naturally identified with a subgroup of U(45), the unitary group in whose fundamental representation would sit all the 45 known Weyl fermions. That is:

$$SMG \subseteq G \subseteq U(45) \tag{12}$$

Clearly, the 45 Weyl states fall into IRs of G each of which must include a number of the IRs of the SMG. It is thus possible to classify the extensions G according to the manner in which the IRs of the SMG fit into the IRs of G *[7]*. We introduce our classification of groups G by first classifying collections of IRs of the SMG, which has 5 different IRs of type $(u\ d)_L$, u_R, d_R, $(\nu_e\ e)_L$ and e_R each occuring 3 times, once for each generation. The different types of collection are:

1. A collection consisting only of *isomorphic* IRs of the SMG (this might be called a "horizontal" or H collection) e.g. $\{u_R, c_R, t_R\}$ or $\{(u\ d)_L, (t\ b)_L\}$. It is this type of collection which forms the IRs of non-abelian horizontal symmetry groups.

2. A collection consisting only of *non-isomorphic* IRs of the SMG (this might be called a "vertical" or V collection) e.g. $\{d^c_L, (\nu_e\ e)_L\}$ or $\{(u\ d)_L, u^c_L, e^c_L\}$. It is this type of collection which forms the IRs of GUTs.

3. A collection consisting of *both* isomorphic and non-isomorphic IRs of the SMG (this might be called a "mixed" or M collection) e.g. $\{d^c_L, (\nu_e\ e)_L, s^c_L, (\nu_\mu\ \mu)_L\}$.

We may now classify the group G according to whether its IRs number among them collections of these 3 types (see Table 1) where, in our notation, numerical subscripts on a group component indicate which generations are transformed non-trivially by an element of that group component. For example, $SU_1(5)$ acts non-trivially (in the usual manner) only on the 1st generation while $SU_{23}(3)$ acts non-trivially (in the usual manner) only on the 2nd and 3rd generations. So a group like $SU_1(5) \times SU_2(5) \times SU_3(5)$ of category (6) has all the fermions grouped into the usual $\bar{5}$ and 10 $SU(5)$ representations but of *different* $SU(5)$ components e.g. $(d^c_L\ e_L - \nu_{eL})$ would transform like $(\bar{5}, 1, 1)$ whereas $(s^c_L\ \mu_L - \nu_{\mu L})$ would transform like $(1, \bar{5}, 1)$ etc. A more complicated example is $SU_1(5) \times SMG_{23} \times SU_{23}(2)_H$ which appears in category (2) and has: the 1st generation states grouped into the usual $SU(5)$ IRs $\bar{5}$ and 10 and transformed only by the $SU_1(5)$ component; and the 2nd and 3rd generation states behaving exactly as in the SM under SMG_{23} but grouped horizontally under $SU_{23}(2)_H$ (in one of the scenarios catalogued in *[8]*).

Table 1 displays all combinations of the 3 different types of collection and gives one or two anomaly-free examples in each case. Note that any group in any of these categories may be extended by the addition of abelian $U(1)_f$

Table 1. Classification of SMG extensions according to the manner in which IRs of the SMG fit into IRs of G. "H", "V" and "M" mean "horizontal", "vertical" and "mixed", while subscript "ij" means that the relevant group component acts only on generations "i" and "j". Some examples of each category are given, and these are gauge and mixed anomaly free so that they may be considered valid *gauge* groups. Note that abelian flavour symmetries may be added on to any group, although the corresponding charges are constrained if anomalies have to be cancelled.

<table>
<tr><td colspan="3" align="center">Type:</td><td rowspan="2" align="center">Anomaly-Free Examples
(involving only 3 generations
where possible)</td></tr>
<tr><td align="center">H</td><td align="center">V</td><td align="center">M</td></tr>
<tr><td rowspan="4" align="center">$\checkmark$</td><td rowspan="2" align="center">$\checkmark$</td><td align="center">$\checkmark$</td><td align="center">(1)   Need ≥ 5 generations
$SU_1(5) \times \{SMG \times SU(2)_H\}_{23} \times \{SU(5) \times SU(2)_H\}_{45}$</td></tr>
<tr><td align="center">$\times$</td><td align="center">(2) $SU_1(5) \times \{SMG \times SU(2)_H\}_{23}$</td></tr>
<tr><td rowspan="2" align="center">$\times$</td><td align="center">$\checkmark$</td><td align="center">(3)   Need ≥ 4 generations
$\{SMG \times SU(2)_H\}_{12} \times \{SU(5) \times SU(2)_H\}_{34}$</td></tr>
<tr><td align="center">$\times$</td><td align="center">(4) $SMG \times SU(2)_H$
$SU(3) \times SU(2) \times U_1(1) \times U_2(1) \times U_3(1) \times SU(2)_H$</td></tr>
<tr><td rowspan="4" align="center">$\times$</td><td rowspan="2" align="center">$\checkmark$</td><td align="center">$\checkmark$</td><td align="center">(5) $SU_1(5) \times \{SU(5) \times SU(2)_H\}_{23}$</td></tr>
<tr><td align="center">$\times$</td><td align="center">(6) $SU(5)$
$SU_1(5) \times SU_2(5) \times SU_3(5)$</td></tr>
<tr><td rowspan="2" align="center">$\times$</td><td align="center">$\checkmark$</td><td align="center">(7) $SU(5) \times SU(2)_H$
$SMG_1 \times \{SU(5) \times SU(2)_H\}_{23}$</td></tr>
<tr><td align="center">$\times$</td><td align="center">(8) SMG
$SMG_1 \times SMG_2 \times SMG_3$</td></tr>
</table>

flavour symmetries. We do not refer to such abelian symmetries as horizontal symmetries as they are not restricted to groups whose IRs constitute horizontal collections. The usual GUT models which have no "extra" fermions appear in Category (6) - this really only means $SU(5)$, but this can be distorted to e.g. $SU_1(5) \times SU_2(5) \times SU_3(5)$. The horizontal symmetry models (e.g.$SMG \times SU(2)_H$, $SMG \times SU(3)_H$) appear in category (4), but again distortions are possible: the SMG part might be widened to any larger subgroup of SMG^3, or the horizontal part might apply to only two of the three generations. Most of the other categories involve some combination of the different types of collection (and some even require more than three generations of fermions in order that no group components act trivially). The exception is category (8), whose IRs are no bigger than those already present in the SM.

We now ask which of the 8 categories of SMG extension in Table 1 can most naturally explain the fermion mass spectrum. No pair of particles from the same family, i.e. of the same electric charge, is degenerate, even order of magnitude wise; consequently it is reasonable to expect that Nature has chosen a model in which the chance of such degeneracy is very low. This argument would seem to disfavour the use of non-abelian horizontal symmetry groups to account for the generation gaps. For example, the positioning of corresponding pairs of states such as $\{e_R, \mu_R\}$ into doublets of $SU(2)_H$ (in the $SMG \times SU(2)_H$ model), or even $\{u_R, c_R, t_R\}$ into triplets of $SU(2)_H$, creates the obvious risk of these states forming degenerate particles. It is difficult to avoid degeneracies among the particles formed by the states in these IRs (without appealing to a hierarchy in the fundamental Yukawa couplings or a finely-tuned cancellation between scalar VEVs - see *[8]* for example). To avoid inviting these degeneracies we should keep isomorphic IRs of the SMG *separate* rather than collected together; the extended group G should thus have no IRs which constitute horizontal or mixed collections of IRs of the SMG. This argument weighs heavily against the extensions of categories (1)–(5) and (7), leaving only the GUTs and category (8) groups. Here we will concentrate on the category (8) models.

The category (8) models are characterised by having:

- the same IRs as exist in the SM.

- the SMG embedded within them as a diagonal subgroup: the subgroup whose elements correspond to identical transformations on each generation.

Since the non-abelian part of the SMG (SMG_{na}) is embedded in the non-abelian part of G (G_{na}) as a diagonal subgroup, it follows *[7]* that G_{na} must be gauged: commuting the SMG_{na} gauge transformations with the (ostensibly global) non-SMG members of G_{na} generates a fully gauged G_{na}. This line of argument for gauging is obviously not applicable to any abelian factors in G, but, as the non-abelian and SM weak hypercharge symmetries are gauged, we consider it reasonable to gauge all of the abelian factors too. We can then further conclude that the "largest" such G which has no net anomalies is $SMG^3 \times U(1)_f^n$. In the rest of this talk we will consider models based on this non-simple gauge group with $n = 1$: the so-called anti-grand unified models.

4 Anti-Grand Unification

The basic ingredient of anti-grand unification is the existence of a fundamental gauge group at the Planck scale M_P containing $N_{gen} = 3$ copies of the Standard Model Group *[9, 10]*:

$$G_{anti} = SMG^3 \equiv SMG_1 \times SMG_2 \times SMG_3 \tag{13}$$

Here SMG_i (i=1,2,3) behaves just like the SMG as far as the ith generation is concerned, but acts trivially on the other 2 generations, i.e. G_{anti} has three times as many generators as the SMG, each fermion generation effectively having its "own" SMG. This SMG^3 group has 9 gauge coupling constants. It is assumed that all these couplings approach a multicritical point in the corresponding lattice gauge theory, where G_{anti} spontaneously breaks down just below M_P to its diagonal subgroup which is to be identified with the usual low energy SMG:

$$G_{anti} \to G_{diag} \equiv SMG \tag{14}$$

Consequently, the running fine structure constants of the SM, $\alpha_i(\mu)$ (i=1,2,3), are predicted to take on values:

$$\alpha_i(M_P) = \frac{\alpha_i^{crit}}{N_{gen}} = \frac{\alpha_i^{crit}}{3} \tag{15}$$

where the critical couplings α_i^{crit} have been estimated using lattice gauge theory Monte Carlo results. Good agreement of (15) with the experimental couplings (extrapolated to M_P using the SM renormalisation group equations) is obtained for the two non-abelian groups $SU(2)_L$ and $SU(3)_c$ *[11]*; but there are as yet unresolved problems in estimating α_1^{crit} for the group $U(1)_Y$ due to the infinite number of invariant subgroups and corresponding phases expected for the U(1) groups in (13). Thus the anti-grand unification model replaces the GUT prediction of the equality of the SM gauge coupling constants at M_{GUT} by the prediction (15) of their actual values close to M_P.

We now turn to the suggestion *[9, 12]* that the broken chiral gauge quantum numbers of the quarks and leptons under the symmetry groups SMG_a (a=1,2,3) could be responsible for the fermion mass hierarchy. Obviously, all of the SM fermions are mass-protected by SMG^3 in much the same way as they are by the SMG. If in the process of the breaking described by (14) some symmetries are only weakly broken, then a hierarchy in the fermion masses would appear in a most natural manner. The SMG^3 model can readily explain the generation mass gaps, but not the splittings within each generation *[12]*. We are thus led to consider extending the gauge group further and, as far as mass suppression is concerned, $SMG^3 \times U(1)_f$ is the only non-trivial anomaly-free extension with no new fermions *[12]*.

5   The $SMG^3 \times U(1)_f$ Model

We first consider if it is possible to obtain satisfactory mass and mixing hierarchies by choosing particular subgroups of SMG^3 to be partially conserved. Obviously the choice is between $SU_a(3)$, $SU_a(2)$, $U_a(1)$ and their products (where, in the notation of (13), the subscript indicates which generation is transformed non-trivially by that particular group component) so we consider each in turn, taking $a = 3$ for illustration.

- for an $SU_3(3)$ all states are singlets except $t_{L,R}$ and $b_{L,R}$ so the only suppressed elements are off-diagonal in the quark matrices (*e.g.* elements linking $t_L \leftrightarrow u_R, c_R$).

- an $SU_3(2)$ is more powerful, with $(t\,b)_L$ and $(\nu_\tau\,\tau)_L$ doublets and all other states singlets. This results, for example, in the elements $t_L \leftrightarrow u_R, c_R, t_R$ being suppressed *i.e.* a column in each of the 3 mass matrices.

- finally, for a partially conserved $U_3(1)$ the 3rd generation states have their usual weak hypercharge quantum numbers while all others have zero charge so the 3rd row and column in each matrix is suppressed, *e.g.* in the up matrix the elements linking $t_L \leftrightarrow u_R, c_R, t_R$ and $t_R \leftrightarrow u_L, c_L$ are suppressed.

None of the above group components is promising as far as mass splitting *within* a generation is concerned. This is because $SU_a(3)$ and $SU_a(2)$ have the same effect on both the quark mass matrices (M_U and M_D) while $U_a(1)$ affects corresponding diagonal elements of M_U and M_D identically (*i.e.* the charge differences are the same for diagonal elements). However, $U_a(1)$ affects off-diagonal elements differently (they have unequal charge differences) and, in order to accommodate the $t - b$ mass splitting, we investigate whether the 3rd generation (and other) masses might have their origin in off-diagonal elements of the mass matrices.

The fundamental Yukawa couplings of the fermions are assumed to be O(1) complex numbers, but otherwise unknown. We therefore determine the modulus of matrix elements ($|M|_{ij}$) up to $O(1)$ numerical factors. It can then be shown [12] that, even assuming all component symmetries are partially conserved, the SMG^3 model inevitably leads to the result:

$$\det M_U \simeq \det M_l \leq \det M_D$$

$$\text{i.e.} \quad m_u m_c m_t \simeq m_e m_\mu m_\tau \leq m_d m_s m_b \tag{16}$$

and so although we can for example accomodate $m_b \ll m_t$ we cannot simultaneously get $m_s \ll m_c$ and $m_u \simeq m_d$.

So we now consider the $SMG^3 \times U(1)_f$ model. The no-anomaly equations for the $U(1)_f$ charges have only one solution carrying new information (*i.e.* which is not a linear combination of the $U_a(1)$ charges (a=1,2,3)) given by:

$$\begin{pmatrix} d_L & u_R & d_R & e_L & e_R \\ s_L & c_R & s_R & \mu_L & \mu_R \\ b_L & t_R & b_R & \tau_L & \tau_R \end{pmatrix} = \begin{pmatrix} 0 & 0 & 0 & 0 & 0 \\ 0 & 1 & -1 & 0 & -1 \\ 0 & -1 & 1 & 0 & 1 \end{pmatrix} \tag{17}$$

There are another two solutions corresponding to permutations of the non-zero charges shown in (17) among the generations, but they are equivalent to this one after simply relabelling the states. Furthermore, any linear combination of the charges of (17) with the $U_a(1)$ charges (a=1,2,3) is also an anomaly-free solution. But the choice of the four generators to span the $U(1)^4$ space should not affect the mass matrix elements and we choose an ansatz which is invariant under such basis changes [12].

Initially we assume that the full gauge group $SMG^3 \times U(1)_f$ is partially conserved and obtain an expression for the mass matrices. The diagonal elements are still identical in all 3 matrices, so m_t must gain its dominant contribution from an off-diagonal element of M_U. Bearing in mind the need for $m_t \gg m_b$, $m_c \gg m_s$ and a quasi-diagonal V_{CKM}, we aim as a first approximation for a parameter choice which yields matrices looking approximately like

$$M_U \simeq \begin{pmatrix} m_u & 0 & 0 \\ 0 & 0 & m_t \\ 0 & m_c & 0 \end{pmatrix}, \; M_D \simeq diag(m_d, m_s, m_b), \; M_l \simeq diag(m_e, m_\mu, m_\tau) \tag{18}$$

As the top quark mass is unsuppressed, we demand $m_t \simeq M_U(2,3) \simeq \langle \phi \rangle_{ws}$. This implies that $SU_3(2)$, $SU_2(3)$, $SU_3(3)$ and the linear combination $4Y_2 - Y_3 + Y_f$ of $U(1)$ generators are not relevant for mass suppression.

We then perform a computer fit with five degrees of freedom to experimental data for the masses and mixing angles, by minimising the function

$$\chi^2 = \sum_{i=1}^{8} [\log f_i - \log \frac{E_i}{m_t}]^2 + \sum_{i=9}^{11} [\log f_i - \log E_i]^2 \tag{19}$$

where

$$E_i = \begin{cases} \text{quark/lepton mass to be fitted} & (i = 1, \ldots, 8) \\ \text{value of } V_{us}, V_{cb}, V_{ub} \text{ to be fitted} & (i = 9, \ldots, 11) \end{cases}$$

Table 2. Results of a χ^2-fit to fermion masses and mixing angles. All masses are running masses evaluated at 1 GeV unless otherwise stated. The third column shows results for $m_t^{phys} = 200\ GeV$ for a case where the fit is biased in favour of obtaining $m_c > m_s$.

Fit Results	$m_t^{phys} = 100\ GeV$	$m_t^{phys} = 200\ GeV$	
		unbiased	biased
χ^2	3.7	5.6	6.9
m_e (MeV)	1.0	1.0	1.0
m_μ (MeV)	120	160	110
m_τ (GeV)	1.4	1.5	1.5
m_d (MeV)	4.9	4.9	4.9
m_s (MeV)	600	790	530
m_b^{phys} (GeV)	5.4	5.5	5.3
m_u (MeV)	4.9	4.9	4.9
m_c (GeV)	0.73	0.53	0.84
V_{us}	0.19	0.22	0.22
V_{cb}	0.016	0.012	0.0048
V_{ub}	0.0030	0.0027	0.0027

This choice of χ^2 fits the E_i assuming that the "theoretical" uncertainty on the model predictions f_i (due to the unknown $O(1)$ Yukawa couplings) is a factor of e which swamps the experimental error on the E_i. The top mass m_t is a free parameter in χ^2 and we perform fits with both $m_t^{phys} = 100\ GeV$ and $200\ GeV$. A good fit to the mass and mixing hierarchies can now be obtained and results are shown in the 1st column of Table 2 for $m_t^{phys} = 100\ GeV$. All the data are fitted within a factor of 2, except for m_s and V_{cb} which are fitted within a factor of 3.

6 Conclusion

The observed values of the quark and lepton masses and the quark mixing angles provide our main experimental clues to the underlying flavour dynamics contained in the physics beyond the SM. Their hierarchical structure strongly suggests the existence of approximately conserved chiral quantum numbers beyond those of the SMG. We presented a classification of the SMG extensions with no new mass-protected particles beyond those of the SM and argued, intuitively, that the category (8) models are particularly suited to study the Yukawa coupling hierarchy problem. The members of this category are characterised by having the same irreducible representations as exist in the SM and by having the SM embedded within them as a diagonal subgroup. In particular the anti-grand unified $SMG^3 \times U(1)_f$ model can naturally reproduce all the qualitative features of the quark-lepton spectrum.

7 Acknowledgement

I should like to thank my collaborators Holger Bech Nielsen and Gerry Lowe for many enjoyable discussions of the fermion mass problem.

References

[1]   C.D. Froggatt & H.B. Nielsen, Nucl. Phys. $\underline{B164}$ (1979) 144;
M. Gell-Mann, P. Ramond & R. Slansky in *Supergravity*, eds. P. van Nieuwen-huizen & D. Z. Freedman (North Holland, New York, 1980) p.315;
T. Yanagida in *Proc. of the Workshop on Unified Theory and Baryon Number in the Universe*, eds. A. Sawada & H. Sugawara (KEK, Tsukuba-Gu, Ibaraki-ken, Japan, 1979) p.95.

[2]   M. Veltman, Acta Phys. Pol. $\underline{B12}$ (1981) 437.

[3]   H. Fritzsch, Phys. Lett. $\underline{B70}$ (1977) 436; Phys. Lett. $\underline{B73}$ (1978) 317;
P. Ramond, R.G. Roberts & G.G. Ross, Nucl. Phys. $\underline{B406}$ (1993) 19.

[4]   C.D. Froggatt & H.B. Nielsen, Nucl. Phys. $\underline{B147}$ (1979) 277.

[5]   S. Dimopoulos, Phys. Lett. $\underline{B129}$ (1983) 417.

[6]   See, for example, the contributions of S. Raby, G.G. Ross and N. Seiberg to these proceedings.

[7]   C.D. Froggatt, G. Lowe & H.B. Nielsen, Nucl. Phys. $\underline{B420}$ (1994) 3.

[8]   D.S. Shaw & R.R. Volkas, Phys. Rev. $\underline{D47}$ (1993) 241.

[9]   C.D. Froggatt & H.B. Nielsen, *Origin of Symmetries* (World Scientific, Singapore, 1991).

[10]   D.L. Bennett, H.B. Nielsen & I. Picek, Phys. Lett. $\underline{B208}$ (1988) 275.

[11]   D.L. Bennett & H.B. Nielsen, *Fitting the Fine Structure Constants by Critical Couplings and Integers*, Proc. of the XXV International Ahrenshoop Symposium on the Theory of Elementary Particles, Gosen, Sept. 1991, NBI Preprint NBI-HE-92-07 (1992);
D.L. Bennett & H.B. Nielsen, *Predictions for Nonabelian Fine Structure Constants from Multicriticality*, NBI Preprint NBI-HE-93-22 (1993).

[12]   C.D. Froggatt, G. Lowe & H.B. Nielsen, Nucl. Phys. $\underline{B414}$ (1994) 579.

The vacuum energy and fermion mass hierarchies[1]

T. GHERGHETTA AND Y. NAMBU
ENRICO FERMI INSTITUTE
AND
DEPARTMENT OF PHYSICS,
UNIVERSITY OF CHICAGO,
CHICAGO, IL 60637

Abstract: Recently Nambu proposed a toy model which generates a mass hierarchy by minimizing the vacuum energy density as a function of Yukawa couplings. We consider the vacuum energy density of the Weinberg-Salam electroweak theory and discuss whether the toy model can be realized for the Standard Model fermions.

1   Introduction

The BCS theory of superconductivity and the Weinberg-Salam electroweak theory are both well known examples of a mass spectrum arising from spontaneous symmetry breaking (SSB) [1]. While the origin of the fermion masses remains a major unsolved problem, it would not be unreasonable to assume that the fermion mass spectrum arises from the SSB of some underlying fundamental symmetry. If this were the case, then one should be able to solve a set of "gap equations" or Schwinger-Dyson equations for the fermion self-energies. In general there may be more than one set of solutions to the nonlinear coupled equations and so a plausible criteria would be to choose that set of solutions which minimizes the vacuum energy density. However, instead of solving a coupled system of gap equations, one can directly obtain the solution by minimizing the vacuum energy density as a function of the Yukawa couplings, where the vacuum expectation value (vev) and other couplings are held fixed. If such a solution were to exist then this would shed new light on the origin of the fermion mass spectrum.

In general the vacuum energy density Ω is a highly divergent quantity and so an ultraviolet cutoff Λ will be used. On dimensional grounds the vacuum energy density will have the form

$$\Omega = \Omega^{(0)}\Lambda^4 + \Omega^{(2)}\Lambda^2 + \Omega^{(4)}\ln\frac{\Lambda}{\mu} \tag{1}$$

where

$$\Omega^{(0)} = F^{(0)}\left(\left\{g_i\ln\frac{\Lambda}{\mu}\right\}, \left\{\frac{m_i}{\Lambda}\right\}, \left\{\frac{\Lambda}{m_i}\right\}, \left\{\frac{m_i}{m_k}\right\}\right) \tag{2}$$

$$\Omega^{(2)} = \sum_i m_i^2 F_i^{(2)} \tag{3}$$

$$\Omega^{(4)} = \sum_{i,k} m_i^2 m_k^2 F_{i,k}^{(4)} \tag{4}$$

---

[1]Talk presented by T. Gherghetta.

and $\{g_i\}$ collectively denote all the couplings. The functions $F_i^{(2)}$ and $F_{i,k}^{(4)}$ depend on the same dimensionless ratios as $F^{(0)}$. Note that in supersymmetric theories the quartic and quadratic terms vanish, $\Omega^{(0)} = \Omega^{(2)} = 0$. The masses $\{m_i\}$ are all assumed to be generated by the Higgs mechanism, so $m_i \simeq f_i v$ where v denotes the vacuum expectation value (vev) of a Higgs field. The vev is the only mass parameter which has a dynamical meaning.

In order to guarantee a minimum solution it will be assumed that the vacuum energy density, Ω is a nonsingular function of the vev and the Yukawa couplings $\{f_i\}$ at each order in perturbation theory. This means eliminating terms like $(\Lambda/m_i)^n$ in (1). A convenient way to achieve this is to impose $\Omega^{(2)} = 0$. This condition is similar to invoking the Veltman condition [2], which requires all quadratically divergent terms in the Higgs self-energy to vanish, if the tree level Higgs mass is zero. Additional motivation for imposing this condition, is the assumption that the low energy physics should be as insensitive as possible to the unknown physics represented by the scale Λ.

The Λ^4 terms in the vacuum energy density are in general not gauge invariant because they receive contributions from the wavefunction renormalization and so pose a potential problem. To understand these terms completely gravity must also be included. We will assume for our purposes that these terms vanish by some underlying supersymmetry or other mechanism at the fundamental level. There should be no problems with gauge invariance for the remaining quadratic and logarithmic terms, because the coefficients depend on physical mass parameters.

Let us now consider the toy model introduced by Nambu [1] which assumes the general conditions noted above. For simplicity consider two flavours with Yukawa couplings f_1 and f_2. Writing $x = f_1^2$ and $y = f_2^2$ the Veltman condition and vacuum energy density to lowest order are

$$x + y = C = constant \tag{5}$$
$$\Omega = -A(x^2 + y^2) + B(x \ln x + y \ln y) = minimum \tag{6}$$

where $A, B > 0$. If $B = 0$ then it is clear that the minimum will occur at the end points $(x, y) = (C, 0)$ or $(0, C)$. However for $B > 0$ the log term dominates for small x and y and the minimum shifts slightly away from the endpoints

$$x \simeq C \qquad y \simeq \exp\left(-\frac{2AC}{B}\right) \tag{7}$$

Provided $AC/B \gg 1$ there will be a mass hierarchy between the two fermion flavours. Clearly the hierarchy results from the logarithmic terms in Ω. Higher order corrections to the conditions (5) and (6) do not destroy this hierarchy.

If one adds a third fermion flavour to the above model $z = f_3^2$ then the minimum solution is found to be $x = y \ll z$. This degeneracy in x and y can be removed by introducing mixing terms of the form $f_i^2 f_k^2 \ln(f_i f_k)$ in the model.

2   The Standard Model

The three flavour Nambu toy model is clearly suggestive of what may be occurring for the top, bottom and tau fermions in the Standard Model. In order to generate terms in the Weinberg-Salam electroweak theory similar to the toy model, we need to compute the quadratic and logarithmic terms in the vacuum energy density to fourth order or two loops, including all counterterms resulting from renormalization.

To make contact with the toy model the renormalization scale parameter μ in each $g_i(\mu)\ln(\Lambda/\mu)$ is set equal to its respective mass $g_i v$. This means that g_i becomes its on-shell value. In particular, the fermion masses and the Yukawa couplings are equivalent for fixed v, and so we may regard them as dynamical variables to be varied.

There are only two 1PI vacuum diagram topologies at two loops as shown in Fig. 1. These are the one point double-loop diagram and the two point diagram. However one does not have to compute every vacuum diagram in the Standard Model with these topologies. Since we want to minimize the vacuum energy density as a function of the Yukawa couplings, we only need to compute the mass dependent part of Ω. This can be obtained by considering

$$\int dm_i \frac{\partial \Omega}{\partial m_i} \tag{8}$$

and noting that the partial derivative is equivalent to attaching an external Higgs leg to the vacuum diagram. Thus the mass dependent quadratic and logarithmic terms in the vacuum energy density are equivalent to computing all Higgs tadpole diagrams.

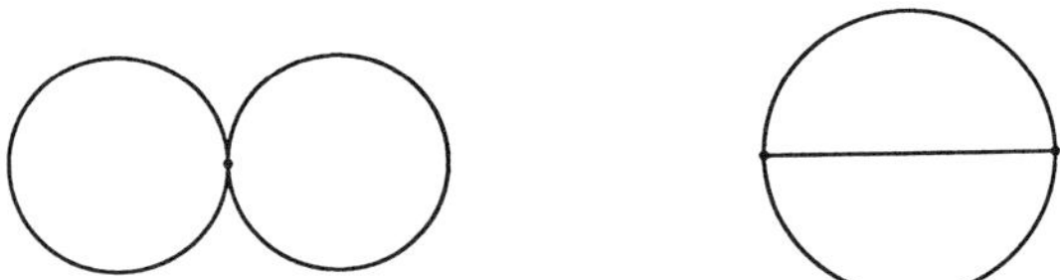

Figure 1. The one point double-loop and two point
vacuum energy diagram topologies.

The two-loop Veltman condition can be obtained by differentiating the coefficients of the pure Λ^2 terms in the vacuum energy density with respect to the vev. These terms arise solely from the two point topology depicted in Fig. 1. This can be understood by noting that off-shell self-energies behave like

$$\Sigma(p) \simeq \ln\frac{\Lambda}{p} \tag{9}$$

and contracting the self-energy external legs gives rise to

$$\int \frac{d^3p}{E} \ln\frac{\Lambda}{p} \simeq \left. p^2 \ln\frac{\Lambda}{p}\right|_0^\Lambda + \int_0^\Lambda dp\, p \simeq \Lambda^2 \tag{10}$$

Note that we are assuming that the tree level Higgs mass parameter is zero so that the vanishing of the quadratic divergences in the vacuum energy density is equivalent to imposing the Veltman condition.

On the other hand vacuum diagrams with the double-loop topology depicted in Fig. 1 give rise to $\Lambda^2 \ln \Lambda/\mu$ terms. This is because a single loop behaves like $\Lambda^2 + \ln \Lambda/\mu$ and the double-loop diagram is just the single loop "squared".

Note also that since we are calculating two loop vacuum diagrams we have to consider all one loop counterterms arising from self-energies. There is no need to consider vertex counterterms because these will be of higher order. Contributions from counterterms are to leading order $\Lambda^2 \ln \Lambda/\mu$ and are important in maintaining gauge invariance.

Recall that in the three flavour toy model the minimum solution had a degeneracy for the two smaller values. This degeneracy may be lifted in the Standard Model by allowing all fermion mixing which would lead to the CKM matrix.

3 Concluding Remarks

The success of Nambu's toy model depends critically on the signs of A and B in (6). Work is still in progress on computing and checking the gauge invariance of the necessary terms in the vacuum energy for the Standard Model and will be reported elsewhere *[3]*. However if Nambu's idea fails for the Standard Model then it may well be useful in other areas. P. Binétruy has already proposed a scheme for a string theory model at this workshop using Nambu's toy model *[4]*.

This work was supported in part by NSF contract PHY-91-23780.

References

[1] Y. Nambu, in *Some New Trends on Fluid Mechanics and Theoretical Physics*, Proceedings of the ICFMTP in Honor of Professor Pei-Yuan Chou's 90th Anniversary, Peking University Press, Beijing, China, 1993, (U. Chicago preprint EFI 92-37).

[2] M. Veltman Acta Phys. Pol. **B12** (1981), 30.

[3] T. Gherghetta and Y. Nambu, in preparation.

[4] P. Binétruy, in *Yukawa Couplings and the Origins of Mass*, Proceedings of the 2nd IFT Workshop, Gainesville, Florida, Feb. 11-13, 1994, this volume.

Comments on Yukawa Couplings and Higgs Bosons from the CMSSM[1] [2]

G.L. KANE
RANDALL LAB OF PHYSICS
UNIVERSITY OF MICHIGAN
ANN ARBOR, MI 48104-1120

Abstract: Although the eventual goal is to fully integrate the fermion masses into the construction of a supersymmetric unified theory, we begin by constructing a minimal supersymmetric Standard Model that is as constrained as possible both theoretically and experimentally. All analysis are done with an accuracy better than that of the data used. It is then possible to say some things about Yukawa couplings, and about Higgs boson masses.

1   The Constrained Minimal Supersymmetric Standard Model (CMSSM)

We take very seriously the idea that the underlying theory will be described by an effective Lagrangian, at a scale of order 10^{16} GeV, that is supersymmetric. Among the reasons are the elegant and satisfying way in which the electroweak symmetry breaking is derivable [1] in such a theory; this already puts constraints on $\tan\beta$ and the soft breaking parameters. Similarly, the unification of the gauge couplings [2] only occurs if the parameters are in certain ranges. In our view another major motivation of the supersymmetric theory is that is provides a natural cold dark matter candidate, for a range of but not all parameters. [3] Thus we want to insist on embedding fermion masses in a theory that implements all these requirements. In addition, of course, all experimental constraints should be included.

On the other hand, we do not want to impose constraints that are too specialized, so that we do not miss good results if we specialize incorrectly. We want to avoid making assumptions that might not be valid, and mislead us about what works. The main question perhaps involves the gauge group at the high scale. For the general analysis [3] we only choose $\sin^2\theta_W = 3/8$ at the high scale, which is true of almost all interesting choices; equivalently, we use the 5/3 factor in the $U(1)$ normalization. Saying more than that requires choosing a gauge group, and we do not yet think there is any compelling reason to choose any particular gauge group. Finally, to fully include fermion masses, we will have to specialize about the gauge group, but we have not yet carried out that work. Our construction of the CMSSM is typical of what a number of groups [4] are doing, and is reported in [3].

Here I want to emphasize that we include a large number of constraints, and do so simultaneously to insure the same model framework is used for all. Also, it is very important to use the appropriate accuracy for the analysis. Some

---

[1] Research supported in part by the United States DoE.
[2] Work in collaboration with C. Kolda, L. Roszkowski, and J. Wells.

examples that illustrate this are

(a) If we use one-loop or two-loop RGE's for running between the electroweak scale and the high scale, with a particular set of other quantities held constant, we find a shift of $\alpha_s\,(M_Z)$ from 0.117 (one-loop) to 0.129 (two-loop). That is a 2σ shift in α_s given the present errors, so it is clearly essential to use two-loop RGE's.

(b) If we use two-loop RGE's and a common superpartner threshold, or separate thresholds for the different superpartners, we find a shift in the unification scale of a factor of about 1.6. That enters squared or even to the fourth power in various constraints.

(c) It is essential to use the full one-loop effective Higgs potential to avoid misleading results.

(d) Perhaps a useful way to illustrate the need for precision is to point out that the two solutions reported in one of the earliest analyses of the present sort are no larger valid when all of the aspects of the analysis are done as accurately as possible. Other sets of parameters do satisfy the conditions, of course, but in general lead to different predictions.

(e) In carrying out the analysis values are used for $\sin^2\theta_W$ and for M_t. If one fixes $\sin^2\theta_W$ at its experimental value and asks for the variation of $\alpha_s\,(M_Z)$ with M_t, one finds that $\delta\alpha_s \simeq -3\%$ as M_t increases over a $\pm 2\sigma$ range. But if one includes the M_t dependence of $\sin^2\theta_W$ [5] and asks the same question, then α_s increases with increasing M_t, about the same amount. Thus there is a full 1σ effect on α_s from including the M_t dependence of $\sin^2\theta_W$. It also increases the value of the unification scale by a factor of about 1.3, which enters squared or to a higher power.

We see that drawing quantitative conclusions about implications of SUSY-GUT frameworks for fermion masses requires a careful numerical analysis at the appropriate level of precision.

Once the constraints are included, we have a set of parameters such as $\tan\beta, M_0, M_{1/2}, A_t$, each allowed to vary over a range. Any (correlated) set is equally consistent with what we know today theoretically and experimentally. To make predictions that are sharper requires more assumptions and is risky.

Of course, nature only chooses one particular set of these "solutions", assuming the whole approach is right. Nevertheless, the best that can be done today at making predictions or drawing conclusions is to calculate for all allowed sets of parameters, all "solutions". In some cases we will make histograms or scatter plots of these solutions. In principle the only firm conclusions are those implied by the boundaries. But if the solutions naturally occur in some region of the quantity being examined that is of interest too.

2   Interpretation, The Value of M_t

One question we can comment on with our solutions is the physical origin of the large value of M_t. Defining $h_t(Q)$ and $h_b(Q)$ to be the t and b Yukawa couplings, with h_{t0} and h_{b0} the values at the unification scale, we can write

$$M_t(Q) = h_t(Q)V \tan\beta / \sqrt{1 + \tan^2\beta},$$

$$M_b(Q) = h_b(Q)V / \sqrt{1 + \tan^2\beta}.$$

Then if $r \equiv (h_t(M_t)/h_b(M_t)) / (h_{t0}/h_{b0})$, we get

$$M_t(M_t)/M_b(M_t) = r \tan\beta \, (h_{t0}/h_{b0}).$$

The left hand side is numerically about 60 (remember that $M_b(M_t) < 3$). The right hand side is a product of three factors, so the factor of 60 could arise as a product with two of them of order unity and the third large, or all of order 4, or many other ways. The factor r measures the running of the Yukawas, and is given by the RGE's. If we neglect the small $U(1)$ terms and the τ Yukawa in the RGE's, then

$$8\pi^2 \left(dh_t/d\ln Q - dh_b/d\ln Q \right) = h_t \left(3h_t^2 + \frac{1}{2}h_b^2 \right) - h_b \left(3h_b^2 + \frac{1}{2}h_t^2 \right). \tag{1}$$

Thus *if* $h_{t0}/h_{b0} \simeq 1$, then the right hand side of this equation vanishes, and h_t and h_b run together. Then the large M_t/M_b is entirely due to a large $\tan\beta$.

What if we allow h_{t0}/h_{b0} to vary? The result is that the running turns out not to be important. All the CMSSM solutions have the property that

$$\tan\beta \, (h_{t0}/h_{b0}) \simeq 60. \tag{2}$$

Thus all theories of fermion masses will need to satisfy this condition, and will not be able to get a range of values from the running.

3   The Mass and Detection of the Lightest SUSY Higgs Boson h

The tree level mass matrix for h°, H° would have a zero eigenvalue if either $B\mu \gg M_Z^2$ or $B\mu \ll M_Z^2$, so effectively the ratio $B\mu/M_Z^2$ determines m_{h°. That gives upper and lower limits on M_{h°, since $B\mu$ cannot get too small because of the electroweak breaking constraint and because of the one-loop contribution to the effective Higgs potential.

As a result, the CMSSM prediction for M_{h° is the histogram shown in Fig. 3.1 for $M_t = 170$ GeV. The natural range is 95 GeV < 125 GeV, though solutions exist up to 136 GeV. Almost none occur below 80 GeV.

LEP2 is initially planned to have $\sqrt{s} \simeq 180$ GeV, so it can search for $m_{h^\circ} \lesssim 80$ GeV. If the energy is increased to 210 GeV, the search is extended to $m_{h^\circ} \lesssim 110$ GeV.

Recently it has been realized *[6]* that if FNAL is upgraded in luminosity to get 10fb^{-1} or more of data, then it can search for the entire range of m_{h° show in Fig. 3.1. The modes $W(\to \ell\nu)h^\circ(\to b\bar{b})$, $Z(\to \ell^+\ell^-)h^\circ(\to b\bar{b})$, and $W(\to jj)h^\circ(\to \tau^+\tau^-)$ are all used.

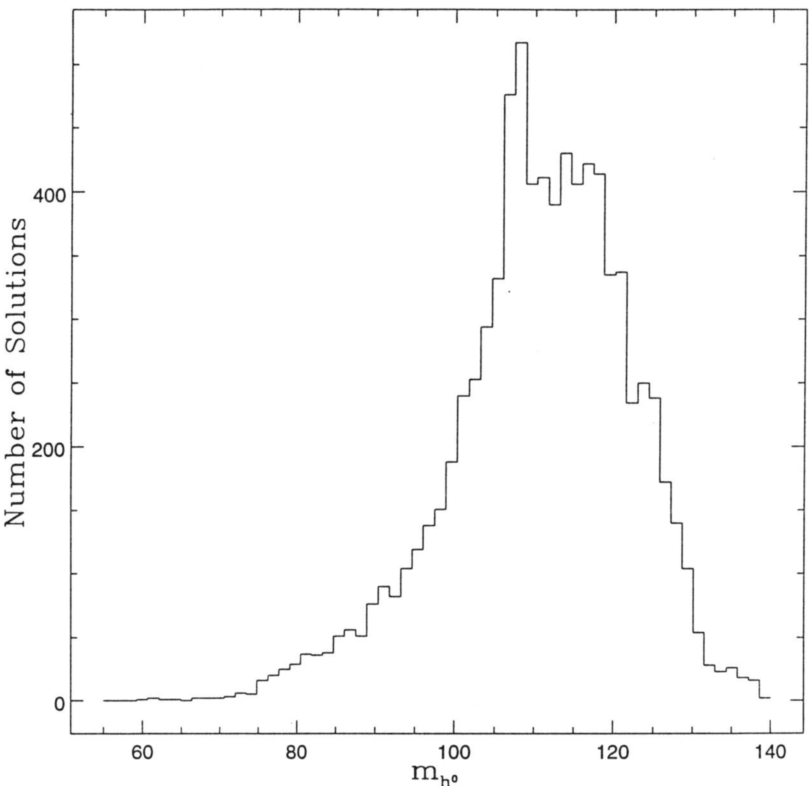

Figure 3.1. The CSSM predictions for m_{h^o}

If these upgrades at LEP and FNAL are not carried out, eventually LHC can carry out the search if sufficiently good detectors are built, and NLC can cover the entire range.

There has been some question about whether h° can be detected over its entire mass range. Fortunately, in the CMSSM it turns out *[3]* that for the entire set of solutions $\sin^2(\beta - \alpha) \simeq 1$. Thus the production cross section for Wh or Zh, proportional to $\sin^2(\beta - \alpha)$, is approximately the same as that for a SM h, over the entire parameter space. Similarly, *[3]* the couplings to t (proportional to $\cos\alpha/\sin\beta$), and to b (proportional to $\sin\alpha/\cos\beta$) are both approximately equal to their SM values since $|\beta - \alpha| \cong \pi/2$. Fig. 3.2 shows $\sin^2(\beta - \alpha)$ for the CMSSM solutions.

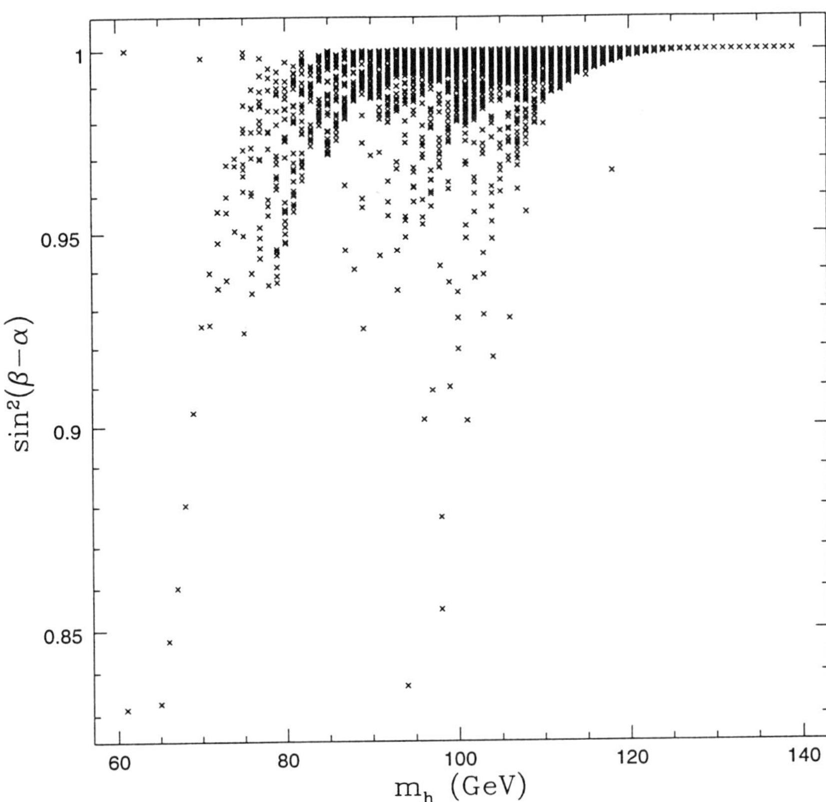

Figure 3.2. The CSSM predictions for $\sin^2\left(\beta - \alpha\right)$, shown vs. m_{h°.

4 Calculable Upper Bound on m_{h°

The discussion of the previous section gave the predictions for m_{h° in the CMSSM. There is a general upper bound [7] on m_{h° in any supersymmetric theory where the couplings remain perturbative up to a high scale. It is somewhat surprising that such a bound is calculable; it's value does not depend on any unknown vev's of the arbitrary Higgs sector, nor on any of the soft SUSY breaking parameters. The value does depend on the gauge group and spectrum, in calculable ways, at about the 10-15% level. Thus m_{h° cannot be pushed up if it is not detected — either a light Higgs boson exists or supersymmetry does not describe nature and explain the physics of the electroweak scale.

The derivation of the upper bound can be summarized fairly simply. It is basically a straightforward analysis. The scalar potential $V = V_F + V_D + V_{SOFT} + V_{MASS}$, for any general Higgs sector. From this the genearl mass matrix $2M_{ij}^2 = \partial^2 V / \partial Z^i \partial Z^j$ can be calculated, where each scalar field is expanded into real and imginary parts, $\varphi^a = Z^{2a} + iZ^{2a+1}, a = 0, 1, 2 \ldots$. The coefficients of powers of Z^i in V depend on the superpotential coefficients, the soft breaking terms, any vev's etc.

The thing to look at is the 2×2 submatrix corresponding to ReH_1°, ReH_2° (these are the two doublets that give mass to the SM particles). One can see that it always has the form

$$m_{ij}^2 = \begin{pmatrix} -J \tan\beta + K & J + L \\ J + L & -J \cot\beta + K' \end{pmatrix} \tag{3}$$

The nice thing is that all the SUSY parameters and vev's that grow or that we don't know are in J, while K, K', L only depend on gauge couplings, $\tan\beta$, and on self-couplings of scalars. Since $det\, M_{ij}^2$ and $Tr\, M_{ij}^2$ are both linear in J, one eigenvalue of M_{ij}^2 is bounded independent of J. A bounded eigenvalue of M_{ij}^2 is also a bound on the smallest eigenvalue of the full M^2 matrix , i.e., on $m_{h^\circ}^2$. The remaining piece to the argument is that the Higgs self-couplings are bounded if the theory is perturbative up to a high scale, so K, K', L are bounded.

5 Comments on $b - \tau$ Unification

A number of people have studied [8] the possibility of setting $h_b = h_\tau$ at the unification scale, or even $h_b = h_\tau = h_t$. We have studied these questions, [9] but have not yet been able to convince ourselves of the meaning of the results. There are several factors. First, or course the Yukawa unification need not occur precisely where the gauge unification does. Second, the Yukawa unification can be somewhat sensitive to effects of high thresholds.

There are two serious practical problems. Since the b and τ Yukawas are not so separated at low energies, their trajectories as the scale increases are not so separated. Thus for a typical set of parameters the τ line and the b band (a band because of uncertainty in the value of m_b) might overlap for 10^9 GeV - 10^{12} GeV, which seems quite far from 10^{16} GeV — but only be different by 20% at 10^{16} GeV. Since uncertainties in α_s, etc. certainly make 10% effects in $m_b(10^{16}$ GeV), it is not obvious how to interpret a 20% deviation.

Second, in the CMSSM we want to study the result for larger $\tan\beta$, and once $\tan\beta > 10$ there are significant effects [10] from loop contributions to m_b, which in principle should be included in the analysis as part of the iterative running procedure. We plan to do this, but we have not yet done it. To study the $b = \tau = t$ case such a procedure is essential.

The $b - \tau$ unification requires either very large or very small $\tan\beta$ (given that $m_t < 190$ GeV), and the $b - \tau - t$ unification can only happen if $\tan\beta$ is large. In both cases one can make a number of testable predictions, mostly for superpartner and h° masses; several talks in this proceedings deal with such analyses.

6 Other Predictions

With the CMSSM one can calculate the expected spectrum of superpartners, $BR\,(b \to s\gamma)$, $BR\,(Z \to b\bar{b})$, Ωh^2, and many other observables. Once superpartners are discovered it will be easy to test this framework, and to see if it is the appropriate framework to try to get new insights into fermion masses. Even now, as data improves it is becoming possible to make progress in constraining the soft-breaking parameters and $\tan\beta$.

References

[1]   For a review, see L.E. Ibanez and G.G. Ross, "Perspectives on Higgs Physics", World Scientific, ed. G.L. Kane, 1993.

[2]   P. Langacker, in the *Proceedings of the PASCOS-90 Symposium*, eds. P. Nath and S. Reucroft (World Scientific, Singapore, 1990); P. Langacker and M.-X. Luo, Phys. Rev. D **44**, 817 (1991).
      U. Amaldi, W. de Boer, and H. Fürstenau, Phys. Lett. B **260**, 447 (1991).

[3]   See the detailed discussion in G.L. Kane, Chris Kolda, Leszek Roszkowski, and James D. Wells, Phys. Rev. **D49** (1994) 6173.

[4]   R. G. Roberts and G. G Ross, Nucl. Phys. **B377**, 571 (1992).
      R. Barbieri and L. J. Hall, Phys. Rev. Lett. **68**, 752 (1992);
      L. J. Hall and U. Sarid, *ibid.* **70**, 2673 (1993).
      P. Langacker and N. Polonsky, Phys. Rev. D **47**, 4028 (1993).
      K. Hagiwara and Y. Yamada, Phys. Rev. Lett. **70**, 709 (1993).
      V. Barger, M. S. Berger, and P. Ohmann, Phys. Rev. D **47**, 1093 (1993).
      H. Arason, D. J. Castaño, B. Keszthelyi, S. Mikaelian, E. J. Piard, P. Ramond, and B. D. Wright, Phys. Rev. D **46**, 3945 (1992).
      H. Arason, D. J. Castaño, B. Keszthelyi, S. Mikaelian, E. J. Piard, P. Ramond, and B. D. Wright, Phys. Rev. D **47**, 232 (1993). See also A. Giveon, L. J. Hall, and U. Sarid, Phys. Lett. B **271**, 138 (1991);
      P. H. Frampton, J. T. Liu, and M. Yamaguchi, *ibid.* **277**, 130 (1992).
      V. Barger, M.S. Berger, P. Ohmann, and R.J.N. Phillips, Madison preprints MAD/PH/755 (April 1993) and MAD/PH/760 (May 1993).
      M. Carena, S. Pokorski, and C. Wagner, Nucl. Phys. **B406**, 59 (1993).
      P. Langacker and N. Polonsky, University of Pennsylvania preprint UPR-0556-T (May 1993).
      R. Arnowitt and P. Nath, Phys. Lett. B **287**, 89 (1992); Phys. Rev. Lett. **69**, 725 (1992); Phys. Lett. B **299**, 58 (1993); Northeastern Univ. preprint

NUB-TH-3048-92.
R. G. Roberts and L. Roszkowski, Phys. Lett. B **309**, 329 (1993).
S. Kelley, J. L. Lopez, D. V. Nanopoulos, H. Pois, and K. Yuan, Nucl. Phys. **B398**, 3 (1993).
J. L. Lopez, D. V. Nanopoulos, and A. Zichichi, Phys. Lett. B **291**, 255 (1992).
M. Drees and M. Nojiri, Phys. Rev. D **47**, 376 (1993).
W. de Boer, R. Ehret, and D. I. Kazakov, IEKP-KA/93-13 (August 1993), talk at the XVI Int. Symposium on Lepton-Photon Interactions, Cornell, August 10-15, 1993.
D. J. Castaño, E. J. Piard, and P. Ramond, University of Florida preprint UFIFT-HEP-93-18 (August 1993).

[5]   P. Langacker and N. Polonsky, Phys. Rev. **D47** (1993) 4028.

[6]   S. Mrenna and G.L. Kane, preprint UM-TH-94-24; see also J. Gunion and T. Han, UCD-94-10 and W. Marciano, A. Stange, and S. Willenbrock, ILL-TH-94-8.

[7]   G. L. Kane, C. Kolda, and J. D. Wells, Phys. Rev. Lett. **70**, 2686 (1993). See also J. R. Espinosa and M. Quirós, Phys. Lett. B **302**, 51 (1993).

[8]   H. Arason, D. J. Castaño, B. Keszthelyi, S. Mikaelian, E. J. Piard, P. Ramond, and B. D. Wright, Phys. Rev. D **47**, 232 (1993). See also A. Giveon, L. J. Hall, and U. Sarid, Phys. Lett. B **271**, 138 (1991);
P. H. Frampton, J. T. Liu, and M. Yamaguchi, *ibid.* **277**, 130 (1992).
V. Barger, M.S. Berger, P. Ohmann, and R.J.N. Phillips, Madison preprints MAD/PH/755 (April 1993) and MAD/PH/760 (May 1993).
M. Carena, S. Pokorski, and C. Wagner, Nucl. Phys. **B406**, 59 (1993).
P. Langacker and N. Polonsky, University of Pennsylvania preprint UPR-0556-T (May 1993).

[9]   Chris Kolda, Leszek Roszkowski, James D. Wells, and G.L. Kane, UM-TH-94-03, to appear in Phys. Rev.

[10]   L. Hall, R. Rattazzi, and U. Sarid, LBL preprint LBL-33997 (June 1993, revised February 1994).
M. Carena, M. Olechowski, S. Pokorski, and C. Wagner, CERN preprint CERN-TH.7163/94 (February 1994).

Some model-independent properties of quark mixing

A. KUSENKO
INSTITUTE FOR THEORETICAL PHYSICS
STATE UNIVERSITY OF NEW YORK AT STONY BROOK
STONY BROOK, NY 11794-3480

Abstract: We discuss some new invariants of quark mixing and show their usefulness with a simple example. We also present some other new tools for analyzing quark mixing.

1 Invariants of quark mixing and their applications

In recent years the increasingly accurate data on quark mixing has stimulated interest in possible predictions concerning quark mass matrices. The parameters of the Cabibbo-Kobayashi-Maskawa mixing matrix V are now known reasonably well [1]. This determination has been made possible partly by the finding that there are only three generations of usual standard-model fermions (with corresponding light or massless neutrinos). Since the diagonalization of the quark matrices in the up and down sectors determines V, one can work back from the knowledge of V to put constraints on the possible forms of (original, non-diagonal) quark mass matrices. However, the data on quark mixing determines these mass matrices only up to an arbitrary unitary similarity transformation. This is a result of the fact that if the up and down quark mass matrices, M_u and M_d, are both acted on by the same unitary operator U_0 according to

$$M_{u,d} \rightarrow U_0 \, M_{u,d} \, U_0^\dagger \tag{1}$$

then the mixing matrix V remains unchanged. There have been many attempts to study specific assumed forms for quark matrices. While this is worthwhile, it is desirable to express the constraints from data on V on the quark mass matrices in an invariant form. In Refs. [2, 3] certain invariant functions of the quark mass matrices I_{pq} were introduced, which are expressed in terms of the quark masses squared and the $|V_{ij}|$:

$$I_{pq} = Tr(H_u^p \, H_d^q) = \sum_{ij} (m_i^{(u)})^{2p} (m_j^{(d)})^{2q} \, |V_{ij}|^2 \tag{2}$$

where $H_q = M_q M_q^\dagger$, and $m_i^{(u)}$ and $m_i^{(u)}$ are the masses of quarks in the "up" and "down" charge sectors.

In this paper we introduce some new invariants of the quark mass matrices (with respect to the transformation (1)) which can be expressed in terms of the measurable quantities only. These new invariants help one simplify the algebraic expressions which relate the elements of the quark mass matrices to the data. This important advantage allows for some new uses of the invariants which we will illustrate with some examples.

We introduce the following new invariants of the transformation (1):

$$K_{pq}(\alpha, \beta) = det(\alpha H_u^p + \beta H_d^q) \tag{3}$$

where $p, q, \alpha, \beta \neq 0$.

The hermitian matrices H_u and H_d can be diagonalized by a unitary similarity transformation:

$$\begin{cases} U_u H_u U_u^\dagger = D_u \\ U_d H_d U_d^\dagger = D_d \end{cases} \tag{4}$$

where $D_q = diag((m_1^{(q)})^2, (m_2^{(q)})^2, (m_3^{(q)})^2)$ are the diagonal matrices of the quark masses squared.

The mixing matrix V can be written then as:

$$V = U_u U_d^\dagger \tag{5}$$

In order to find an expression for K_{pq} in terms of U_{ij} we will need the following

Theorem 1.1. *If A and B are two 3×3 matrices such that $det(A) \neq 0$ and $det(B) \neq 0$ then the following relation holds:*

$$det(A + B) = det(A) + det(B) + det(A)\, Tr(A^{-1}B) + det(B)\, Tr(AB^{-1}) \tag{6}$$

Proof. We denote the elements of matrices A and B by A_{ij} and B_{ij} correspondingly. Their co-factors (which are equal to the corresponding minors, up to sign) will be written as $\hat{A}_{ij}$ and $\hat{B}_{ij}$. Then each determinant may be decomposed in a sum (Laplace expansion):

$$det(A) = \sum_i A_{ij} \hat{A}_{ij} = \sum_j A_{ij} \hat{A}_{ij}$$

$$det(B) = \sum_i B_{ij} \hat{B}_{ij} = \sum_j B_{ij} \hat{B}_{ij}$$

By definition, the determinant of a 3×3 matrix is a sum of $3! = 6$ terms:

$$det(A + B) = \sum (-1)^r (A_{1k_1} + B_{1k_1})(A_{2k_2} + B_{2k_2})(A_{3k_3} + B_{3k_3}) \tag{7}$$

where r is the sign of the permutation $\left(\begin{smallmatrix} 1 & 2 & 3 \\ k_1 & k_2 & k_3 \end{smallmatrix}\right)$.

The terms in the sum (7) which contain only the elements of A can be arranged as $det(A)$. Similarly, the terms containing only B's give $det(B)$. The terms containing one element of A multiplied by two elements of B, or visa versa, can be rewritten as:

$$\sum_{i,j} (A_{ij} \hat{B}_{ij} + B_{ij} \hat{A}_{ij}) \tag{8}$$

We can now use an identity:

$$(A^{-1})_{ij} = \frac{1}{det(A)}\hat{A}_{ji}$$

to rewrite (8) as:

$$\sum_{i,j}(A_{ij}\hat{B}_{ij} + B_{ij}\hat{A}_{ij}) = det(B)\sum_{ij}A_{ij}(B^{-1})_{ji} + det(A)\sum_{ij}(A^{-1})_{ji}B_{ij} =$$

$$det(A)\,Tr(A^{-1}B) + det(B)\,Tr(AB^{-1})$$

Altogether we get

$$det(A+B) = det(A) + det(B) + det(A)\,Tr(A^{-1}B) + det(B)\,Tr(AB^{-1})$$

which is the statement of Theorem 1.1. This completes the proof. ∎

Theorem 1.1 may be easily generalized to the case of 2×2 matrices, in which case the last two terms in (6) are equal and correspond to a redundant counting of the same terms in a sum similar to (7). Thus for the 2×2 matrices we get:

$$det(A+B) = det(A) + det(B) + det(A)\,Tr(A^{-1}B) \equiv$$
$$det(A) + det(B) + det(B)\,Tr(AB^{-1}) \tag{9}$$

The immediate consequence of Theorem 1.1 and equation (2) is the following relation:

$$K_{pq}(\alpha,\beta) \equiv det(\alpha H_u^p + \beta H_d^q) =$$
$$\alpha^3\,(m_1^{(u)}m_2^{(u)}m_3^{(u)})^{2p}\,[\,1 + (\beta/\alpha)\sum_{ij}[(m_j^{(d)})^{2q}/(m_i^{(u)})^{2p}]\,U_{ij}\,] + \tag{10}$$
$$\beta^3\,(m_1^{(d)}m_2^{(d)}m_3^{(d)})^{2q}\,[\,1 + (\alpha/\beta)\sum_{ij}[(m_j^{(u)})^{2p}/(m_i^{(d)})^{2q}]\,U_{ij}\,]$$

We also notice that

$$K_{pq}(\alpha,\pm\beta) = \alpha^3(m_1^{(u)}m_2^{(u)}m_3^{(u)})^{2p}(1 \pm (\beta/\alpha)I_{(-p)\,q})$$
$$\pm\beta^3(m_1^{(d)}m_2^{(d)}m_3^{(d)})^{2q}(1 \pm (\alpha/\beta)I_{p\,(-q)}) \tag{11}$$

Any four independent invariants from the set $\{I_{pq}, K_{pq}\}$ contain all the physical information about the CKM matrix.

If the mass matrices are assumed to be hermitian, one can introduce a similar set of invariants:

$$\tilde{I}_{pq} = Tr(\,M_u^p\,M_d^q\,)$$
$$\tilde{K}_{pq}(\alpha,\beta) = det(\alpha M_u^p + \beta M_d^q) \tag{12}$$

The formulae, similar to (2), (10) and (11), will also hold for $\tilde{I}, \tilde{K}$:

$$\tilde{I}_{pq} = Tr(M_u^p\, M_d^q) = \sum_{ij} (m_i^{(u)})^p (m_j^{(d)})^q\, |V_{ij}|^2 \tag{13}$$

$$\tilde{K}_{pq}(\alpha, \beta) \equiv det(\alpha M_u^p + \beta M_d^q) =$$
$$\alpha^3\, (m_1^{(u)} m_2^{(u)} m_3^{(u)})^p\, [\, 1 + (\beta/\alpha) \sum_{ij} [(m_j^{(d)})^q/(m_i^{(u)})^p]\, U_{ij}\,] + \tag{14}$$
$$\beta^3\, (m_1^{(d)} m_2^{(d)} m_3^{(d)})^q\, [\, 1 + (\alpha/\beta) \sum_{ij} [(m_j^{(u)})^p/(m_i^{(d)})^q]\, U_{ij}\,]$$

The odd powers of the mass eigenvalues may appear in some of the $\tilde{I}$ and $\tilde{K}$-type invariants. As usual, the signs of the fermion masses are ambiguous. However, for a given model, different choices of signs will in general result in different predictions for the CKM matrix. This is because in general the mass matrices do not commute with all of the diagonal matrices of the form $diag(\pm 1, \pm 1, \pm 1)$, where the $+$ and $-$ signs are chosen arbitrarily but so as to not get the plus or minus identity. We discussed the effect of the sign choices on the CKM mixing parameters on the example of some particular model [4]. Once the choice of signs for the fermion eigenvalues (determined in practice by the best fit to the data) is made, there is no further sign-related ambiguity in the $\tilde{I}$ and $\tilde{K}$-type invariants.

Now we would like to illustrate the usefulness of the invariants discussed above. As an example, we will take the model proposed in [4]. This model gives predictions for the low energy data on fermion masses and mixing which are in reasonable agreement with experiment. This model is formulated in the context of an SO(10) supersymmetric grand unified theory. The model has the following Yukawa matrices at the GUT scale:

$$M_u = \begin{pmatrix} 0 & A_u & 0 \\ A_u & B_u & 0 \\ 0 & 0 & C_u \end{pmatrix} \tag{15}$$

$$M_d = \begin{pmatrix} 0 & A_d e^{i\phi} & 0 \\ A_d e^{-i\phi} & B_d e^{i\theta} & B_d \\ 0 & B_d & C_d \end{pmatrix} \tag{16}$$

$$M_e = \begin{pmatrix} 0 & A_d e^{i\phi} & 0 \\ A_d e^{-i\phi} & -3B_d e^{i\theta} & -3B_d \\ 0 & -3B_d & C_d \end{pmatrix} \tag{17}$$

The low energy effective theory below GUT thresholds is assumed to be the Minimal Supersymmetric Standard Model (MSSM). After the renormalization effects are taken into account, the model agrees with the data on fermion mixing for a broad range of the t-quark mass, $m_t = 150...190\ GeV$. In these fits the phase θ is relatively unimportant and can be taken to be zero.

The values of the parameters $|A_q|$ and C_q ($q = u, d$), in (15) and (16) are simply related to the masses of quarks. On the contrary, the phase ϕ, which must have a nonzero value in order for the model to agree with experiment, depends on both the masses of quarks and their mixing parameters. It is a common procedure to diagonalize the matrices M_u and M_d numerically (or by means of the Taylor series expansion in terms of the quark mass ratios) and compare the corresponding CKM matrix to the data. On the other hand, the method of invariants discussed above offers a simpler and more elegant solution: the phase ϕ can be expressed analytically in terms of the measurable quantities only. Because the mass matrices in (15) and (16) are hermitian for $\theta = 0$, one can use a $\tilde{K}$-type invariant. On one hand, the allowed range for the value of $\tilde{K}_{11}(1,1)$ is known in terms of the measurable quantities from (14). On the other hand,

$$
\begin{aligned}
\tilde{K}_{11}(1,1) &= det(M_u + M_d) = \\
&= |A_u + A_d|\,(C_u + C_d) = \\
&= (2|A_u A_d|cos(\phi) + |A_u|^2 + |A_d|^2)(C_u + C_d)
\end{aligned}
\tag{18}
$$

And therefore

$$
cos(\phi) \;=\; \frac{1}{2\,|A_u|\,|A_d|}\;(\tilde{K}_{11}(1,1)/(C_u + C_d) \;-\; |A_u|^2 \;-\; |A_d|^2)
\tag{19}
$$

The right-hand side of (19) is known in terms of the quark masses and the CKM mixing parameters. Therefore, one can use the relation (19) to evaluate the phase ϕ without having to explicitly diagonalize the mass matrices.

For this model and others, it is important to have a general methodology to determine (1) how many unremovable phases there are in quark mass matrices and (2) which elements of these matrices can be rephased to be real. We have presented a general method to answer both of these questions [5]. The analogous questions for lepton mass matrices are more complicated, but we have also given a general answer to them [6].

In summary, we have introduced and studied some new invariants of the quark mass matrices which are model- and weak basis-independent. The identities (10) and (2), as well as (13) and (14), involving these invariants, provide important constraints on the possible forms of quark mass matrices M_u and M_d since they directly relate the elements of these matrices to the measurable parameters $|V_{ij}|^2$ and quark masses, and thereby enable one to avoid the explicit calculation of the eigenvectors of H_u and H_d.

2   A method to generate families of viable quark mass matrices

In this section we would like to present some useful mathematical tools, first proposed in [7], for the study of quark mass matrices and the connection with quark mixing. Our method was recently applied in [8] to generate the families of acceptable solutions for the fermion mass matrices.

The idea of the method is to utilize the well-known experimental fact that the CKM matrix is close to the 3×3 identity matrix: $|V_{ij}| \approx \delta_{ij}$. We make use of this fact by writing V as

$$V = e^{i\alpha H} \tag{20}$$

where H is some hermitian matrix and α is a real number. One may choose H to have its dominant (largest, in absolute value) eigenvalue to be 1. Then for α consistent with the data [1], we find that $|\alpha| \approx 0.3$. We can now expand V in the powers of α:

$$V = 1 + i\alpha H - \frac{1}{2}\alpha^2 H^2 + ... + \frac{1}{n!}(i\alpha H)^n + ... \tag{21}$$

It was shown in Ref. [7] that for any practical purposes it is sufficient to consider the first and the second order in α to match the precision to which the relevant quantities (CKM parameters, quark masses, etc.) are known.

The matrix H, corresponding to a given matrix V with distinct eigenvalues v_i, can be easily computed using the Sylvester's theorem:

$$i\alpha H = \sum_{k=1}^{3} ln(v_k) \frac{\prod_{i\neq k}(V - v_i \times 1)}{\prod_{i\neq k}(v_k - v_i)} \tag{22}$$

Then the usual system of equations:

$$\begin{cases} U_{u,L} M_u U_{u,R}^\dagger = diag(m_u, m_c, m_t) \equiv D_u \\ U_{d,L} M_d U_{d,R}^\dagger = diag(m_d, m_s, m_b) \equiv D_d \\ U_{u,L} U_{d,L}^\dagger = V \end{cases} \tag{23}$$

is satisfied for the family of solutions:

$$\begin{cases} M_u = U_u^\dagger D_u U_u = \\ D_u + i\alpha x[D_u, H] - \frac{1}{2}\alpha^2 x^2[[D_u, H], H] + ... \\ \\ M_d = U_d^\dagger D_d U_d = \\ D_d + i\alpha(x-1)[D_d; H] - \frac{1}{2}\alpha^2(x-1)^2[[D_d, H], H] + ... \end{cases} \tag{24}$$

depending on some arbitrary parameter x. (It is assumed that $|x|$, as well as $|1 - x|$, is sufficiently small to preserve the convergence of the series.)

To summarize, in this presentation we have discussed two different approaches to analyzing quark mixing. First, we introduced some new invariants of quark mixing and showed the usefulness of our method of invariants on a simple example. Then we have also discussed some new mathematical tools which allow one to generate the mass matrices consistent with the data.

The author would like to thank Professor Robert Shrock for many helpful discussions and comments.

References

[1]   Particle Data Group, *Review of Particle Properties*, Phys. Rev. **D 45** (1992) III.65; J. Rosner, talk at this conference; S. Stone, talk at this conference.

[2]   C. Jarlskog, Phys. Rev. **D 35** (1987) 1685. (See also C. Jarlskog, Phys. Rev. Lett. **55** (1985) 1039.)

[3]   G. C. Branco and L. Lavoura, Phys. Lett. **B 208** (1988) 123.

[4]   A. Kusenko and R. Shrock, Phys. Rev. **D 49**, (1994) 4962 (hep-ph/9307344).

[5]   A. Kusenko and R. Shrock, Phys. Rev. **D 50**, R30 (1994) (hep-ph/9310307).

[6]   A. Kusenko and R. Shrock, Phys. Lett. **B 323** (1994) 18 (hep-ph/9311307).

[7]   A. Kusenko, Phys. Lett. **B 284** (1992) 390.

[8]   C. H. Albright and S. Nandi, FERMILAB-PUB-93-316-T; FERMILAB-PUB-94-061-T (hep-ph/9403268); C. H. Albright, talk at this conference.

Masses of the Light Quarks

H. Leutwyler
Institute for Theoretical Physics,
University of Bern
Sidlerstr. 5,
CH-3012 Bern, Switzerland
HLEUTWYLER@ITP.UNIBE.CH

Abstract: The magnitude of m_u, m_d and m_s is discussed on the basis of Chiral Perturbation Theory. In particular, the claim that $m_u = 0$ leads to a coherent picture for the low energy structure of QCD is examined in detail. It is pointed out that this picture leads to violent flavour asymmetries in the matrix elements of the scalar and pseudoscalar operators, which are in conflict with the hypothesis that the light quark masses may be treated as perturbations.

1   Effective low energy theory of QCD

At low energies, the behaviour of scattering amplitudes or current matrix elements can be described in terms of a *Taylor series expansion* in powers of the momenta. The electromagnetic form factor of the pion, e.g., may be exanded in powers of the momentum transfer t. In this case, the first two Taylor coefficients are related to the total charge of the particle and to the mean square radius of the charge distribution, respectively,

$$f_{\pi^+}(t) = 1 + \tfrac{1}{6}\langle r^2\rangle_{\pi^+}\, t + O(t^2) \ . \tag{1}$$

Scattering lengths and effective ranges are analogous low energy constants occurring in the Taylor series expansion of scattering amplitudes.

For the straightforward expansion in powers of the momenta to hold it is essential that the theory does not contain massless particles. The exchange of photons, e.g., gives rise to Coulomb scattering, described by an amplitude of the form $e^2/(p' - p)^2$ which does not admit a Taylor series expansion. Now, QCD does not contain massless particles, but it does contain very light ones: pions. The occurrence of light particles gives rise to singularities in the low energy domain which limit the range of validity of the Taylor series representation. The form factor $f_{\pi^+}(t)$, e.g., contains a cut starting at $t = 4M_\pi^2$, such that the formula (1) provides an adequate representation only for $t \ll 4M_\pi^2$. To extend this representation to larger momenta, one needs to account for the singularities generated by the pions. This can be done, because the reason why M_π is so small is understood: the pions are the Goldstone bosons of a hidden, approximate symmetry. The low energy singularities generated by the remaining members of the pseudoscalar octet $(K^\pm, K^0, \bar{K}^0, \eta)$ can be dealt with in the same manner, exploiting the fact that the Hamiltonian of QCD is approximately invariant under $\mathrm{SU}(3)_{\mathrm{R}} \times \mathrm{SU}(3)_{\mathrm{L}}$. If the three light quark flavours u, d, s, were massless, this symmetry would be an exact one. In reality, chiral symmetry is broken by the quark mass term ocurring in the QCD Hamiltonian

$$H_{\text{QCD}} = H_0 + H_1 \ , \qquad H_1 = \int d^3x \{ m_u \bar{u}u + m_d \bar{d}d + m_s \bar{s}s \} \ .$$

For yet unknown reasons, the masses m_u, m_d, m_s however happen to be small, such that H_1 can be treated as a perturbation. First order perturbation theory shows that the expansion of the square of the pion mass in powers of m_u, m_d, m_s starts with

$$M_{\pi^+}^2 = (m_u + m_d)B\{1 + O(m_u, m_d, m_s)\} \ , \tag{2}$$

while, for the kaon, the leading term contains the mass of the strange quark,

$$M_{K^+}^2 = (m_u + m_s)B + \ldots \qquad M_{K^0}^2 = (m_d + m_s)B + \ldots \tag{3}$$

This explains why the pseudoscalar octet contains the eight lightest hadrons and why the mass pattern of this multiplet very strongly breaks eightfold way symmetry: M_π^2, M_K^2 and M_η^2 are proportional to combinations of quark masses, which are small but very different from one another, $m_s \gg m_d > m_u$. For all other multiplets of SU(3), the main contribution to the mass is given by the eigenvalue of H_0 and is of order Λ_{QCD}, while H_1 merely generates a correction which splits the multiplet, the state with the largest matrix element of $\bar{s}s$ ending up at the top.

The effective field theory combines the expansion in powers of momenta with the expansion in powers of m_u, m_d, m_s. The resulting new improved Taylor series, which explicitly accounts for the singularities generated by the Goldstone bosons, is referred to as chiral perturbation theory (χPT). It provides a solid mathematical basis for what used to be called the "PCAC hypothesis" [1, 2, 3, 4].

It does not appear to be possible to account for the singularities generated by the next heavier bound states, the vector mesons, in an equally satisfactory manner. The mass of the ρ-meson is of the order of the scale of QCD and cannot consistently be treated as a small quantity. Although the vector meson dominance hypothesis does lead to valid estimates (an example is given below), a coherent framework which treats these estimates as leading terms of a systematic approximation scheme is not in sight.

The effective low energy theory replaces the quark and gluon fields of QCD by a set of pseudoscalar fields describing the degrees of freedom of the Goldstone bosons π, K, η. It is convenient to collect these fields in a 3×3 matrix $U(x) \in$ SU(3). Accordingly, the Lagrangian of QCD is replaced by an effective Lagrangian, which only involves the field $U(x)$ and its derivatives. The most remarkable point here is that this procedure does not mutilate the theory: if the effective Lagrangian is chosen properly, the effective theory is mathematically equivalent to QCD [1, 4].

On the level of the effective Lagrangian, the combined expansion introduced above amounts to an expansion in powers of derivatives and powers of the quark mass matrix

$$m = \begin{pmatrix} m_u & & \\ & m_d & \\ & & m_s \end{pmatrix}$$

Lorentz invariance and chiral symmetry very strongly constrain the form of the terms occurring in this expansion. Counting m like two powers of momenta, the expansion starts at $O(p^2)$ and only contains even terms

$$\mathcal{L}_{eff} = \mathcal{L}^2_{eff} + \mathcal{L}^4_{eff} + \mathcal{L}^6_{eff} + \dots$$

The leading contribution is of the form

$$\mathcal{L}^2_{eff} = \tfrac{1}{4}F^2\text{tr}\{\partial_\mu U^+ \partial^\mu U\} + \tfrac{1}{2}F^2 B\,\text{tr}\{m(U + U^+)\} \qquad (4)$$

and involves two independent coupling constants: the pion decay constant F and the constant B occurring in the mass formulae (2) and (3). The expression (4) represents a compact summary of the soft pion theorems established in the 1960's: the leading terms in the chiral expansion of the scattering amplitudes and current matrix elements are given by the tree graphs of this Lagrangian.

At order p^4, the effective Lagrangian contains terms with four derivatives such as

$$\mathcal{L}^4_{eff} = L_1[\text{tr}\{\partial_\mu U^+ \partial^\mu U\}]^2 + \dots$$

as well as terms with one or two powers of m. Altogether, ten coupling constants occur [3], denoted $L_1, \dots, L_{10}$. Four of these are needed to specify the scattering matrix to first nonleading order. The terms of order m^2 in the meson mass formulae (2) and (3) involve another three of these constants. The remaining three couplings concern current matrix elements.

As an illustration, consider again the e.m. form factor $f_{\pi+}(t)$. To order p^2, the chiral representation reads [3]

$$f_{\pi+}(t) = 1 + \frac{t}{F^2}\{2L_9 + 2\phi_\pi(t) + \phi_K(t)\} + O(t^2, tm) \qquad (5)$$

In this example, the leading term (tree graph of $\mathcal{L}^2_{eff}$) is trivial, because $f_{\pi+}(0)$ represents the charge of the particle. At order p^2, there are two contributions: the term linear in t arises from a tree graph of $\mathcal{L}^4_{eff}$ and involves the coupling constant L_9, while the functions $\phi_\pi(t)$ and $\phi_K(t)$ originate in one loop graphs generated by $\mathcal{L}^2_{eff}$. The loop integrals contain a logarithmic divergence which is absorbed in a renormalization of L_9 – the net result for $f_{\pi+}(t)$ is independent of the regularization used. The representation (5) shows how the straightforward Taylor series (1) is modified by the singularites due to $\pi\pi$ and $K\bar{K}$ intermediate states. At the order of the chiral expansion we are considering here, these singularities are described by the one loop integrals $\phi_\pi(t), \phi_K(t)$ which contain cuts starting at $t = 4M_\pi^2$ and $t = 4M_K^2$, respectively. The result (5) also shows that chiral symmetry does not determine the pion charge radius: its magnitude depends on the value of the coupling constant L_9 – the effective Lagrangian is consistent with chiral symmetry for any value of the coupling constants. The symmetry, however, *relates* different observables. The slope of the K_{l_3} form factor $f_+(t)$, e.g., is also fixed by L_9. The experimental value of this slope [5], $\lambda_+ = 0.030$, can therefore be used to first determine the magnitude of L_9 and then to calculate the pion charge radius. This gives $\langle r^2 \rangle_{\pi+} = 0.42$ fm^2, to be compared with the experimental result, 0.44 fm^2 [6].

2 Scale of the loop expansion

One of the main problems encountered in the effective Lagrangian approach is the occurrence of an entire fauna of effective coupling constants. If these constants are treated as totally arbitrary parameters, the predictive power of the method is equal to zero – as a bare minimum, an estimate of their order of magnitude is needed.

Let me first drop the masses of the light quarks and send the heavy ones to infinity. In this limit, QCD is a theoretician's paradise: a theory without adjustable dimensionless parameters. In particular, the effective coupling constants $F, B, L_1, L_2, \ldots$ are calculable — the available, admittedly crude evaluations of F and B on the lattice demonstrate that the calculation is even feasible. As discussed above, the coupling constants $L_1, L_2, \ldots$ are renormalized by the logarithmic divergences occurring in the one loop graphs. This property sheds considerable light on the structure of the chiral expansion and provides a rough estimate for the order of magnitude of the effective coupling constants [7]. The point is that the contributions generated by the loop graphs are smaller than the leading (tree graph) contribution only for momenta in the range $\mid p \mid \lesssim \Lambda_\chi$, where [8]

$$\Lambda_\chi \equiv 4\pi F / \sqrt{N_f} \tag{6}$$

is the scale occurring in the coefficient of the logarithmic divergence (N_f is the number of light quark flavours). This indicates that the derivative expansion is an expansion in powers of $(p/\Lambda_\chi)^2$ with coefficients of order one. The stability argument also applies to the expansion in powers of m_u, m_d and m_s, indicating that the relevant expansion parameter is given by $(M_\pi/\Lambda_\chi)^2$ and $(M_K/\Lambda_\chi)^2$, respectively.

3 Low lying excited states

A more quantitative picture can be obtained along the following lines. Consider again the e.m. form factor of the pion and compare the chiral representation (5) with the dispersion relation

$$f_{\pi^+}(t) = \frac{1}{\pi} \int_{4M_\pi^2}^{\infty} \frac{dt'}{t' - t} \mathrm{Im} f_{\pi^+}(t') \ .$$

In this relation, the contributions ϕ_π, ϕ_K from the one loop graphs of χPT correspond to $\pi\pi$ and $K\bar{K}$ intermediate states. To leading order in the chiral expansion, the corresponding imaginary parts are slowly rising functions of t. The most prominent contribution on the r.h.s., however, stems from the region of the ρ-resonance which nearly saturates the integral: the vector meson dominance formula, $f_{\pi^+}(t) = (1 - t/M_\rho^2)^{-1}$, which results if all other contributions are dropped, provides a perfectly decent representation of the form factor for small values of t. In particular, this formula predicts $\langle r^2 \rangle_{\pi^+} = 0.39 \text{ fm}^2$, in satisfactory agreement with observation (0.44 fm^2). This implies that the effective coupling constant L_9 is approximately given by [3]

$$L_9 = \frac{F^2}{2M_\rho^2} \ . \tag{7}$$

In the channel under consideration, the pole due to ρ exchange thus represents the dominating low energy singularity – the $\pi\pi$ and $K\bar{K}$ cuts merely generate a small correction. More generally, the validity of the vector meson dominance formula shows that, for the e.m. form factor, the scale of the derivative expansion is set by $M_\rho = 770$ MeV.

Analogous estimates can be given for all effective coupling constants at order p^4, saturating suitable dispersion relations with contributions from resonances [9, 10], e.g.

$$L_5 = \frac{F^2}{4M_S^2} \ , \qquad L_7 = -\frac{F^2}{48M_{\eta'}^2} \ , \tag{8}$$

where $M_S \simeq 980$ MeV and $M_{\eta'} = 958$ MeV are the masses of the scalar octet and pseudoscalar singlet, respectively. In all those cases where direct phenomenological information is available, these estimates do remarkably well. I conclude that the observed low energy structure is dominated by the poles and cuts generated by the lightest particles – hardly a surprise.

4 Magnitude of the effective coupling constants

The effective theory is constructed on the asymptotic states of QCD. In the sector with zero baryon number, charm, beauty, ... , the Goldstone bosons form a complete set of such states, all other mesons being unstable against decay into these (strictly speaking, the η occurs among the asymptotic states only for $m_d = m_u$; it must be included among the degrees of freedom of the effective theory, nevertheless, because the masses of the light quarks are treated as a perturbation — in massless QCD, the poles generated by the exchange of this particle occur at $p = 0$). The Goldstone degrees of freedom are explicitly accounted for in the effective theory — they represent the dynamical variables. All other levels manifest themselves only indirectly, through the values of the effective coupling constants. In particular, low lying levels such as the ρ generate relatively small energy denominators, giving rise to relatively large contributions to some of these coupling constants.

In some channels, the scale of the chiral expansion is set by M_ρ, in others by the masses of the scalar or pseudoscalar resonances occurring around 1 GeV. This confirms the rough estimate (6). The cuts generated by Goldstone pairs are significant in some cases and are negligible in others, depending on the numerical value of the relevant Clebsch-Gordan coefficient. If this coefficient turns out to be large, the coupling constant in question is sensitive to the renormalization scale used in the loop graphs. The corresponding pole dominance formula is then somewhat fuzzy, because the prediction depends on how the resonance is split from the continuum underneath it.

The quantitative estimates of the effective couplings given above explain why it is justified to treat m_s as a perturbation. At order p^4, the symmetry breaking part of the effective Lagrangian is determined by the constants $L_4, \ldots, L_8$. These constants are immune to the low energy singularities generated by spin 1 resonances, but are affected by the exchange of scalar or pseudoscalar particles. Their magnitude is therefore determined by the scale

$M_S \simeq M_{\eta'} \simeq 1$ GeV. Accordingly, the expansion in powers of m_s is controlled by the parameter $(M_K/M_S)^2 \simeq \frac{1}{4}$. The asymmetry in the decay constants, e.g., is given by [10]

$$\frac{F_K}{F_\pi} = 1 + \frac{M_K^2 - M_\pi^2}{M_S^2} + \chi\text{logs} + O(m^2) \ , \tag{9}$$

where the term "χlogs" stands for the chiral logarithms generated by intermediate states with two Goldstone bosons. This shows that the breaking of the chiral and eightfold way symmetries is controlled by the mass ratio of the Goldstone bosons to the non-Goldstone states of spin zero. In χPT, the observation that the Goldstones are the lightest hadrons thus acquires quantitative significance.

5 Light quark masses

A crude estimate for the order of magnitude of the light quark masses was given shortly after the discovery of QCD [11] :

$$m_u \simeq 4\,\text{MeV} \ , \qquad m_d \simeq 6\,\text{MeV} \ , \qquad m_s \simeq 125 - 150\,\text{MeV} \ .$$

Many papers dealing with the issue have appared since then. Weinberg [12] pointed out that χPT leads to an improved estimate for the ratios $m_u : m_d : m_s$. Using the Dashen theorem [13] to account for e.m. self energies, the lowest order mass formulae given in eqs. (2) and (3) imply [12]

$$\frac{m_u}{m_d} = \frac{M_{K^+}^2 - M_{K^0}^2 + 2M_{\pi^0}^2 - M_{\pi^+}^2}{M_{K^0}^2 - M_{K^+}^2 + M_{\pi^+}^2}$$
$$\frac{m_s}{m_d} = \frac{M_{K^0}^2 + M_{K^+}^2 - M_{\pi^+}^2}{M_{K^0}^2 - M_{K^+}^2 + M_{\pi^+}^2} \ . \tag{10}$$

Numerically, this gives

$$\frac{m_u}{m_d} = 0.55 \ , \qquad \frac{m_s}{m_d} = 20 \ . \tag{11}$$

The higher order terms in the chiral expansion generate corrections to the mass formulae (10), controlled by the parameter $(M_K/\Lambda_\chi)^2$. Roughly, the numerical result (11) should therefore hold to within 20 or 30%.

The corrections of $O(p^4)$ to the above mass formulae were worked out some time ago [3]. In particular, it was shown that these corrections drop out when taking the double ratio

$$Q^2 \equiv \frac{M_K^2}{M_\pi^2} \cdot \frac{M_K^2 - M_\pi^2}{M_{K^0}^2 - M_{K^+}^2} \ . \tag{12}$$

The observed values of the meson masses thus provide a tight constraint on one particular ratio of quark masses,

$$Q^2 = \frac{m_s^2 - \hat{m}^2}{m_d^2 - m_u^2}\{1 + O(m^2)\} \ , \tag{13}$$

with $\hat{m} = \frac{1}{2}(m_u + m_d)$. The constraint may be visualized by plotting the ratio m_s/m_d versus m_u/m_d [14]. Dropping the corrections of $O(m^2) = O(M_K^4/\Lambda_\chi^4)$, the resulting curve takes the form of an ellipse,

$$\left(\frac{m_u}{m_d}\right)^2 + \frac{1}{Q^2}\left(\frac{m_s}{m_d}\right)^2 = 1 \ , \tag{14}$$

with Q as major semi-axis (the term $\hat{m}^2/m_s^2$ has been discarded, as it is numerically very small). The meson masses occurring in the double ratio (12) refer to pure QCD. Using the Dashen theorem to correct for the e.m. self energies, one obtains $Q = 24.1$. For this value of the semi-axis, the ellipse passes through the point specified by Weinberg's mass ratios, eq. (11).

6   Corrections to the Dashen Theorem

The Dashen theorem is subject to corrections from higher order terms in the chiral expansion. As usual, there are two categories of contributions: loop graphs of order e^2m and terms of the same order from the derivative expansion of the effective e.m. Lagrangian.

The Clebsch-Gordan coefficients occurring in the loop graphs are known to be large, indicating that two-particle intermediate states generate sizeable corrections; the corresponding chiral logarithms tend to increase the e.m. contribution to the kaon mass difference [15]. The numerical result depends on the scale used when evaluating the logarithms. In fact, taken by themselves, chiral logs are unsafe at any scale.

The magnitude of the contributions from the terms of order e^2m occurring in the effective Lagrangian is estimated in two recent papers by Donoghue et al. [16] and Bijnens [17]. Although the framework used is rather different, they come up with the same conclusion: the corrections are large, increasing the value $(M_{K^+} - M_{K^0})_{e.m.} = 1.3$ MeV predicted by Dashen to 2.3 MeV (Donoghue et al.) and 2.6 MeV (Bijnens), respectively.

In the present context, the main point is that even large corrections of this size only lead to a small change in the value of Q, because the mass difference between K^+ and K^0 is predominantly due to $m_d > m_u$: the value $Q = 24.1$ (Dashen) is lowered to $Q = 22.1$ (Donoghue et al.) and $Q = 21.6$ (Bijnens), respectively. Expressed in terms of the error bars attached to the value of Q in [10], a doubling of the electromagnetic self-energy modifies the value of Q by $1\frac{1}{2}\sigma$.

The decay $\eta \to 3\pi$ provides an independent measurement of Q: writing the decay rate in the form $\Gamma_{\eta \to \pi^+\pi^-\pi^0} = \Gamma_0/Q^4$, χPT predicts the value of Γ_0 in a parameter free manner [3]. Although the calculation accounts for all corrections of $O(p^4)$, the numerical accuracy is rather modest because, due to strong final state interaction effects, the calculated corrections are quite large. Since the quantity Q enters in the fourth power, the value $\Gamma_{\eta \to \pi^+\pi^-\pi^0} = 283 \pm 28$ eV given by the particle data group [5] still yields a decent measurement: $Q = 20.6 \pm 1.7$. The fact that this result is significantly smaller than the value predicted with the Dashen theorem represents an old puzzle. The problem disappears if the e.m. contribution to the kaon mass difference is significantly larger than indicated by the Dashen theorem *[18, 19, 20]*. In particular, the values of Q obtained in *[16, 17]* are consistent with the one from η decay.

The theoretical uncertainties in the η decay amplitude could be reduced. The calculation referred to above only accounts for the corrections of order p^4 and includes final state interaction effects and $\eta\eta'$ mixing only to that order. A dispersive analysis along the lines indicated by Khuri and Treiman *[21]*, which uses the χPT predictions only for the subtraction constants would likely lead to a more accurate estimate of the major semi-axis *[22]*.

7   The ratio m_u/m_d

Chiral perturbation theory thus fixes one of the two quark mass ratios in terms of the other, to within small uncertainties. The ratios themselves, i.e., the position on the ellipse, are a more subtle issue. Kaplan and Manohar *[14]* pointed out that the corrections to the lowest order result (11) for m_u/m_d cannot be determined on purely phenomenological grounds. They argued that these corrections might be large and that the u-quark might actually be massless. This possibility is of particular interest, because the strong CP problem would then disappear. Several authors *[23]* have given arguments in favour of $m_u = 0$.

Let me show the picture this reasoning leads to. The lowest order mass formulae (2), (3) imply that the ratio m_u/m_d determines the K^0/K^+ mass difference, the scale being set by M_π:

$$M_{K^0}^2 - M_{K^+}^2 = \frac{m_d - m_u}{m_u + m_d} \cdot M_\pi^2 + \ldots \tag{15}$$

The formula holds up to corrections from higher order terms in the chiral expansion and up to e.m. contributions. It is illustrated in figure 1, taken from *[24]*. The upper curve corresponds to the value $(M_{K^+} - M_{K^0})_{e.m.} = 1.3\,\text{MeV}$, which follows from the Dashen theorem. The correction of Donoghue et al. *[16]* shifts the result by 1 MeV (lower curve). The horizontal line is the experimental value. Hence the corrections from the higher order terms must generate the contributions shown by the arrows. In particular, if m_u is assumed to vanish, the lowest order mass formula predicts a mass difference which exceeds the observed value by a factor of four. The disaster can only be blamed on the "corrections" from the higher order terms. It is evident that, under such circumstances, it does not make sense to truncate the expansion at first nonleading order and to fool around with the numerics of the effective coupling constants occurring therein. The conclusion to draw from the assumption $m_u = 0$ is that χPT is unable to account for the masses of the Goldstone bosons. The fact that it happens to work remarkably well in other cases must then be accidental. I prefer to conclude that the assumption $m_u = 0$ is not tenable.

8   Phenomenological ambiguities

The preceding presentation is an extract of a rapporteur talk given in 1992 *[24]*. I thought that this coffee was rather cold by now and dealt with it only briefly in the manner described above, concentrating the remainder of the talk on issues which I expected to be more interesting to the audience, such as the role of winding number and the mass of the η' *[25, 26]*. The vigorous

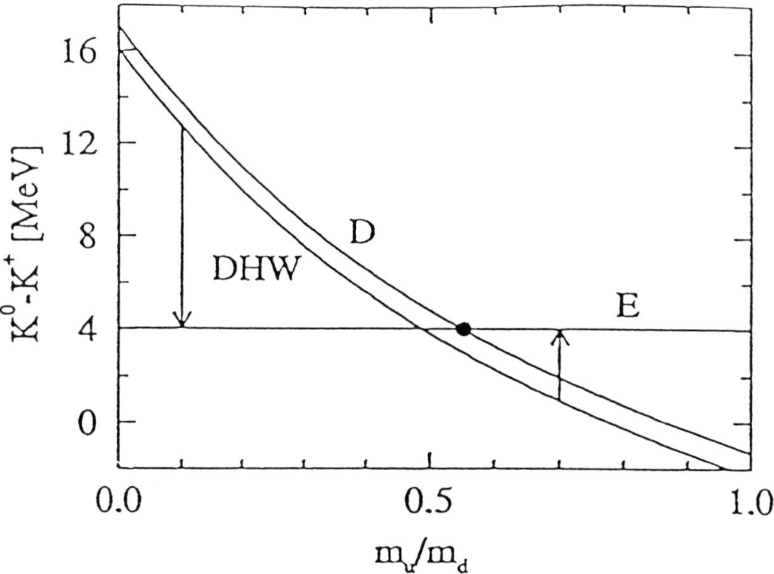

Figure 1: Sensitivity of the first order prediction for the kaon mass difference to m_u/m_d. The two curves differ in the estimate used for the e.m. self energies. E is the experimental value. The dot corresponds to Weinberg's leading order result.

discussion triggered by the claim that $m_u = 0$ is firmly ruled out by now and the remarkable statements made by some of the participants indicate, however, that not everyone fully shares this view of the matter. I conclude that the arguments given previously do not suffice and elaborate a little further.

To set up notation, I first briefly review the observation of Kaplan and Manohar [14], who pointed out that the matrix

$$m' = \alpha_1 m + \alpha_2 (m^+)^{-1} \det m$$

transforms in the same manner as m. For a real, diagonal mass matrix, the transformation amounts to

$$m'_u = \alpha_1 m_u + \alpha_2 m_d m_s \quad (\text{cycl. } u \to d \to s \to u) \ . \tag{16}$$

Symmetry alone does therefore not distinguish m' from m. If $\mathcal{L}_{eff}(U, \partial U, \ldots, m)$ is an effective Lagrangian consistent with chiral symmetry, so is $\mathcal{L}_{eff}(U, \partial U, \ldots, m')$.

Since only the product Bm enters the Lagrangian, α_1 merely changes the value of the constant B. The term proportional to α_2 is a correction of order m^2 which, upon insertion in $\mathcal{L}^2_{eff}$ generates a contribution to $\mathcal{L}^4_{eff}$. The contribution can again be removed by changing some of the coupling constants:

$$B' = B/\alpha_1 \ , \quad L'_6 = L_6 - \alpha \ , \quad L'_7 = L_7 - \alpha \ , \quad L'_8 = L_8 + 2\alpha \ , \tag{17}$$

with $\alpha = \alpha_2 F^2/32\alpha_1 B$. The effective Lagrangian is therefore invariant under a simultaneous change of the quark mass matrix and of the coupling constants. Accordingly, the meson masses and scattering amplitudes which one calculates with this Lagrangian remain the same. The elliptic constraint considered above, e.g., is invariant under the operation (up to terms of order $(m_u - m_d)^2/m_s^2$, which were neglected). The chiral representation for the Green functions of the vector and axial currents are also invariant. Since there is experimental information only about masses, scattering amplitudes and matrix elements of the electromagnetic or weak currents and since the chiral representation for these does not distinguish m, B, L_i from m', B', L'_i, phenomenology does not allow one to determine the magnitude of the constants B, L_6, L_7, L_8.

We are not dealing with a hidden symmetry of QCD here — this theory is not invariant under the change (16) of the quark masses. In particular, the matrix elements of the scalar and pseudoscalar operators are modified. Consider, e.g., the vacuum-to-pion matrix element of the pseudoscalar density. The Ward identity for the axial current implies that this matrix element is given by

$$\langle 0| \, \bar{d} i \gamma_5 u | \pi^+ \rangle = \sqrt{2} \, F_{\pi^+} M^2_{\pi^+} / (m_u + m_d) \tag{18}$$

The relation is exact, except for electroweak corrections. It involves the physical quark masses and is not invariant under the above transformation. For the ratio of the K and π matrix elements, the effective Lagrangian yields the following low energy representation:

$$\frac{\langle 0| \, \bar{s} i \gamma_5 u | K^+ \rangle}{\langle 0| \, \bar{d} i \gamma_5 u | \pi^+ \rangle} = 1 + \frac{4(M^2_{K^+} - M^2_{\pi^+})}{F^2}(4L_8 - L_5) + \chi \text{logs} + O(m^2) \ .$$

The relation is analogous to the formula for the ratio of the corresponding matrix elements of the axial currents,

$$\frac{F_{K+}}{F_{\pi+}} = 1 + \frac{4(M_{K+}^2 - M_{\pi+}^2)}{F^2} L_5 + \chi\text{logs} + O(m^2) \ .$$

While the ratio of decay constants is invariant under the KM transformation, the one of the pseudoscalar densities is not. The main difference between the two cases is that nature is kind enough to provide us with a weak interaction probe, testing the matrix elements of the vector and axial currents at low energies, while a probe which would test those of the scalar and pseudoscalar currents is not available — the Higgs particle is too heavy for this purpose. The observed rates of the decays $K \to \mu\nu$ and $\pi \to \mu\nu$ imply $F_{K+}/F_{\pi+} = 1.22$ and thereby permit a phenomenological determination of the effective coupling constant L_5, while L_8 cannot be determined on purely phenomenological grounds.

If the electroweak interactions were not available to probe the low energy structure of QCD, the coupling constant L_5 would also count amoung those quantities for which direct phenomenological information is absent, for the following reason. The field

$$U' = U\{1 + \alpha_3(U^\dagger m - m^\dagger U) - \tfrac{1}{3}\alpha_3 \text{tr}(U^\dagger m - m^\dagger U) + O(m^2)\} \qquad (19)$$

transforms in the same manner as U. For a constant quark mass matrix, the change of variables $U \to U'$ is equivalent to a change of the effective coupling constants:

$$L_5' = L_5 - 6\bar{\alpha} \ , \quad L_7' = L_7 + \bar{\alpha} \ , \quad L_8' - 3\bar{\alpha} \ , \quad \bar{\alpha} = F^2\alpha_3/24B \ . \qquad (20)$$

Hence the χPT results for the masses of the Goldstone bosons and for their scattering amplitudes only involve those combinations of coupling constants which are invariant under this operation. The only difference to the transformation considered above is that (17) leaves the matrix elements of the vector and axial currents invariant, while (20) does not.

The ambiguity pointed out by Kaplan and Manohar does not indicate that the effective Lagrangian possesses any symmetries beyond those of the QCD Lagrangian. It does not concern QCD as such, but originates in the fact that the electromagnetic and weak interactions happen to probe the low energy structure of the system exclusively through vector and axial currents.

9 Additive renormalization of m and all that

One of the claims repeatedly made in the discussion is that the quark masses occurring in the effective Lagrangian must be distinguished from those entering the Lagrangian of QCD. Indeed, this problem does arise within early formulations of the effective Lagrangian technique, which exclusively dealt with the properties of the Goldstone bosons on the mass shell. There, the structure of the effective Lagrangian was inferred from *global* symmetry arguments and the only place where the quark mass matrix entered was through the transformation law of the term which explicitly breaks the symmetries of the QCD Hamiltonian.

The framework used in current work on χPT, however, does not rely on global symmetry arguments, but identifies the quark masses through their contributions to the Ward identities, which express the symmetries of the underlying theory in *local* form. The method was introduced ten years ago *[2, 4]*. It is not limited to on-shell matrix elements, but also specifies the low energy expansion of the Green functions formed with the vector, axial, scalar and pseudoscalar currents. The Ward identities involve the physical quark masses, not some effective low energy version thereof. *For the effective theory to reproduce the low energy structure of the Green functions, the mass matrix occurring in the effective Lagrangian must be identified with the physical one.* Likewise, the magnitude of the effective coupling constants L_6, L_7 and L_8 is not an inherently ambiguous issue; the effective theory reproduces the low energy expansion of QCD for precisely one value of these constants.

There is a distinction between the currents associated with chiral symmetry and the scalar or pseudoscalar densities in that the latter pick up *multiplicative* renormalization, contragredient to the renormalization of the quark mass matrix. It is crucial that the renormalization used is mass independent, such that the Green functions of the scalar and pseudoscalar operators are given by the response of the effective action to a local change in the quark mass matrix. In the ratio of matrix elements considered above, the renormalization scale then drops out — this ratio is a perfectly well-defined pure number, which may be estimated, e.g., by means of QCD sum rule techniques. In fact, the sum rule for the two-point functions of the pseudoscalar currents is one of the main sources of information available today for the absolute magnitude of the light quark masses. Once numerical evaluations on a lattice reach sufficiently small quark masses, this method will allow a more precise determination of low energy matrix elements involving the scalar and pseudoscalar currents.

Some authors *[27]* have pointed out that *instantons* offer a physical interpretation for the KM-ambiguity. In the field of an instanton, the Dirac operator develops discrete zero modes. As pointed out by t'Hooft *[28]*, these modes undergo an effective interaction which indeed generates a self-energy contribution for the up-quark proportional to $m_d m_s$. The problem with this picture is that the spectrum of the Dirac operator on which it is based is fictitious. Banks and Casher *[29]* have shown that the spontaneous breakdown of chiral symmetry requires the small eigenvalues of the Dirac operator to be distributed uniformly: the level density $\rho(\lambda)$ approaches the value $\rho(0) = |\langle 0| \bar{q}q |0\rangle|/\pi$ when $\lambda \to 0$. For the model described in *[27]*, the spectrum instead contains an isolated zero mode, such that the level density develops a peak there, $\rho(\lambda) \propto \delta(\lambda)$. The peak may generate large flavour symmetry breaking effects, but at the same time, it shows that the model is in conflict with spontaneous chiral symmetry breakdown. The calculations carried out in *[27]* are based on the assumption that a framework which is unable to account for the leading term in the effective Lagrangian (the one related to the constant $B \leftrightarrow \langle 0| \bar{q}q |0\rangle$) can be trusted when analyzing the higher order terms, i.e., the corrections.

There is some progress in understanding the level structure in a picture which describes the vacuum as a dense collection of instantons *[26]*. In that framework, the zero modes of the individual instantons merge into a continuum, such that the fictitious peak mentioned above disappears. For models with a decent spectrum, the symmetry breaking effects should be of reasonable size.

10   The $K^0 - K^+$ mass difference

As pointed out in [24], the lowest order χPT formula for the $K^0 - K^+$ mass difference disagrees with observation by a factor of four — if m_u is set equal to zero. Since I did not find this puzzle discussed in the literature [23], I do it myself and first try to guess the reasoning put forward by the defence.

Any lawyer worth a fraction of his fee would immediately point out that the problem disappears if the quark masses occurring in the lowest order formula are replaced by suitable KM-transforms m'_u, m'_d, m'_s. In particular, one may take $\alpha_1 = 1$ and fix α_2 in such a manner that the ratio of the primed masses agrees with the Weinberg ratios. This amounts to the statement that the chiral perturbation series should be reordered, replacing the expansion in powers of the physical quark masses by an expansion in the primed ones. With the above specification of these, the leading term of the reordered series agrees with the experimental value of the $K^0 - K^+$ mass difference, by construction. So, the first conclusion to draw from $m_u = 0$ is that the expansion of the meson masses in powers of the physical quark masses does not make sense — χPT must be replaced by a reordered series. Why does this not take care of the problem ?

The point is that the operation very strongly distorts the matrix elements of the scalar operators, which are obtained from the effective action by expanding in powers of the physical mass matrix, not by some hand made variant thereof. Consider, e.g., the scalar form factors

$$\langle K^+ | \bar{u}u - \bar{d}d | K^+ \rangle = S_{K^+}(t) \ , \qquad \langle \pi^+ | \bar{u}u - \bar{s}s | \pi^+ \rangle = S_{\pi^+}(t) \ .$$

Since the operation $s \leftrightarrow d$ takes a K^+ into a π^+, the two form factors coincide in the SU(3) limit. The difference $S_{K^+}(t) - S_{\pi^+}(t)$ is a symmetry breaking effect of order $m_s - m_d$. According to the Feynman-Hellmann theorem [30], the values of these form factors at $t = 0$ are related to the derivatives of the meson masses with respect to the quark masses. The kaon matrix element of the operator $\bar{u}u$, e.g., represents the first derivative of $M_{K^+}^2$ with respect to m_u, such that

$$S_{K^+}(0) = \left(\frac{\partial}{\partial m_u} - \frac{\partial}{\partial m_d} \right) M_{K^+}^2 \ , \qquad S_{\pi^+}(0) = \left(\frac{\partial}{\partial m_u} - \frac{\partial}{\partial m_s} \right) M_{\pi^+}^2 \ .$$

The expansion of the meson masses in powers of m_u, m_d, m_s starts with [3]

$$M_{K^+}^2 = (m_u + m_s)B\{1 + (m_u + m_s)K_3 + (m_u + m_d + m_s)K_4 + \dots \}$$
$$M_{\pi^+}^2 = (m_u + m_d)B\{1 + (m_u + m_d)K_3 + (m_u + m_d + m_s)K_4 + \dots \} \ ,$$

where chiral logarithms and terms of $O(m^3)$ are dropped. Evaluating the derivatives, one readily establishes the low energy theorem

$$r \equiv \frac{S_{K^+}(0)}{S_{\pi^+}(0)} = \left(\frac{m_s - m_u}{m_d - m_u} \cdot \frac{M_{K^0}^2 - M_{K^+}^2}{M_{K^0}^2 - M_{\pi^+}^2} \right)^2 \{1 + O(m^2)\} \ , \qquad (21)$$

valid up to chiral logarithms (in the present case, these are proportional to M_π^2 and therefore tiny, of order 1%). The theorem is of the same character as the one which leads to the elliptic constraint: the corrections are of second order in the quark masses. The relevant combination of quark masses which occurs here, however, fails to be invariant under the KM-transformation; the relation only holds if the masses entering the ratio $(m_s - m_u)/(m_d - m_u)$ are identified with the physical ones.

Suppose now that m_u vanishes. The formula then predicts that the kaon matrix element is about three times smaller than the pion matrix element (the precise value of r depends on the electromagnetic corrections: using the value $(M_{K^0} - M_{K^+})_{\rm QCD} = 5.3\,{\rm MeV}$, which follows from the Dashen theorem, I obtain $r = 0.30$, while the values $(M_{K^0} - M_{K^+})_{\rm QCD} = 6.3\,{\rm MeV}$ and $6.6\,{\rm MeV}$ found by Donoghue et al. [16] and Bijnens [17] lead to $r = 0.36$ and $r = 0.38$, respectively). So, $m_u = 0$ leads to the prediction that the evaluation of the above matrix elements with sum rule or lattice techniques will reveal extraordinarily strong flavour symmetry breaking effects.

Massless up-quarks usually come in a luxurious wrapping with plausible well-known statements of general nature. Indeed, there is no logical contradiction with the proposal that (i) the expansion of the meson masses in powers of the physical quark masses fails and (ii) a suitable reordering of the series leads to meaningful results. Also, the prediction that the matrix elements of the scalar and pseudoscalar operators exhibit dramatic flavour symmetry breaking effects is not in conflict with phenomenology. Although theoretical tools such as sum rules lead to the conclusion that the picture does not make sense, these are more fragile than solid experimental facts. It is conceivable that the present example represents an exception to Einstein's statement: "Raffiniert ist der Herrgott, aber boshaft ist er nicht." Even then, the literature on massless up-quarks leaves much to be desired, as it does not address any of the consequences for the low energy structure of QCD, which are rather bizarre.

11 Does the quark condensate vanish ?

Let me finally turn to an entirely different picture [31], which I do not believe either, but find conceptually more interesting. The picture may be motivated by an analogy with spontaneous magnetization. There, spontaneous symmetry breakdown occurs in two quite different modes: ferromagnets and antiferromagnets. For the former, the magnetization develops a nonzero expectation value, while for the latter, this does not happen. In either case, the symmetry is spontaneosly broken (for a discussion of the phenomenon within the effective Lagrangian framework, see [32]). The example illustrates that operators which the symmetry allows to pick up an expectation value may, but need not do so.

The standard low energy analysis assumes that the quark condensate is the leading order parameter of the spontaneously broken symmetry, such that the Goldstone boson masses are proportional to the square root of the quark mass. The proportionality coefficient follows from the relation of Gell-Mann, Oakes and Renner [33]

$$F_\pi^2 M_\pi^2 = (m_u + m_d)|\langle 0|\, \bar{u}u\, |0\rangle| + \ldots$$

The simplest version of the question addressed in the papers quoted above is whether the quark condensate indeed tends to a nonzero limit when the quark masses are turned off, like the magnetization of a ferromagnet, or whether it tends to zero, as it is the case for the magnetization of an antiferromagnet. If the second option were realized in nature, the pion mass would not be determined by the quark condensate, but by the terms of order m^2, which the Gell-Mann-Oakes-Renner formula neglects. More generally, one may envisage a situation, where the quark condensate is different from zero, but small, such that, at the physical value of the quark masses, the terms of order m and m^2 both yield a significant contribution. In the literature, this framework is referred to as "Generalized χPT" [31]. The "generalized" and "ordinary" chiral perturbation theories only differ in the manner in which the symmetry breaking effects are treated.

The main problem with GχPT is that much of the predictive power of the standard framework is then lost. The most prominent example is the Gell-Mann-Okubo formula for the octet of Goldstone bosons, which does not follow within GχPT. As mentioned above, this prediction of the standard framework is satisfied remarkably well, thus supporting that picture, but it is not unfair to say that this may be a coincidence, like the $\Delta I = \frac{1}{2}$-rule, whose explanation within the Standard Model is also considerably more complex than the rule itself. The original motivation for the study of the generalized setting was the discrepancy between theory and experiment with regard to the πN σ-term, which indeed represented quite a fountain, continuously spitting out new ideas, one more strange than the other. In the meantime, this fountain has run dry, because the discrepancy was shown to be due to an inadequate treatment of a form factor, which enters the relation between the scattering amplitude and the σ-term matrix element [34]. ¿From my point of view, the generalized szenario is still of some interest, mainly as a kinematical framework, which parametrizes the vicinity of the χPT predictions and allows one to judge the significance of the experimental information concerning these: the standard framework represents a special case of the generalized one. Although this is not very practical in general, one may even put the cart before the horse and claim that the predictions of χPT are based on extra assumptions, which the generalized setting does not make and which should be tested experimentally. Indeed, χPT leads to very strong predictions concerning, e.g., the scattering lengths of $\pi\pi$ scattering [35], which are sensitive to the explicit breaking of chiral symmetry generated by the quark masses and which it is important to test experimentally.

12   Summary and Conclusion

It is difficult for me to summarize the present knowledge concerning the masses of the light quarks in an unbiased manner. The following conclusions unavoidably reflect my own views and mainly rely on work done together with Gasser. These views are dyed with the prejudice that a straightforward expansion of the matrix elements in powers of the light quark masses makes sense without reordering the series and that the Gell-Mann-Oakes-Renner relation is not spoiled by higher order effects.

1. Concerning the absolute magnitude of the quark masses, the most re-

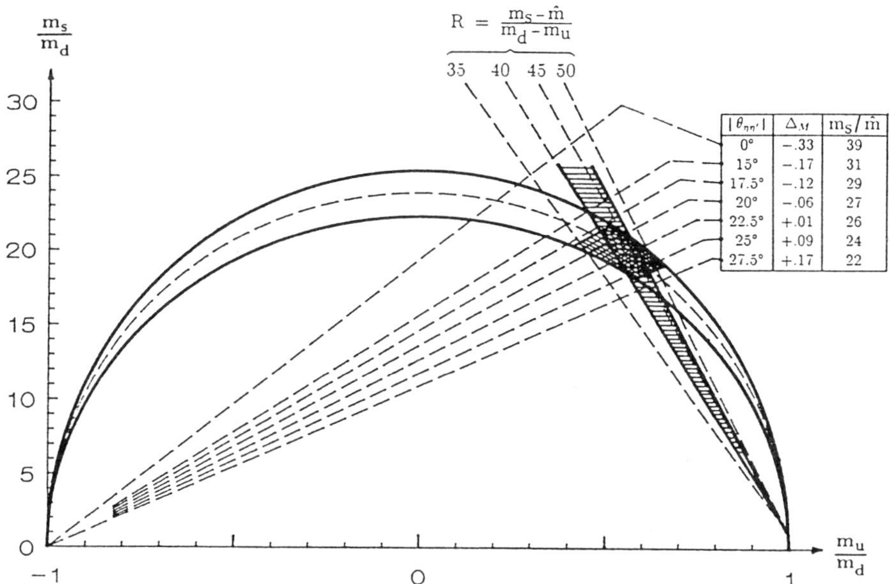

$\|\theta_{\eta\eta'}\|$	Δ_M	$m_s/\hat{m}$
0°	−.33	39
15°	−.17	31
17.5°	−.12	29
20°	−.06	27
22.5°	+.01	26
25°	+.09	24
27.5°	+.17	22

Figure 2: The elliptic band shows the range permitted by the low energy theorem (14). The cross-hatched area is the intersection of this band with the sector selected by the phenomenology of $\eta\eta'$-mixing. The shaded wedge shows the constraint imposed on the ratio R by the mass splittings in the baryon octet.

liable estimates are based on evaluations of QCD sum rules [36]. With the experience gained from the various applications of these, the evaluation leads to a remarkably stable result, the visible noise generated by uncertainties of the input being quite small [37]. The main uncertainty stems from the systematic error of the method, which it is hard to narrow down. The results do not indicate a significant change as compared to the estimates given in [38]. In the long run, lattice evaluations should allow a more accurate determination, but further progress with light dynamical fermions is required before the numbers obtained can be taken at face value.

2. Chiral perturbation theory constrains the ratios m_s/m_d and m_u/m_d to the ellipse specified in equation (14) and displayed in figure 2 (taken from [10]). The magnitude of the semiaxis Q is known to an accuracy of 10%. The value $Q = 24$ shown in the figure relies on the Dashen Theorem. The estimates given in [16, 17] indicate that this theorem receives large corrections from higher order terms, reducing the value of Q by about 10%.

3. The position on the ellipse is best discussed in terms of the ratio

$$R = \frac{m_s - \hat{m}}{m_d - m_u} \, ,$$

which represents the relative magnitude of SU(3) breaking compared to isospin breaking. If m_u were to vanish, this ratio would be related to the semiaxis by $R \simeq Q \simeq 24$ and the ratio of the strange quark mass to $\hat{m} = \frac{1}{2}(m_u + m_d)$ would be given by $m_s/\hat{m} \simeq 2Q \simeq 48$. The quark mass estimates obtained from QCD sum rules and from numerical evaluations on a lattice are in flat disagreement with this pattern, but confirm the standard picture, where $m_s/\hat{m} \simeq Q \simeq 24$.

4. The value of R was investigated much before QCD had been discovered. For a long time, it was taken for granted that isospin breaking is an electromagnetic effect. The early literature on the subject is reviewed in [38]. In 1981, Gasser [39] analyzed the nonanalytic terms occurring in the chiral expansion of the octets of meson and baryon masses and demonstrated that these remove the discrepancies which had affected previous determinations of the ratio R. A thorough discussion of the available information concerning this ratio [38] showed that the data on $\Sigma^+ - \Sigma^-$, $\Xi^0 - \Xi^-$, $K^0 - K^+$ and $\rho\omega$-mixing allow four independent determinations of R (in the case of the proton-neutron mass difference, the higher order corrections turn out to be large, so that this source of information does not significantly affect the overall result). Within the noise visible in the calculation, estimated at 10-15%, the results turned out to be mutually consistent. Adding the uncertainties of the individual values quadratically, the analysis implies $R = 43.5 \pm 2.2$. Discarding the information from $K^0 - K^+$ and from $\rho\omega$-mixing, one instead obtains $R = 43.5 \pm 3.2$, which corresponds to the range depicted as a shaded wedge in figure 2.

5. The above error bars rely on the prejudice that the matrix elements of the operators $\bar{u}u$, $\bar{d}d$, $\bar{s}s$ do not exhibit strong SU(3) breaking effects. These were estimated on the basis of the same rough argument, which identifies the quark mass difference $m_s - \hat{m}$ with the mass splitting within the SU(3) multiplets [40]. When analyzing the meson mass spectrum to second order in the chiral expansion [3], we realized, of course, that the mass spectrum of the Goldstone bosons does not suffice to independently determine the ratios Q and R, but the algebraic nature of the ambiguity was pointed out only later, by Kaplan and Manohar [14]. In the opinion of some authors, this ambiguity appears to represent the most interesting aspect of the matter, indicating that the notion of quark mass is fuzzy at low energies, due to an *additive renormalization problem*, generated by *nonperturbative effects*, such as *instantons*, etc. The detailed discussion given above is another attempt at fighting these dangerous ogres, which the innocent observer might take for windmills (earlier attempts are described in [41, 10, 24]). Gradually, Rosinante is getting sick and tired.

6. One of the consequences we did infer from the second order analysis is that the $\eta\eta'$ mixing angle cannot be determined with the Gell-Mann-Okubo formula ($\theta_{\eta\eta'} = -10°$), because there are other effects of the same algebraic order of magnitude, originating in $\mathcal{L}^4_{eff}$. Using the information about the value of R discussed above, we obtained $\theta_{\eta\eta'} = -20° \pm 4°$. The phenomenological analysis of the decays $\eta \to \gamma\gamma$ and $\eta' \to \gamma\gamma$, performed independently a year later [42], neatly confirmed this prediction.

7. As noted in [3], the result for the coupling constant L_7, which follows from the above value of R, is consistent with the η' dominance formula (8). This indicates that the sum rule used in the derivation of that formula is nearly

saturated by the η' intermediate state also in the real world, not only in the large N_c limit, where the formula is exact.

8. Turning the argument around, the phenomenology on the mixing angle available today [43] may be used to determine the quark mass ratios. This is indicated by the dashed straight lines in figure 2. Since the data agree with the prediction, it is clear that the information about the mass ratios obtained in this manner is consistent with the one extracted from isospin breaking in the baryon octet. Note, however, that the phenomenological analysis of the η and η' decays relies on large N_c arguments. Although the resulting picture yields a coherent understanding of quite a few processes [44, 45], the validity of these arguments needs to be examined more carefully to arrive at firm conclusions [46].

9. The branching ratio $\Gamma(\psi' \to \psi\pi^0)/\Gamma(\psi' \to \psi\eta)$ provides further information about R. Using the lowest order formula, the data available at the time gave $R = 28^{+7}_{-4}$ [38]. The formula is valid up to SU(3) breaking effects; according to the general rule of thumb, these are expected to be of order 20 or 30%. Since we saw no way of evaluating these, the information derived from this branching ratio did not significantly affect our analysis and was discarded. Meanwhile, the data are slightly more precise [5]; the lowest order formula now yields $R = 31 \pm 4$. Furthermore, Donoghue and Wyler [47] have investigated the second order corrections, using the multipole expansion, which relates the relevant matrix elements to $\langle 0| G_{\mu\nu}\tilde{G}^{\mu\nu}|\pi^0\rangle$, $\langle 0| G_{\mu\nu}\tilde{G}^{\mu\nu}|\eta\rangle$. As it is the case with the sum rule used to estimate L_7, these matrix elements involve pseudoscalar operators and do thus not suffer from the KM-ambiguity. The calculation yields remarkably small SU(3)-breaking effects, such that the value obtained for R remains close to 31 ± 4.

The problems faced by this attempt at estimating R are listed in [24]. In particular, as discussed in detail in [48], it is not clear that the relevant matrix elements may be replaced by those of the operator $G_{\mu\nu}\tilde{G}^{\mu\nu}$. The calculation is of interest, because it is independent of other determinations, but at the present level of theoretical understanding, it is subject to considerable uncertainties. In view of these, it appears to me that the numerical result is perfectly consistent with the value $R = 43.4$, which follows from the Weinberg ratios.

As evidenced by the example of $\eta\eta'$ mixing and quite a few others, the estimates of the effective coupling constants which follow from the hypothesis that the low energy structure is dominated by the singularities due to the lowest lying levels lead to a rather predictive framework, which until now has passed all experimental tests. It would be of interest to estimate the symmetry breaking effects in the branching ratio $\Gamma(\psi' \to \psi\pi^0)/\Gamma(\psi' \to \psi\eta)$ on the basis of this hypothesis and to see whether they indeed increase the value of R obtained with the lowest order formula. If not, there would be a problem with the simple picture I am advocating here.

10. There are examples, where the higher order corrections turn out to be large, such as $\eta \to 3\pi$ or scalar form factors. In all cases I know of, one can put the finger on the culprit responsible for the enhancement. Invariably, the problem arises, because the perturbations generated by the quark mass term

are enhanced by small energy denominators. In particular, the enhancement due to strong final state interactions in the S-waves often generates sizeable corrections, which are perfectly well understood — χPT itself predicts how large the S-wave phase shifts are near threshold and that they rapidly grow with energy. Once the origin of the phenomenon is understood, one may take it into account and arrive at a reliable low energy representation, even if the straightforward chiral expansion thereof contains relatively large contributions from higher order terms.

11. In the case of the mass formulae, the low energy singularities do not generate a significant enhancement of the higher order contributions. Accordingly, the estimates of the effective coupling constants mentioned above imply that the Weinberg ratios only receive small corrections. There is some evidence for a 10% decrease in Q, related to the e.m. self energies: The Dashen theorem is suspected to receive large corrections, because some of the low energy singularities do generate large symmetry breaking effects in that case.

12. Although the above hypothesis is the most natural setting for an effective theory I can think of, it evidently involves assumptions which go beyond pure symmetry and phenomenology. One may dismiss this hypothesis and the estimates for R obtained with it, requiring only that the mass term of the light quarks, which breaks the chiral symmetry of the QCD Hamiltonian, may be treated as a perturbation. With $m_u = 0$, the lowest order formula for $M_{K^0} - M_{K^+}$ is off by a factor of four, in flat contradiction with this requirement. The disaster can be avoided only if the factor $(m_d - m_u)/(m_d + m_u)$ accounts for at least half of the discrepancy, $(m_d - m_u)/(m_d + m_u) < \frac{1}{2}$. This alone yields $m_u/m_d > \frac{1}{3}$.

13. Accordingly, if m_u is assumed to be equal to zero, χPT must be thrown overboard. The expansion in powers of the quark masses does then not make sense and the matrix elements of the scalar and pseudoscalar operators must exhibit extraordinarily strong SU(3) breaking effects. These cannot be explained with the low energy singularities listed in the particle data tables. As far as I know, the instanton model of Choi et al. *[27]* is the only theoretical scenario proposed to substantiate the claim that QCD may give rise to strong flavour symmetry breakings of this type. There, the occurrence of discrete zero modes does indeed produce such effects. The same feature, however, is the reason why the model is inconsistent with spontaneously broken chiral symmetry. Dilute instanton coffee is undrinkable. Bold discussion remarks may warm it up a little, but that merely intensifies the unacceptable flavour characteristics of the brewing.

I am indebted to Hans Bijnens, Jürg Gasser, Marina Nielsen and Daniel Wyler for valuable comments.

References

[1]   S. Weinberg, *Physica* A96 (1979) 327.

[2]   J. Gasser and H. Leutwyler, *Ann. Phys. (N.Y.)* 158 (1984) 142.

[3]   J. Gasser and H. Leutwyler, *Nucl. Phys.* B250 (1985) 465, 517, 539.

[4] The foundations of the method are discussed in detail in
H. Leutwyler, *Ann. Phys. (N.Y.)* 235 (1994) 165.

[5] Review of Particle Properties, *Phys. Rev.* D45 (1992).

[6] S.R. Amendolia et al., *Nucl. Phys.* B277 (1986) 168.

[7] H. Georgi, *Weak Interactions and Modern Particle Theory*
(Benjamin/Cummings, Menlo Park, 1984);
H. Georgi and A. Manohar, *Nucl. Phys.* B234 (1984) 189.

[8] M. Soldate and R. Sundrum, *Nucl. Phys.* B340 (1990) 1;
R.S. Chivukula, M.J. Dugan and M. Golden, *Phys. Rev.* D47 (1993) 2930.

[9] G. Ecker et al., *Nucl. Phys.* B321 (1989) 311; *Phys. Lett.* B223 (1989) 425.

[10] H. Leutwyler, *Nucl. Phys.* B337 (1990) 108.

[11] H. Leutwyler, *Phys. Lett.* B48 (1974) 431; *Nucl. Phys.* B76 (1974) 413;
J. Gasser and H. Leutwyler, *Nucl. Phys.* B94 (1975) 269.

[12] S. Weinberg, in *A Festschrift for I.I. Rabi*, ed. L. Motz (New York Acad. Sci., 1977) p. 185.

[13] R. Dashen, *Phys. Rev.* 183 (1969) 1245.

[14] D. B. Kaplan and A. V. Manohar, *Phys. Rev. Lett.* 56 (1986) 2004.

[15] P. Langacker and H. Pagels, *Phys. Rev.* D8 (1973) 4620;
K. Maltman and D. Kotchan, *Mod. Phys. Lett.* A5 (1990) 2457;
G. Stephenson, K. Maltman and T. Goldman, *Phys. Rev.* D43 (1991) 860.

[16] J. Donoghue, B. Holstein and D. Wyler, *Phys. Rev.* D47 (1993) 2089.

[17] J. Bijnens, *Phys. Lett.* B306 (1993) 343.

[18] J. Donoghue, B. Holsten and D. Wyler, *Phys. Rev. Lett.* 69 (1992) 3444.

[19] J. Donoghue, Lectures given at the Theoretical Advanced Study Institute (TASI),
Boulder, Colorado (1993).

[20] D. Wyler, Proc. XVI Kazimierz Meeting on Elementary Particle Physics, eds.
Z. Ajduk et al., World Scientific (1994).

[21] N. Khuri and S. Treiman, *Phys. Rev.* 119 (1960) 1115;
C. Roiesnel and T.N. Truong, *Nucl. Phys.* B187 (1981) 293.

[22] A. V. Anisovich, "Dispersion relation technique for three-pion system and the
P-wave interaction in $\eta \to 3\pi$ decay", preprint Petersburg Nuclear Physics
Institute, Gatchina TH-62-1993/1931;
J. Kambor, C. Wiesendanger and D. Wyler, in preparation;
A. V. Anisovich and H. Leutwyler, in preparation.

[23] For a review, see T. Banks, Y. Nir and N. Seiberg, in these Proceedings.

[24] H. Leutwyler, Proc. XXVI Int. Conf. on High Energy Physics, Dallas, Aug.
1992, ed. J. R. Sanford, AIP Conf. Proc. No. 272, (1993).

[25] H. Leutwyler and A. Smilga, *Phys. Rev.* D 46 (1992) 5607;
A. Smilga and J. Stern, *Phys. Lett.* B318 (1993) 531.

[26] E. Shuryak and J. Verbaarschot, *Nucl. Phys.* B 410 (1993) 37;
E. Shuryak and J. Verbaarschot, *Nucl. Phys.* A 560 (1993) 306;
J. Verbaarschot and I. Zahed, *Phys. Rev. Lett.* 70 (1993) 3852;
J. Verbaarschot, *Nucl. Phys.* B 426 (1994) 559; B 427 (1994) 534;

Phys. Rev. Lett. 72 (1994) 2531; *Phys. Lett.* B 329 (1994) 351;
A. Smilga and J. Verbaarschot, "Spectral sum sules and finite volume partition function in gauge theories with real and pseudoreal fermions", preprint Univ. Minnesota, TPI-MINN-94/10-T (1994).

[27]   K. Choi, C.W. Kim and W.K. Sze, *Phys. Rev. Lett.* 61 (1988) 794;
K. Choi and C.W. Kim, *Phys. Rev.* D40 (1989) 890;
K. Choi, *Nucl. Phys.* B383 (1992) 58; *Phys. Lett.* B292 (1992) 159.

[28]   G. t'Hooft, *Phys. Rev.* D14 (1976) 3432.

[29]   T. Banks and A. Casher, *Nucl. Phys.* B169 (1980) 103.

[30]   H. Hellman, "Einführung in die Quantenchemie", Deuticke, Leipzig (1937);
R. P. Feynman, *Phys. Rev.* 66 (1939) 340.

[31]   For a recent exposition of GχPT, see
J. Stern, H. Sazdjan and N. H. Fuchs, *Phys. Rev.* D47 (1993) 3814;
M. Knecht et al., *Phys. Lett.* B313 (1993) 229.
The early literature is reviewed in
M. D. Scadron, Rep. Prog. Phys. 44 (1981) 213.

[32]   S. Randjbar-Daemi, A. Salam and J. Strathdee, *Phys. Rev.* B48 (1993) 3190;
H. Leutwyler, *Phys. Rev.* D49 (1994) 3033.

[33]   M. Gell-Mann, R. J. Oakes and B. Renner, *Phys. Rev.* 175 (1968) 2195.

[34]   J. Gasser, H. Leutwyler and M. Sainio, *Phys. Lett.* B253 (1991) 260.

[35]   J. Gasser and H. Leutwyler, *Phys. Lett.* B125 (1983) 321, 325.

[36]   A. I. Vainshtein et al., *Sov. J. Nucl. Phys.* 27 (1978) 274;
B. L. Ioffe, *Nucl. Phys.* B188 (1981) 317; B191 (1981) 591(E).

[37]   The literature may be traced with the following papers:
C. Adami, E. G. Drukarev and B. L. Ioffe *Phys. Rev.* D48 (1993) 2304;
V. L. Eletsky and B. L. Ioffe, *Phys. Rev.* D48 (1993) 1441;
C. A. Dominguez, C. Van Gend and N. Paver, *Phys. Lett.* B253 (1991) 241;
S. Narison, *Phys. Lett.* B216 (1989) 191;
C. A. Dominguez and E. de Rafael, *Ann. Phys.* 174 (1986) 372;
X. Jin, M. Nielsen and J. Pasupathy, "Calculation of $\langle p|\bar{u}u-\bar{d}d|p\rangle$ from QCD sum rule and the neutron-proton mass difference", hep-ph/9405202.

[38]   J. Gasser and H. Leutwyler, *Phys. Reports* 87 (1982) 77.

[39]   J. Gasser, *Ann. Phys. (N.Y.)* 136 (1981) 62.

[40]   The chiral logarithms occurring in the expansion are worked out in
R. F. Lebed and M. A. Luty, *Phys. Lett.* B 329 (1994) 479.

[41]   M. de Cervantes, "Don Quixote", part I, chapter 8, Madrid (1605).

[42]   J. F. Donoghue, B. R. Holstein and Y. C. R. Lin, *Phys. Rev. Lett.* 55 (1985) 2766.

[43]   F. Gilman and R. Kauffmann, *Phys. Rev.* D36 (1987) 2761;
Riazuddin and Fayyazuddin, *Phys. Rev.* D37 (1988) 149;
ASP Collaboration, N. A. Roe et al., *Phys. Rev.* D41 (1990) 17.

[44]   J. Bijnens, A. Bramon and F. Cornet, *Z. Phys.* C46 (1990) 599.

[45]   The DAFNE Physics Handbook, eds. L. Maiani, G. Pancheri and N. Paver, INFN–Frascati (1992).

[46] G. M. Shore and G. Veneziano, *Nucl. Phys.* B381 (1992) 3.

[47] J. Donoghue and D. Wyler, *Phys. Rev.* D45 (1992) 892.

[48] K. Gottfried, *Phys. Rev. Lett.* 40 (1978) 598;
Y. P. Tung and T. N. Yan, *Phys. Rev.* D41 (1990) 155;
M. Luty and R. Sundrum, *Phys. Lett.* B312 (1993) 205.

The Fermion Mass Problem
in Grand Unified String Theories
with $SU(3)$ Gauge Family Symmetry

A.A. MASLIKOV
INSTITUTE FOR HIGH ENERGY PHYSICS
142284 PROTVINO, MOSCOW REGION, RUSSIA

I.A. NAUMOV
MOSCOW STATE UNIVERSITY, DEPARTMENT OF PHYSICS
117 234 MOSCOW, RUSSIA

G.G. VOLKOV
INFN SEZIONE DI PADOVA AND
DIPARTIMENTO DI FISICA UNIVERSITÀ DI PADOVA
VIA MARZOLO 8, 35100 PADUA, ITALY.
INSTITUTE FOR HIGH ENERGY PHYSICS
142284 PROTVINO, MOSCOW REGION, RUSSIA
VOLKOV_G@MX.IHEP.SU

Abstract: In the framework of four dimensional heterotic superstring with free fermions we investigate the rank eight Grand Unified String Theories (GUST) which contain the $SU(3)_H$-gauge family symmetry. GUSTs of this type accomodate naturally three fermion families presently observed and, moreover, can describe fermion mass spectrum without high dimensional representations of conventional unification groups. We explicitly construct GUST with gauge symmetry $G = SU(5) \times U(1) \times (SU(3) \times U(1))_H \subset SO(16)$ in free complex fermion formulation. As the GUSTs originating from Kac-Moody algebras (KMA) contain only low-dimensional representations it is usually difficult to break the gauge symmetry. We solve this problem taking for the observable gauge symmetry the diagonal subgroup G^{sym} of rank 16 group $G \times G \subset SO(16) \times SO(16) \subset E(8) \times E(8)$. Such a construction effectively corresponds to a level two KMA, and therefore some higher dimensional representations of the diagonal subgroup appear. This (due to $G \times G$- tensor Higgs fields) allows one to break GUST symmetry down to the $SU(3^c) \times U(1)_{em}$. In this approach the observed electromagnetic charge Q^{em} can be viewed as a sum of two Q^I- and Q^{II} charges of each G- group. In this case below the scale where $G \times G$ breaks down to G^{symm} the spectrum does not contain particles with exotic fractional charges. In these GUST there has to exist "superweak" light chiral matter.

1 Introduction

Since a couple of years superstring theories and particularily the heterotic string theory, provide the efficient way to construct the Grand Unified Superstring Theories ($GUST$) of all known interactions. Despite of that it is still difficult to construct unique and fully realistic low energy models resulting after decoupling of massive string modes. This is because it is only in 10-dimensional space-time that there exist just two consistent (invariant under reparametrization, superconformal, modular, Lorentz and SUSY transformations) theories with gauge symmetries $E(8) \times E(8)$ or $spin(32)/Z_2$ [1, 2] which after compactification of the six extra space coordinates (into the Calabi-Yau [3, 4] manifolds or into the orbifolds) can be used for constructing GUSTs. Unfortunately, the process of compactification to four dimensions is not unique and the number of possible low energy models is very large. On the other hand, starting the construction of the theory directly in 4-dimensional space-time requires including a considerable number of free bosons or fermions into the internal string sector of the heterotic superstring [5, 6, 7]. This leads to as large internal symmetry group as e.g. rank 22 group. The way of breaking this primordial symmetry is again not unique and leads to a huge number of possible models, each of them giving different low energy predictions.

On the other side, due to the presence of the affine Kac-Moody algebra (KMA) $\hat{g}$ (which is a 2-dimensional manifestation of gauge symmetries of the string itself) on the world sheet, string constructions yield definite predictions as to what representation of the symmetry group can be used for low energy models building. Therefore the following long-standing questions have the chance to be answered in this kind of unification schemes:

1. How are the chiral matter fermions assigned to the multiplets of the unifying group?

2. How is the GUT gauge symmetry breaking realized?

3. What is the origin and the form of the fermion mass matrices?

The first of these problems is, of course, closely connected with the quantization of the electromagnetic charge of matter fields. In addition, string constructions can shed some light on the questions about the number of generation and possible existence of mirror fermions which remain unanswered in conventional GUTs.

In this note starting from the rank 16 grand unified gauge group (which is the minimal rank allowed in strings [6, 10]) of the form $G \times G$ and making use of the KMA which select the possible gauge group representations we construct the string model based on the diagonal subgroup $G^{symm} \subset G \times G \subset SO(16) \times SO(16)$ ($\subset E_8 \times E_8$). We discuss and consider $G^{symm} = SU(5) \times U(1) \times (SU(3) \times U(1))_H \subset SO(16)$ where the factor $(SU(3) \times U(1))_H$ is interpreted as the horizontal gauge family symmetry. We explain how the unifying gauge symmetry can be broken down to the Standard Model group. Furthermore, the horizontal interaction predicted in our model can give alternative description of the fermion mass matrices without invoking high dimensional Higgs representations. In contrast to other GUST constructions, our model does not contain particles with exotic fractional electric charges. This important virtue of the model is due to the symmetric construction of the electromagnetic charge Q_{em} from Q^I and Q^{II} – the two electric charges of each of the $U(5)$ groups.

2   World-Sheet Kac-Moody Algebra And Main Features of Rank Eight GUST

Let's begin with a short review of the KMA results $[8, 9]$. The KMA $\hat{g}$ allow to grade the representations R of the gauge group by a level number x (a non negative integer) and by a conformal weight $h(R)$.

An irreducible representation of the affine algebra $\hat{g}$ is characterized by the vacuum representation of the algebra g and the value of the central term k, which is connected with the level number by the relation $x = 2k/\psi^2$ where ψ is the highest root. The value of the level number of the KMA determines the possible highest weight unitary representation which are present in the spectrum, in the following way:

$$x = \frac{2k}{\psi^2} \geq \sum_{i=1}^{rank\ g} n_i m_i, \tag{1}$$

where the sets of non-negative integers $\{m_i = m_1, ..., m_r\}$ and $\{n_i = n_1, ..., n_r\}$ define the highest root and the highest weight of a representation R respectively $[8, 9, 10]$.

In fact, the KMA on the level one is realized in the 4-dimensional heterotic superstring theories with free world sheet fermions $[6, 7, 10]$. For these models the level of KMA coincides with the Dynkin index of representation M to which free fermions are assigned:

$$x = x_M = \frac{Q_M}{\psi^2} \frac{dim M}{dim g} \tag{2}$$

(Q_M is a quadratic Casimir eigenvalue of representation M) and equals one in cases when real fermions form vector representation M of $SO(2N)$, or when the world sheet fermions are complex and M is the fundamental representation of $U(N)$ $[8, 9]$.

Thus, in strings with KMA on the level one realized on the world-sheet, only very restricted set of unitary representations can arise in the spectrum $[8, 9, 10]$:

1. singlet and totally antisymmetric tensor representations of $SU(N)$ groups, for which $m_i = (1, ..., 1)$;

2. singlet, vector and spinor representations of $SO(2N)$ groups, with $m_i = (1, 2, 2, ...2, 1, 1)$;

3. singlet, $\underline{27}$, and $\underline{\bar{27}}$-plets of $E(6)$, corresponding to $m_i = (1, 2, 2, 3, 2, 1)$;

4. singlet of $E(8)$, with $m_i = (2, 3, 4, 6, 5, 4, 3, 2)$.

Therefore only these representations can be used to incorporate matter and Higgs fields in GUSTs.

For example, to describe chiral matter fermions in GUST with the gauge symmetry group $SU(5) \times U(1) \subset SO(10)$ the following sum of the level-one complex representations: $\underline{1}(-5/2) + \underline{5}(+3/2) + \underline{10}(-1/2) = \underline{16}$ can be used. On the other side, as real representations of $SU(5) \times U(1) \subset SO(10)$, from which Higgs fields can arise, one can take for example $\underline{5} + \underline{\bar{5}}$ representations arising from real representation $\underline{10}$ of $SO(10)$. Also, real Higgs representations like $\underline{10}(-1/2) + \underline{\bar{10}}(+1/2)$ of $SU(5) \times U(1)$ originating from $\underline{16} + \underline{\bar{16}}$ of $SO(10)$, which has been used in ref. *[10]* for further symmetry breaking, are allowed.

Another example is provided by the decomposition of $SO(16)$ representations under $SU(8) \times U(1) \subset SO(16)$. Here only singlet, $\underline{16}$, $\underline{128}$ and $\underline{128}'$ representations of $SO(16)$ are allowed by the KMA ($\underline{128}$ and $\underline{128}'$ are the two nonequivalent, real spinor representations). From the item 2. we can obtain the following $SU(8) \times U(1)$ representations: singlet, $\underline{8} + \underline{\bar{8}}$ $(= \underline{16})$, $\underline{8} + \underline{56} + \underline{\bar{56}} + \underline{\bar{8}}$ $(= \underline{128})$ and $\underline{1} + \underline{28} + \underline{70} + +\underline{\bar{28}} + \underline{\bar{1}}$ $(= \underline{128}')$.

However, as we will demonstrate, in each of the string sectors the requirement of the modular invariance (one of the GSO projections) necessarily eliminates either $\underline{128}$ or $\underline{128}'$. It is therefore important that, in order to incorporate chiral matter in the model, only one spinor representation is sufficient. Moreover, if one wants to solve the chirality problem applying further GSO projections (which break the gauge symmetry also the representation $\underline{\bar{10}}$ which otherwise, together with $\underline{10}$, could form real Higgs representation, disappears from this sector. Therefore, the existence of $\underline{\bar{10}}_{-1/2} + \underline{10}_{1/2}$, needed for breaking $SU(5) \times U(1)$ is incompatible with the possible solution of the chirality problem for the family matter fields.

Thus, in the rank eight group $SU(8) \times U(1) \subset SO(16)$ with Higgs representations from the level-one KMA only, one can not arrange for further symmetry breaking. Moreover, construction of the realistic fermion mass matrices seems to be impossible. In old-fashioned GUTs (see e.g. *[11]*), not originating from strings, the representations of level-two (which are not the representations of level-one) were commonly used to solve these problems.

The way out from this difficulty is based on the following important observations. Firstly, all higher-dimensional representations of (simple laced) groups like $SU(N)$, $SO(2N)$ or $E(6)$, which belong to the level-two of the KMA (according to the equation 1), appear in the direct product of the level- one representations:

$$R_G(x = 2) \subset R_G(x = 1) \times R_G{}'(x = 1). \tag{3}$$

For example, the level-two representations of $SU(5)$: $\underline{15}$, $\underline{24}$, $\underline{40}$, $\underline{45}$, $\underline{50}$, $\underline{75}$ will appear in the direct products of: $\underline{5} \times \underline{5}$, $\underline{5} \times \underline{\bar{5}}$, $\underline{5} \times \underline{10}$, etc. respectively. In the case of $SO(10)$ the level two representations $\underline{45}$, $\underline{54}$, $\underline{120}$, $\underline{126}$, $\underline{210}$, $\underline{144}$ can be obtained by the suitable direct products: $\underline{10} \times \underline{10}$, $\underline{\bar{16}} \times \underline{10}$, $\underline{16} \times \underline{16}$, $\underline{\bar{16}} \times \underline{16}$. The level-two representations $\underline{78}$, $\underline{351}$, $\underline{351}'$ $\underline{650}$ of $E(6)$ are factors of the decomposition of the direct products of: $\underline{\bar{27}} \times \underline{27}$ or $\underline{27} \times \underline{27}$. The only exception from this rule is the the $E(8)$ group, two level-two representations ($\underline{248}$ and $\underline{3875}$) of which cannot be constructed as a product of level-one representations *[12]*.

Secondly, the diagonal (symmetric) subgroup G^{symm} of $G \times G$ effectively corresponds to the level-two KMA $g(x = 1) \oplus g(x = 1)$ because taking the $G \times G$ representations in the form (R_G, R_G') of the $G \times G$, where R_G and R_G' belong to the level-one of G, one obtains representations of the form $R_G \times R_G'$ when one considers only the diagonal subgroup of $G \times G$. This observation is crucial, because such a construction allows one to obtain level-two representations. (This construction has implicitly been used in *[13]* where we have constructed some examples of GUST with gauge symmetry realized as a diagonal subgroup of direct product of two rank eight groups $U(8) \times U(8) \subset SO(16) \times SO(16)$.)

In strings, however, not all level-two representations can be obtained in that way because, as we will demonstrate, some of them become massive (with masses of order of the Planck scale). The condition ensuring that in the string spectrum states transforming as a representation R are massless reads:

$$h(R) = \frac{Q_R}{2k + Q_{ADJ}} = \frac{Q_R}{2Q_M} \leq 1, \tag{4}$$

where Q_i is the quadratic Casimir invariant of the corresponding representations, and M has been already defined before (see eq. 2).

This condition, when combined with 1, gives a restriction at the rank of GUT's group $(r \leq 8)$, whose representations can accomodate chiral matter fields. For example, for $G = SO(16)$ or $E(6) \times SU(3)$, representations $\underline{128}$, $(\underline{27}, \underline{3})$ $(h(\underline{128}) = 1, h(\underline{27}, \underline{3}) = 1)$ respectively, satisfy both conditions. Obviously, these (important for incorporation of chiral matter) representations will exist at the level-two KMA of the symmetric subgroup of the group $G \times G$.

In general, condition 4 severely constrains massless string states transforming as $(R_G(x = 1), R_G{}'(x = 1))$ of the direct product $G \times G$. For example, for $SU(8) \times SU(8)$ and for $SU(5) \times SU(5)$ constructed from $SU(8) \times SU(8)$ only representations of the form

$$R_{N,N} = \big((\underline{N}, \underline{N}) + h.c.\big), \quad \big((\underline{N}, \underline{\bar{N}}) + h.c.\big); \tag{5}$$

with $h(R_{N,N}) = (N-1)/N$, where $N = 8$ or 5 respectively can be massless *[13]*. For $SO(2N) \times SO(2N)$ massless states are contained only in representations

$$R_{v,v} = (\underline{2N}, \underline{2N}) \tag{6}$$

with $h(R_{v,v}) = 1$. Thus, for the GUSTs based on a diagonal subgroup $G^{symm} \subset G \times G$, G^{symm} - high dimensional representations, which are embedded in $R_G(x = 1) \times R{}'_G(x = 1)$ are also severely constrained by the condition 4.

For spontaneous breaking of $G \times G$ gauge symmetry down to G^{symm} (rank $G^{symm} = $ rank G) one can use the direct product of representations $R_G(x = 1) \times R_G(x = 1)$, where $R_G(x = 1)$ is the fundamental representation of $G = SU(N)$ or vector representation of $G = SO(2N)$. Furthemore, $G^{symm} \subset G \times G$ can subsequently be broken down to a smaller dimension gauge group (of the same rank as G^{symm}) through the VEVs of the adjoint representations which can appear as a result of $G \times G$ breaking. Alternatively, the real Higgs superfields 5 or 6 can directly break the $G \times G$ gauge symmetry down to a $G_1^{symm} \subset G^{symm}$ (rank $G_1^{symm} \leq $ rank G^{symm}). For example when $G = SU(5) \times U(1)$ or $SO(10) \times U(1)$ $G \times G$ can directly be broken in that way down to $SU(3^c) \times G_{EW}^I \times G_{EW}^{II} \times ....$

The above examples show clearly, that within the framework of GUSTs with the KMA one can get interesting gauge symmetry breaking chains including the realistic ones provided $G \times G$ gauge symmetry group is considered. However the lack of the higher dimensional representations (which are forbidden by 4) on the level-two KMA prevents the construction of the realistic fermion mass matrices. That is why we consider an extended grand unified string model of rank eight.

There are good physical reasons for including the horizontal $SU(3)_H$ group into the unification scheme. Firstly, this group naturally accommodates three fermion families presently observed (explaining their origin) and, secondly, can provide correct and economical description of the fermion mass spectrum and mixing without invoking high dimensional representation of conventional $SU(5)$, $SO(10)$ or $E(6)$ gauge groups. Construction of a string model (GUST) containing the horizontal gauge symmetry provides additional, strong motivation to this idea. Moreover, the fact that in GUSTs high dimensional representations are forbidden by the KMA is a very welcome feature in this context.

All this leads us naturally to consider possible forms for horizontal symmetry G_H, and G_H quantum number assignments for quarks (anti-quarks) and leptons (anti-leptons) which can be realized within GUST's framework. To include the horizontal interactions with three known generations in the ordinary GUST it is natural to consider rank eight gauge symmetry. In this paper we concentrate on $SO(16)$ which is a maximal subgroup of $E(8)$ and which contains the rank eight subgroup $SO(10) \times (U(1) \times SU(3))_H$. We will be, therefore, concerned with the following chain (see Fig. 1):

$$E(8) \longrightarrow SO(16) \longrightarrow SO(10) \times (U(1) \times SU(3))_H \longrightarrow$$
$$\longrightarrow SU(5) \times U(1)_{Y_5} \times (SU(3) \times U(1))_H.$$

According to this scheme one can get $SU(3)_H \times U(1)_H$ gauge family symmetry with $N_g = 3+1$ (there are also other possibilities as eg. $E(6) \times SU(3)_H \subset E(8)$ $N_g = 3$ generations can be obtained due to the second way of $E(8)$ gauge symmetry breaking via $E(6) \times SU(3)_H$, see Fig. 1), where the possible, additional, fourth massive matter superfield is a singlet of $SU(3)_H$ and transforms as $\underline{16}$ under the $SO(10)$ group. The full chiral $SO(10) \times SU(3) \times U(1)$ matter multiplets can be constructed from $SU(8) \times U(1)$–multiplets

$$(\underline{8} + \underline{56} + \underline{\bar{8}} + \underline{\bar{56}}) = \underline{128} \tag{7}$$

of $SO(16)$. In the 4-dimensional heterotic superstring with free complex world sheet fermions, in the spectrum of the Ramond sector there can appear also representations which are factors in the decomposition of $\underline{128}'$. In particular, $SU(5)$-decouplets $(\underline{10} + \underline{\bar{10}})$ from $(\underline{28} + \underline{\bar{28}})$ of $SU(8)$. However their $U(1)_5$ hypercharge prevent using them for $SU(5) \times U(1)_5$–symmetry breaking. Thus, in this approach we have only singlet and $(\underline{5} + \underline{\bar{5}})$ Higgs fields which can break the grand unified $SU(5) \times U(1)$ gauge symmetry. Therefore it is necessary (as we already explained) to construct rank eight GUST based on a diagonal subgroup $G^{symm} \subset G \times G$ primordial symmetry group, where in each rank eight group G the Higgs fields will appear only in singlets and in the fundamental representations as in (see 5).

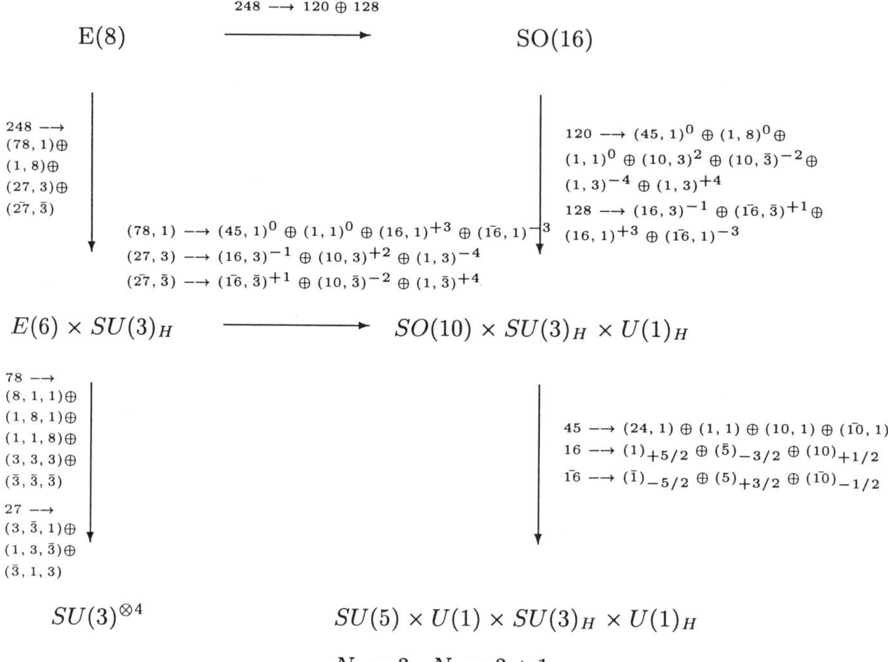

Figure 1. The possible ways of E(8) gauge symmetry breaking leading to the 3+1 or 3 generations.

A comment concerning $U(1)$ factors can be made here. Since the available $SU(5) \times U(1)$ decouplets have non-zero hypercharges with respect to $U(1)_5$ and $U(1)_H$, these $U(1)$ factors may remain unbroken down to the low energies in the model considered which seems to be very interesting.

3   GUST Construction with $SU(3) \times U(1)$ Gauge Family Symmetry in Free Fermion Formulation

In heterotic string theories [1, 2] $(N = 1\,SUSY)_{LEFT}$ $(N = 0\,SUSY)_{RIGHT}$ $\oplus \mathcal{M}_{c_L;c_R}$ with $d \leq 10$, the conformal anomalies of the space-time sector are canceled by the conformal anomalies of the internal sector $\mathcal{M}_{c_L;c_R}$, where $c_L = 15 - 3d/2$ and $c_R = 26 - d$ are the conformal anomalies in the left- and right–moving string sectors respectively.

In the fermionic formulation of the four-dimensional heterotic string theory in addition to the two transverse bosonic coordinates X_μ, $\bar{X}_\mu$ and their left-moving superpartners ψ_μ, the internal sector $\mathcal{M}_{c_L;c_R}$ contains 44 right-moving $(c_R = 22)$ and 18 left-moving $(c_L = 9)$ real fermions. The model is completely defined by a set Ξ of spin boundary conditions for all these world-sheet fermions. In a diagonal basis the vectors of Ξ are determined by the values of phases $\alpha(f)$

Table 1. Basis of the boundary conditions for all world-sheet fermions.

Vectors	$\psi_{1,2}$	$\chi_{1,..,6}$	$y_{1,...,6}$	$\omega_{1,...,6}$	$\bar{\varphi}_{1,...,12}$	$\bar{\Psi}_{1,...,8}$	$\bar{\Phi}_{1,...,8}$
b_1	11	111111	111111	111111	1^{12}	1^8	1^8
b_2	11	111111	000000	000000	0^{12}	$1/2^8$	0^8
b_3	11	111100	000011	000000	$0^4 1^8$	0^8	1^8
$b_4 = S$	11	110000	001100	000011	0^{12}	0^8	0^8
b_5	11	001100	000000	110011	1^{12}	$1/4^5 - 3/4^3$	$-1/4^5 3/4^3$
b_6	11	110000	000011	001100	$1^2 0^4 1^6$	1^8	0^8

$\in (-1,1]$ fermions f acquire $(f \longrightarrow -\exp(i\pi\alpha(f))f)$ when parallel transported around the string. To construct the GUST according to the scheme outlined at the end of the previous section we consider the set Ξ which forms the $Z_2 \times Z_4 \times Z_2 \times Z_2 \times Z_8 \times Z_2$ group under addition (mod 2) and is generated by the basis with six elements $B = b_1, b_2, b_3, b_4 \equiv S, b_5, b_6$ given in Table 1.

Following [7] we construct the canonical basis in such a way that the vector $\bar{1}$, which belongs to Ξ, is the first element b_1 of the basis. The basis vector $b_4 = S$ is the generator of supersymmetry [10] responsible for the conservation of the space-time $SUSY$.

We have chosen a basis in which all left movers $(\psi_\mu; \chi_i, y_i, \omega_i; i = 1, ...6)$ (on which the world sheet supersymmetry is realized nonlinearly) as well as 12 right movers $(\bar{\varphi}_k; k = 1, ...12)$ are real whereas $(8 + 8)$ right movers $\bar{\Psi}_A$, $\bar{\Phi}_M$ are complex. Right- and left-moving real fermions can be used for breaking G^{comp} symmetry [6, 10]. In order to have a possibility to reduce the rank of the compactified group G^{comp}, we have to select the spin boundary conditions for the maximal possible number, $N_{LR} = 12$, of left-moving, $\chi_{3,4,5,6}$, $y_{1,2,5,6}$, $\omega_{1,2,3,4}$, and right-moving, $\bar{\phi}^{1,...12}$ $(\bar{\phi}^p = \bar{\varphi}_p, p = 1, ...12)$ real fermions. The KMA based on 16 complex right moving fermions gives rise to the "observable" gauge group, G^{obs}, with:

$$rank(G^{obs}) \leq 16. \tag{8}$$

In our approach the basis vector b_2 is constructed as a complex vector with the $1/2$ spin-boundary conditions for the right-moving fermions $\bar{\Psi}_A$, $A = 1, ...8$. Initially it generates chiral matter fields in the $\underline{8} + \underline{56} + \underline{\bar{56}} + \underline{\bar{8}}$ representations of $SU(8) \times U(1)$, which subsequently are decomposed under $SU(5) \times U(1) \times SU(3) \times U(1)$ to which $SU(8) \times U(1)$ gets broken by applying the b_5 GSO projection.

Generalized GSO projection coefficients are originally defined up to fifteen signs some of which, are fixed by the supersymmetry conditions. Below, in Table 2, we present a set of numbers

$$\gamma \begin{bmatrix} b_i \\ b_j \end{bmatrix} = \frac{1}{i\pi} \log \mathcal{C} \begin{bmatrix} b_i \\ b_j \end{bmatrix}.$$

which we use as basis for our GSO projections.

The physical states in the Hilbert space of a given sector α are obtained acting on the vacuum $|0>_\alpha$ with the bosonic and fermionic operators with

Table 2. The choice of the GSO basis $\gamma \begin{bmatrix} b_i \\ b_j \end{bmatrix}$, b_i. i numbers rows and j – columns.

	b_1	b_2	b_3	b_4	b_5	b_6
b_1	0	1	1	1	1	0
b_2	1	1/2	0	0	1/4	1
b_3	1	−1/2	0	0	1/2	0
b_4	1	1	1	1	1	1
b_5	0	1	0	0	−1/2	0
b_6	0	0	0	0	1	1

frequencies

$$n(f) = 1/2 + 1/2\alpha(f), \quad n(f^*) = 1/2 - 1/2\alpha(f^*) \tag{9}$$

and subsequently applying the generalized GSO projections. The physical states satisfy the Virasoro condition:

$$M_L^2 = -1/2 + 1/8\,(\alpha_L \cdot \alpha_L) + N_L = -1 + 1/8\,(\alpha_R \cdot \alpha_R) + N_R = M_R^2, \tag{10}$$

where $\alpha = (\alpha_L, \alpha_R)$ is a sector in the set Ξ, $N_L = \sum_L(frequencies)$ and $N_R = \sum_R(frequencies)$.

We keep the same sign convention for the fermion number operator F as in [7, 10]. For complex fermions we have $F_\alpha(f) = 1$, $F_\alpha(f^*) = -1$ with the exception of the periodic fermions for which we get $F_{\alpha=1}(f) = -1/2(1 - \gamma_{5f})$, where $\gamma_{5f}|\Omega> = |\Omega>$, $\gamma_{5f}b_o^+|\Omega> = -b_o^+|\Omega>$.

The full Hilbert space of the string theory is constructed as a direct sum of different sectors $\sum_i m_i b_i$, $(m_i = 0, 1, .., N_i)$, where the integers N_i define additive groups $Z(b_i)$ of the basis vectors b_i. The generalized GSO projection leaves in sectors α those states, whose b_i-fermion number satisfies:

$$\exp(i\pi b_i F_\alpha) = \delta_\alpha \mathcal{C}^* \begin{bmatrix} \alpha \\ b_i \end{bmatrix}, \tag{11}$$

where the space-time phase $\delta_\alpha = \exp(i\pi\alpha(\psi_\mu))$ is equal -1 for the Ramond sector and $+1$ for the Neveu-Schwarz sector. In our case of the $Z_2^4 \times Z_4 \times Z_8$ model, we initially have 256×2 sectors. After applying the GSO-projections we get only 49×2 sectors containing massless states, which depending on the vacuum energy values, E_L^{vac} and E_R^{vac}, can be naturally divided into some classes [13] and which determine the GUST representations.

The NS sector contains the $N = 1$ supergravity multiplet as well as gauge fields (with their $N = 1$ superpartners) of the observable and compactified gauge groups. For example, massless string states yielding gauge fields of the group

$$[SU(5) \times U(1)]_{GUT} \times [SU(3) \times U(1)]_H. \tag{12}$$

corresponding to $U(8)^I$ are constructed as follows:

$$\psi_{1/2}^\mu|\Omega>_L \otimes \Psi_{1/2}^a \Psi_{1/2}^{b*}|\Omega>_R, \quad a, b = 1, ..., 5 ;$$

$$\psi_{1/2}^\mu|\Omega>_L \otimes \Psi_{1/2}^i \Psi_{1/2}^{j*}|\Omega>_R, \quad i, j = 1, 2, 3 , \tag{13}$$

N°	b_1,b_2,b_3,b_4,b_5,b_6	SO_{hid}	$U(5)^I$	$U(3)^I$	$U(5)^{II}$	$U(3)^{II}$	$\tilde{Y}_5^I$	$\tilde{Y}_3^I$	$\tilde{Y}_5^{II}$	$\tilde{Y}_3^{II}$
1	RNS		5	$\bar{3}$	1	1	−1	−1	0	0
			1	1	5	$\bar{3}$	0	0	−1	−1
	0 2 0 1 2(6) 0		5	1	5	1	−1	0	−1	0
			1	3	1	3	0	1	0	1
			5	1	1	3	−1	0	0	1
			1	3	5	1	0	1	−1	0
2	0 1 0 0 0 0		1	3	1	1	5/2	−1/2	0	0
			$\bar{5}$	3	1	1	−3/2	−1/2	0	0
			10	1	1	1	1/2	3/2	0	0
	0 3 0 0 0 0		1	1	1	1	5/2	3/2	0	0
			$\bar{5}$	1	1	1	−3/2	3/2	0	0
			10	3	1	1	1/2	−1/2	0	0
3	0 0 1 1 3 0	$-_1\,\pm_2$	1	1	1	3	0	−3/2	0	−1/2
	0 0 1 1 7 0	$-_1\,\pm_2$	1	$\bar{3}$	1	1	0	1/2	0	3/2
	0 2 1 1 3 0	$+_1\,\pm_2$	1	$\bar{3}$	1	3	0	1/2	0	−1/2
	0 0 1 1 7 0	$+_1\,\pm_2$	1	1	1	1	0	−3/2	0	−3/2
4	1 1 1 0 1 1	$\mp_1\,\pm_3$	1	1	1	$\bar{3}$	0	−3/2	0	1/2
	1 1 1 0 5 1	$\mp_1\,\pm_3$	1	$\bar{3}$	1	1	0	1/2	0	−3/2
	1 3 1 0 1 1	$\pm_1\,\pm_3$	1	$\bar{3}$	1	$\bar{3}$	0	1/2	0	1/2
	1 3 1 0 5 1	$\pm_1\,\pm_3$	1	1	1	1	0	−3/2	0	−3/2
5	0 1(3) 1 0 2(6) 1	$-_1\,\pm_3$	1	$3(\bar{3})$	1	1	$\pm 5/4$	$\pm 1/4$	$\pm 5/4$	$\mp 3/4$
		$+_1\,\pm_3$	$5(\bar{5})$	1	1	1	$\pm 1/4$	$\mp 3/4$	$\pm 5/4$	$\mp 3/4$
	0 1(3) 1 0 4 1	$-_1\,\pm_3$	1	1	1	$3(\bar{3})$	$\pm 5/4$	$\mp 3/4$	$\pm 5/4$	$\pm 1/4$
		$+_1\,\pm_3$	1	1	$5(\bar{5})$	1	$\pm 5/4$	$\mp 3/4$	$\pm 1/4$	$\mp 3/4$
6	1 2 0 0 3(5) 1	$\pm_1\,-_4$	1	1	1	1	$\pm 5/4$	$\pm 3/4$	$\mp 5/4$	$\mp 3/4$
	1 1(3) 0 1 5(3) 1	$+_1\,\mp_4$	1	1	1	1	$\pm 5/4$	$\pm 3/4$	$\pm 5/4$	$\pm 3/4$
	0 0 1 0 2(6) 0	$\mp_3\,+_4$	1	1	1	1	$\pm 5/4$	$\mp 3/4$	$\pm 5/4$	$\mp 3/4$

Table 3. The list of quantum numbers of the fermionic states with positive chirality.

Similar states arise also for $U(8)^{II}$ generated by $\bar{\Phi}_M$ complex fermions. In the compactified $SO(12)$ sector there remains $SO(2)_1 \times SO(2)_2 \times SO(2)_3 \times SO(6)_4$ gauge group.

In NS sector the b_3 GSO projection leaves $(5,\bar{3}) + (\bar{5},3)$ Higgs superfields:

$$\chi_{1/2}^{1,2}|\Omega>_L \otimes \Psi_{1/2}^a \Psi_{1/2}^{i*};\ \Psi_{1/2}^{a*}\Psi_{1/2}^i|\Omega>_R \quad \text{and exchange } \Psi \longrightarrow \Phi \qquad (14)$$

In Table 3 states in row No 1 only are vectorlike. Chirality under hidden $SO(2)^3 \times SO(6)_4$ is denoted as $\pm_i$. Lower signs in rows No 5, 6 correspond to sectors with components given in brackets.

Four $(3_H + 1_H)$ generations of chiral matter fields from $(SU(5) \times SU(3))_I$ group forming $SO(10)$–multiplets $(\underline{1},\underline{3}) + (\underline{\bar{5}},\underline{3}) + (\underline{10},\underline{3})$; $(\underline{1},\underline{1}) + (\underline{\bar{5}},\underline{1}) + (\underline{\bar{10}},\underline{1})$ are contained in b_2 and $3b_2$ sectors. Applying b_3 GSO projection to the $3b_2$ sector yields the following massless states:

$$b^+_{\psi_{12}} b^+_{\chi_{34}} b^+_{\chi_{56}} |\Omega>_L \otimes \left\{ \Psi^{i*}_{3/4} \; , \quad \Psi^a_{1/4}\Psi^b_{1/4}\Psi^c_{1/4}, \quad \Psi^a_{1/4}\Psi^i_{1/4}\Psi^j_{1/4} \right\} |\Omega>_R,$$

$$b^+_{\chi_{12}} b^+_{\chi_{34}} b^+_{\chi_{56}} |\Omega>_L \otimes \left\{ \Psi^{a*}_{3/4} \; , \quad \Psi^a_{1/4}\Psi^b_{1/4}\Psi^i_{1/4}, \quad \Psi^i_{1/4}\Psi^j_{1/4}\Psi^k_{1/4} \right\} |\Omega>_R (15)$$

with the space-time chirality $\gamma_{5\psi_{12}} = -1$ and $\gamma_{5\psi_{12}} = 1$, respectively. In these formulae the Ramond creation operators $b^+_{\psi_{1,2}}$ and $b^+_{\chi_{\alpha,\beta}}$ of the zero modes are built of a pair of real fermions (as indicated by double indices): $\chi_{\alpha,\beta}$, $(\alpha,\beta) = (1,2)$, $(3,4)$, $(5,6)$. Here, as in 13 and 14 indices take values $a, b = 1,...,5$ and $i, j = 1,2,3$ respectively.

We stress that without using the b_3 projection we would get matter super-multiplets belonging to real representations only i.e. "mirror" particles would remain in the spectrum. The b_6 projection instead, eliminates all chiral matter superfields from $U(8)^{II}$ group.

Since the matter fields form the chiral multiplets of $SO(10)$, it is possible to write down $U(1)_{Y_5}$–hypercharges of massless states. In order to construct the right electromagnetic charges for matter fields we must define the hypercharges operators for the observable $U(8)^I$ group as

$$Y_5 = \int_0^\pi d\sigma \sum_a \Psi^{*a}\Psi^a, \quad Y_3 = \int_0^\pi d\sigma \sum_i \Psi^{*i}\Psi^i \qquad (16)$$

and analogously for the $U(8)^{II}$ group.

Then the following orthogonal combinations

$$\tilde{Y}_5 = \frac{1}{4}(Y_5 + 5Y_3), \quad \tilde{Y}_3 = \frac{1}{4}(Y_3 - 3Y_5), \qquad (17)$$

play the role of the hypercharge operators of $U(1)_{Y_5}$ and $U(1)_{Y_H}$ groups, respectively. Below we shall write the hypercharges as $(\tilde{Y}_5^I, \tilde{Y}_3^I, \tilde{Y}_5^{II}, \tilde{Y}_3^{II})$.

In the next section we discuss the problem of rank eight GUST gauge symmetry breaking. The matter is that according to the results of section 2 the Higgs fields $(\underline{10}_{1/2} + \underline{\overline{10}}_{-1/2}$ do not appear. As an illustration we can consider the GUST construction involving $SO(10)$ as GUT gauge group. We consider the set Ξ' consists of six vectors $B = b_1, b_2, b_3, b_4 \equiv S, b_5, b_6$ given in Table 4.

Corresponding GSO projections leading to $G^{comp} \times (SO(10) \times SU(4))^{\times 2}$ are given in Table 5.

The condition of generation chirality in this model results in choice of Higgs fields as a vector representations of $SO(10)$ ($\underline{16} + \underline{\overline{16}}$ are absent). According to conclusion 6 the only Higgs fields $(\underline{10}, \underline{1}; \underline{10}, \underline{1})$ of $(SO(10) \times SU(4))^{\times 2}$ appear in model which can be used for GUT gauge symmetry.

Table 4. Basis of the boundary conditions for basis Ξ'.

Vectors	$\psi_{1,2}$	$\chi_{1,\dots,6}$	$y_{1,\dots,6}$	$\omega_{1,\dots,6}$	$\bar{\varphi}_{1,\dots,12}$	$\Psi_{1,\dots,8}$	$\Phi_{1,\dots,8}$
b_1	11	111111	111111	111111	1^{12}	1^8	1^8
b_2	11	111111	000000	000000	0^{12}	$1^5 1/3^3$	0^8
b_3	11	000000	111111	000000	$0^8 1^4$	$0^5 1^3$	$0^5 1^3$
$b_4 = S$	11	110000	001100	000011	0^{12}	0^8	0^8
b_5	11	111111	000000	000000	0^{12}	0^8	$1^5 1/3^3$
b_6	11	001100	110000	000011	110011111100	1^8	0^8

Table 5. The choice of the GSO basis. $\gamma \begin{bmatrix} b_i \\ b_j \end{bmatrix}$, b_i. i numbers rows and j – columns.

	b_1	b_2	b_3	b_4	b_5	b_6
b_1	0	1	0	0	1	0
b_2	0	2/3	1	1	1	1
b_3	0	1	0	1	1	1
b_4	0	0	0	0	0	0
b_5	0	1	1	1	2/3	0
b_6	0	1	0	1	1	1

4 Gauge Symmetry Breaking and GUST Spectrum

In the first GUST construction presented in the preceding section there exists a possibility to break the GUST group $(U(5) \times U(3)) \times (U(5) \times U(3))$ down to the symmetric group by the ordinary Higgs mechanism [13]:

$$G^I \times G^{II} \to G^{symm} \to \dots \tag{18}$$

To achieve such breaking one can use nonzero vacuum expectation values of the tensor Higgs fields (see Table 3, row No 1), contained in the $2b_2 + 2(6)b_5(+S)$ sectors which transform under the $(SU(5) \times U(1) \times SU(3) \times U(1))^{symm}$ group in the following way:

$$\begin{aligned}
(\underline{5}, \underline{1}; \underline{5}, \underline{1})_{(-1,0;-1,0)} &\to (\underline{24}, \underline{1})_{(0,0)} + (\underline{1}, \underline{1})_{(0,0)}; \\
(\underline{1}, \underline{3}; \underline{1}, \underline{3})_{(0,1;0,1)} &\to (\underline{1}, \underline{8})_{(0,0)} + (\underline{1}, \underline{1})_{(0,0)},
\end{aligned} \tag{19}$$

$$\begin{aligned}
(\underline{5}, \underline{1}; \underline{1}, \underline{3})_{(-1,0;0,1)} &\to (\underline{\bar{5}}, \underline{3})_{(1,1)}; \\
(\underline{1}, \underline{3}; \underline{5}, \underline{1})_{(0,1;-1,0)} &\to (\underline{5}, \underline{\bar{3}})_{(-1,-1)}.
\end{aligned} \tag{20}$$

The diagonal vacuum expectation values for the Higgs fields 19 break the GUST group $(U(5) \times U(3))^I \times (U(5) \times U(3))^{II}$ down to the "skew"-symmetric group with the generators $\triangle_{symm}$ of the form:

$$\triangle_{symm}(t) = \theta(t) \times 1 + 1 \times t, \tag{21}$$

where $\theta(t) = -t^*$. The corresponding hypercharge of the symmetric group reads:

$$\bar{Y} = \tilde{Y}^{II} - \tilde{Y}^I. \tag{22}$$

Similarily, for the electromagnetic charge we get:

$$Q_{em} = Q^{II} - Q^{I} =$$
$$= (T_5^{II} - T_5^{I}) + \frac{2}{5}(\tilde{Y}_5^{II} - \tilde{Y}_5^{I}) = \bar{T}_5 + \frac{2}{5}\bar{Y}_5, \tag{23}$$

where $T_5 = diag(\frac{1}{15}, \frac{1}{15}, \frac{1}{15}, \frac{2}{5}, -\frac{3}{5})$. Note, that this charge quantization does not lead to exotic states with fractional electromagnetic charges (e.g. $Q_{em} = \pm 1/2, \pm 1/6$).

Thus, in the breaking scheme 21 it is possible to avoid colour singlet states with fractional electromagnetic charges, to achieve desired GUT breaking and moreover to get the usual value for the weak mixing angle $\sin^2\theta_W = 3/8$ at the unification scale (see [14]).

Adjoint representations which appear on the *rhs* of 19 can be used for further breaking of the symmetric group. This can lead to the final physical symmetry

$$(SU(3^c) \times SU(2_{EW}) \times U(1_Y) \times U(1)') \times (SU(3_H) \times U(1_H)) \tag{24}$$

with low-energy gauge symmetry of the quark – lepton generations with an additional $U(1)'$–factor.

Note, that using the same Higgs fields as in 19, there exists also another, interesting way of breaking the $G^I \times G^{II}$ gauge symmetry:

$$G^I \times G^{II} \to SU(3^c) \times SU(2)_{EW}^I \times SU(2)_{EW}^{II} \times U(1_{\bar{Y}}) \times$$
$$\times SU(3_H)^I \times SU(3_H)^{II} \times U(1_{\bar{Y}_H}) \to .... \tag{25}$$

It is attractive because it naturally solves the Higgs doublet–triplet mass splitting problem with rather low energy scale of GUST symmetry breaking.

In turn, the Higgs fields $\hat{h}_{(\Gamma,N)}$ from the NS sector

$$(\underline{5}, \underline{\bar{3}})_{(-1,-1)} + (\underline{\bar{5}}, \underline{3})_{(1,1)} \tag{26}$$

N=2 SUSY :	V	$=$	$(1, \frac{1}{2})$	$+$	$(\frac{1}{2}, 0)$	$SU(8)$
$\Downarrow$						$\Downarrow$
N=1 SUSY :	$V_{N=2}$	$\to$	$V_{N=1}$	$+$	$S_{N=1}$	$SU(5) \times SU(3) \times U(1)$

	J=1	J=1/2	J=1/2	J=0
$E_{vac} = [-\frac{1}{2}; -1]$ NS sector	(63)	—	—	(63)
$E_{vac} = [0; -1]$ SUSY sector	—	(63) × 2	(63) × 2	—
	Gauge multiplets			

$$\Downarrow \quad b_5 \quad \text{GSO} \quad \text{projection}$$

	J=1	J=1/2	J=1/2	J=0
$E_{vac} = [-\frac{1}{2}; -1]$ NS sector	(24,1)+(1,1)+(1,8)	—	—	(5,$\bar{3}$)+($\bar{5}$,3)
$E_{vac} = [0; -1]$ SUSY sector	—	((24,1)+(1,1)+(1,8))× 2	((5,$\bar{3}$)+($\bar{5}$,3))× 2	—
	Gauge multiplets		Higgs multiplets	

originate from N=2 SUSY vector representation $\underline{63}$ of $SU(8)^I$ (or $SU(8)^{II}$) by applying b_5 GSO projection (see Fig. 2). These Higgs fields (and fields 20) can be used for constructing chiral fermion (see Table 3, row No 2) mass matrices.

The b spin boundary conditions (Tabl.1) generate chiral matter and Higgs fields with the $GUST$ gauge symmetry $G_{comp} \times (G^I \times G^{II})_{obs}$ (where $G_{comp} = U(1)^3 \times SO(6)$ and $G^{I,II}$ have been already defined). The chiral matter spectrum, which we denote $\hat{\Psi}_{(\Gamma,N)}$ with ($\Gamma = \underline{1},\underline{5},\underline{10}$; $N = \underline{3},\underline{1}$), consists of $N_g = 3_H + 1_H$ families. See Table 3, row No 2 for the $((SU(5) \times U(1)) \times (SU(3) \times U(1))_H)^{symm}$ quantum numbers.

The $SU(3_H)$ anomalies of the matter fields (row No 2) are naturally canceled by the chiral "horizontal" superfields forming two sets: $\hat{\Psi}^H_{(1,N;1,N)}$ and $\hat{\Phi}^H_{(1,N;1,N)}$, $\Gamma = \underline{1}$, $N = \underline{1},\underline{3}$, (with both $SO(2)_2$ chiralities, see Table 3, row No 3, 4).

The horizontal fields (No 3, 4) compensate all $SU(3)^I$ anomalies introduced by the chiral matter spectrum (No 2) of the $(U(5) \times U(3))^I$ group (due to b_6 GSO projection the chiral fields of the $(U(5) \times U(3))^{II}$ group disappear from the final string spectrum). Performing the decomposition of fields (No 3, 4) under $(SU(5) \times SU(3))^{symm}$ we get (among other) three "horizontal" fields:

$$(\underline{1},\underline{\bar{3}})_{(0,-1)}, \quad (\underline{1},\underline{1})_{(0,-3)}, \quad (\underline{1},\underline{\bar{6}})_{(0,1)}, \tag{27}$$

coming from $\hat{\Psi}^H_{(1,3;1,1)}$, (or $\hat{\Psi}^H_{(1,1;1,\bar{3})}$), $\hat{\Psi}^H_{(1,1;1,1)}$ and $\hat{\Psi}^H_{(1,3;1,\bar{3})}$ respectively which make the low energy spectrum of the resulting model 25 $SU(3_H)^{symm}$- anomaly free. The other fields arising from (rows No 3, 4, Table 3) form anomaly-free representations of $(SU(3_H) \times U(1_H))^{symm}$:

$$2(\underline{1},\underline{1})_{(0,0)}, \quad (\underline{1},\underline{\bar{3}})_{(0,-1(2))} + (\underline{1},\underline{3})_{(0,1(-2))}, \quad (\underline{1},\underline{8})_{(0,0)}. \tag{28}$$

The superfields $\hat{\phi}_{(\Gamma,N)} + h.c.$, where ($\Gamma = \underline{1},\underline{5}$; $N = \underline{1},\underline{3}$), from the Table 3, row No 5 forming representations of $(U(5) \times U(3))^{I,II}$ have either Q^I or Q^{II} exotic fractional charges. Due to the strong G^{comp} gauge forces these fields may develop the double scalar condensate $< \hat{\phi}\hat{\phi} >$, which can also serve for $U(5) \times U(5)$ gauge symmetry breaking. For example, the composite condensate $< \hat{\phi}_{(5,1;1,1)}\hat{\phi}_{(1,1;\bar{5},1)} >$ can break the $U(5) \times U(5)$ gauge symmetry down to the symmetric diagonal subgroup with generators of the form

$$\triangle_{symm}(t) = t \times 1 + 1 \times t, \tag{29}$$

so for the electromagnetic charges we would have the form

$$Q_{em} = Q^{II} + Q^I. \tag{30}$$

leading again to no exotic, fractionally charged states in the low-energy string spectrum.

The superfields which transform nontrivially under the compactified group $G^{comp} = SO(6) \times SO(2)^{\times 3}$, (denoted as $\hat{\sigma} + h.c.$), and which are singlets of $(SU(5) \times SU(3)) \times (SU(5) \times SU(3))$, arise in three sectors, see Table 3, row No

6. The superfields $\hat{\sigma}$ form the spinor representations $\underline{4} + \underline{\bar{4}}$ of $SO(6)$ and they are also spinors of one of the $SO(2)$ groups. They have following hypercharges $\tilde{Y}_5^{I,II}$, $\tilde{Y}_3^{I,II}$:

$$\tilde{Y} = (5/4, \mp 3/4; \; 5/4, \mp 3/4), \quad \tilde{Y} = (5/4, \; 3/4; -5/4, -3/4). \tag{31}$$

With respect to the diagonal G^{symm} group with generators given by 21 or 29, the fields $\hat{\sigma}$ from sets a), b) or the set c), are of zero hypercharges and can, therefore, be used for breaking the $SO(6) \times SO(2)^{\times 3}$ group.

Note, that for the fields $\hat{\phi}$ and for the fields $\hat{\sigma}$ any other electromagnetic charge quantization diffrent than 23 or 30 would lead to "quarks" and "leptons" with the exotic fractional charges, for example, for the $\underline{5}$- and $\underline{1}$- multiplets according to the values of hypercharges (see eqs.31)the generator Q^{II} (or Q^{I}) has the eigenvalues

$(\pm 1/6, \pm 1/6, \pm 1/6, \pm 1/2, \mp 1/2)$ or $\pm 1/2$, respectively.

5   Discussion and Prospects

In this work we have considered the rank eight GUST with gauge symmetry $G = SU(5) \times U(1) \times (SU(3) \times U(1))_H \subset SO(16)$ which contains $N_g = \underline{3}_H + \underline{1}_H$ families. GUSTs originating from level-one KMA contain only low-dimensional representations of the unification group. It is, therefore, difficult to break the gauge symmetry. In order to solve this problem we have considered as the observable gauge symmetry the diagonal subgroup G^{sym} of rank 16 group $G \times G \subset SO(16) \times SO(16) \; (\subset E(8) \times E(8))$. This construction allows to break the GUST symmetry to the low energy gauge group, which includes the G_H family gauge symmetry. The virtue of the $GUST^{symm}$ considered is that its low-energy spectrum does not contain particles with exotic charges. This GUST which is based on the maximal invariant $SO(16)$ subgroup of E(8) leads to the $3_H + 1_H$, therefore it will be interesting to consider GUST construction which is based on the $E(6) \times SU(3)_H \subset E(8)$ gauge symmetry *[16]* and which predicts only three quark–lepton families.

The ability to correctly describe the fermion masses and mixings will, of course, connstitute the decisive criterion for selection of a model of this kind. Therefore, within our approach one has to

1. study the possible nature of the G_H horizontal gauge symmetry ($N_g = 3_H$ or $3_H + 1_H$),

2. investigate the possible cases for G_H-quantum numbers for quarks (anti-quarks) and leptons (anti-leptons), i.e. whether one can obtain vector-like or axial-like structure (or even chiral $G_{HL} \times G_{HR}$ structure) for the horizontal interactions.

3. the structure of the sector of the matter fields which are needed for the $SU(3)_H$ anomaly cancelation (chiral neutral "horizontal" or "mirror" fermions),

4. write down all possible renormalizable and relevant non-renormalizable contributions to the superpotential W and their consequences for fermion mass matrices.

All these questions are currently under investigation *[16]*. Here we restrict ourselves to some general remarks only.

With the chiral matter and "horizontal" Higgs fields available in the model constructed in this paper, the possible form of the renormalizable (trilinear) part of the superpotential responsible for fermion mass matrices is well restricted by the gauge symmetry:

$$W_1 = g\sqrt{2}\left[\hat{\Psi}_{(1,3)}\hat{\Psi}_{(\bar{5},1)}\hat{h}_{(5,\bar{3})} + \hat{\Psi}_{(1,1)}\hat{\Psi}_{(\bar{5},3)}\hat{h}_{(5,\bar{3})} + \right.$$

$$\left. +\hat{\Psi}_{(10,3)}\hat{\Psi}_{(\bar{5},3)}\hat{h}_{(\bar{5},3)} + \hat{\Psi}_{(10,3)}\hat{\Psi}_{(10,1)}\hat{h}_{(5,\bar{3})}\right] \tag{32}$$

From the above form of the Yukawa couplings follows that two (chiral) generations 27 have to be very light (comparing to M_W scale). The construction of realistic quarks and leptons mass matrices depends, of course, on the nature of the horizontal interactions. In the construction described in Sec.3 there is a freedom of choosing spin boundary conditions for $N_{LR}=12$ left and right fermions in the basis vectors b_3, b_5, b_6,..., which in the Ramond sector $2b_2$, may yield another Higgs fields, denoted as $\tilde{h}_{(\Gamma,N)}$ and transforming as $(\underline{5},\underline{3})_{(-1,1)} + (\underline{\bar{5}},\underline{\bar{3}})_{(1,-1)} \subset \underline{28} + \underline{\bar{28}}$ of $SU(8)$. Using these Higgs fields we get the following alternative form of the renormalizable part of the superpotential W:

$$W'_1 = g\sqrt{2}\left[\hat{\Psi}_{(1,3)}\hat{\Psi}_{(\bar{5},3)}\tilde{h}_{(5,3)} + \hat{\Psi}_{(10,1)}\hat{\Psi}_{(\bar{5},3)}\tilde{h}_{(\bar{5},3)} + \right.$$

$$\left. +\hat{\Psi}_{(10,3)}\hat{\Psi}_{(10,3)}\tilde{h}_{(5,3)} + \hat{\Psi}_{(10,3)}\hat{\Psi}_{(\bar{5},1)}\tilde{h}_{(\bar{5},3)}\right] \tag{33}$$

To construct the realistic fermion mass matrices one has to use also the Higgs fields 19,20 and (Table 3, No 5) and also to take into account all relevant non-renormalizable contributions *[10]*.

The Higgs fields 19 can be used for constructing Yukawa couplings of the horizontal superfields (No 3 and 4). The most general contribution of these fields to the superpotential is:

$$W_2 = g\sqrt{2}\left[\hat{\Phi}^H_{(1,1;1,\bar{3})}\hat{\Phi}^H_{(1,\bar{3};1,1)}\hat{\Phi}_{(1,3;1,3)} + \hat{\Phi}^H_{(1,1;1,1)}\hat{\Phi}^H_{(1,\bar{3};1,\bar{3})}\hat{\Phi}_{(1,3;1,3)} + \right.$$

$$+\hat{\Phi}^H_{(1,\bar{3};1,\bar{3})}\hat{\Phi}^H_{(1,\bar{3};1,\bar{3})}\hat{\Phi}_{(1,\bar{3};1,\bar{3})} + \hat{\Psi}^H_{(1,3;1,1)}\hat{\Psi}^H_{(1,3;1,\bar{3})}\hat{\Phi}_{(1,3;1,3)} +$$

$$\left. +\hat{\Psi}^H_{(1,1;1,\bar{3})}\hat{\Psi}^H_{(1,3;1,\bar{3})}\hat{\Phi}_{(1,\bar{3};1,\bar{3})}\right] \tag{34}$$

From this expression it follows that some of the horizontal fields in 28 (No 3, 4) remain massless at the tree-level. This is a remarkable prediction: fields 28 interact with the ordinary chiral matter fields only through the $U(1_H)$ and $SU(3_H)$ gauge boson and therefore are very interesting in the context of the experimental searches for the new gauge bosons.

The superfields $\hat{\Phi}^H_{(1,3;1,1)}$ and $\hat{\Psi}^H_{(1,\bar{3};1,1)}$ (see No 3, 4) can be used to construct the non-renormalizable contributions to the superpotential W. For example, the term

$$\Delta W_1 = \frac{cg^3}{M_{Pl}^2} \Psi_{(10,1)} \Psi_{(10,1)} \Phi_{(1,3;5,1)} \hat{\Phi}^H_{(1,3;1,1)} \hat{\Phi}^H_{(1,3;1,1)} \tag{35}$$

can give contribution to the mass to the fourth generation down–type quark ($c = \mathcal{O}(1)$, see [10]). To get a reasonable value of the mass for this quark we must arrange for the $SU(3_H)^I$ gauge symmetry breaking at the energy scale near the Plank scale, i.e. $< \hat{\Phi}^H_{(1,3;1,1)} > = < \hat{\Phi}^H_{(1,\bar{3};1,1)} >\sim M_{Pl}$. In this case one can get the $SU(3_H)^{II}$ -family gauge group with a low energy breaking symmetry scale. Finally, we remark that the Higgs sector of our GUST allows for conservation of the G_H gauge family symmetry down to the low energies ($\sim \mathcal{O}(1TeV)$ [15]). Thus we can expect at this energy region new interesting physics (new gauge bosons, new chiral matter fermions, superweak-like CP–violation in K,- B,- D-meson decays with $\delta_{KM} < 10^{-4}$ [15]).

ACKNOWLEDGMENTS.   One of us (G.V) would like to thank INFN for financial support and the staff of the Physics Departments in Padova and Trieste, especially, Profs. C. Voci and N. Paver, for warm hospitality during his stay in Padova. It is a great pleasure for him to thank Profs. G. Costa, L. Fellin, J. Ellis, A. Masiero, D. Nanopoulos, N. Paver and M. Tonin for useful discussions and for the help and support. Many interesting discussions and help from P. Chankowski and S. Sergeev during this work are also acknowledged. Finally, he would like to express his gratitude to Prof. P. Drigo and her colleagues of the Medical School of the University of Padova.

This publication was made possible in part by travel grants No 1103/1 and 1103/3 and, by Grant No RMP000 from the International Science Foundation.

References

[1]   M.B.Green and J.H.Schwarz *Phys.Lett.B* **149** (1984),117.

[2]   D.J.Gross, J.A.Harvey, E.Martinec and R.Rohm, *Nucl.Phys.* **B256** (1985), 253.

[3]   P.Candelas, G.T.Horowitz, A.Strominger and E.Witten, *Nucl.Phys.* **B258** (1985), 46.

[4]   E.Witten, *Nucl.Phys.* **B268** (1986), 79.

[5]   K.S. Narain, *Phys.Lett* **B169** (1986), 246.
      W. Lerche, D. Lust, A.N. Schellekens, *Nucl.Phys.* **B287** (1987), 477.

[6]   H.Kawai, D. Lewellen and S.-H.H.Tye, *Phys.Rev.Lett.* **57** (1986),1832; *Nucl.Phys.* **288B** (1987) 1.

[7]   I.Antoniadis, C.Bachas, C.Kounnas, *Nucl.Phys.* **B289** (1988), 87.
      I.Antoniadis, C.Bachas, *Nucl.Phys.* **B298** (1988), 586.

[8]   P.Goddard, and D.Olive, *Int.J.Mod.Phys.* **A1** (1986), 1.

[9]   L. Alvarez-Gaume, G.Sierra and C.Gomez, *"Topics in Conformal Field Theory", in "Physics and Mathematics of Strings"* **World Scientific, Singapore** (1990), 16-184.

[10]   I.Antoniadis, J.Ellis, J.S.Hagelin and D.V.Nanopoulos, *Phys.Lett.* **B194** (1987), 231;**B208** (1988),209;
       J.Ellis, J.I. Lopez, D.V. Nanopoulos, *Phys.Lett.* **B245** (1990), 375.
       S.Kalara, J.I. Lopez, D.V. Nanopoulos, *Nucl.Phys.* **B353** (1991), 650.

[11] P. Langacker, *Phys.Rep.* **72** (1981),185 - 385.

[12] E.B. Vinberg, A.L. Onichik, *"Seminar on Group Lee"*, in Russian **Nauka, Moscow** (1988),
R. Slansky, *Phys.Rep.* **79** (1981),1–128.

[13] S.M. Sergeev and G.G. Volkov, **preprint DFPD/TH/51** (1992),*Rus. Nucl. Phys.* **v.57** (1994), No 1;

[14] A. Schellekens, *Phys. Lett.* **B237** (1990), 363;

[15] A.N.Amaglobeli, A.R.Kereselidze, A.G.Liparteliani, G.G.Volkov, *Phys. Lett.* **B237** (1990), 417;
A.A.Maslikov and G.G.Volkov, *Towards a Low Energy Quark-Lepton Family with Non-Abelian Gauge Symmetry*, **ENSLAPP-A-382/92**, (1992),
V.A.Monich, B.V.Struminsky, G.G.Volkov, *Phys. Lett.* **B104** (1981), 382.

[16] G.G.Volkov in preparation.

Approximate Flavor Symmetries

ANDRIJA RAŠIN [1] [2]
DEPARTMENT OF PHYSICS
UNIVERSITY OF CALIFORNIA, BERKELEY, CA 94720
AND THEORETICAL PHYSICS GROUP, LAWRENCE BERKELEY LABORATORY
UNIVERSITY OF CALIFORNIA, BERKELEY, CA 94720
LBL-35530;UCB-PTH-94/09

Abstract: We discuss the idea of approximate flavor symmetries. Relations
between approximate flavor symmetries and natural flavor conservation
and democracy models is explored. Implications for neutrino physics are
also discussed.

1 Introduction

In the Standard Model, Yukawa couplings λ are defined as couplings between
fermions and Higgs scalars:

$$\mathcal{L}_Y = (\lambda_{ij}^{U^a} \bar{Q}_i U_j \frac{\tilde{H}_a}{\sqrt{2}} + \lambda_{ij}^{D^a} \bar{Q}_i D_j \frac{H_a}{\sqrt{2}} + \lambda_{ij}^{E^a} \bar{L}_i E_j \frac{H_a}{\sqrt{2}} + h.c.), \tag{1}$$

where Q_i and L_i are $SU(2)$ doublet quarks and leptons while U_i , D_i and E_i
are $SU(2)$ singlets and $i = 1, 2, 3$ is a generation label. H_a are Higgs SU(2)
doublets, and $a = 1, \ldots, n$, where n is the number of Higgs doublets.

The Standard Model can easily be extended to accommodate neutrino
masses. We can introduce higher dimensional operators $\frac{1}{M} H H L_i L_j$ which give
Majorana neutrino masses. We can also add SU(2) singlet neutrinos which can
also have Majorana masses or Dirac masses similar to (1). Neutrino mass is
then in general described by a 6×6 Yukawa matrix.

As opposed to, e.g. couplings of fermions and vector bosons, Yukawa couplings in the Standard Model are not constrained nor related to each other
by any symmetry principle; i.e., they enter the Lagrangian as arbitrary complex numbers. These numbers are then only fixed by experiment, namely by
measuring fermion masses and mixing angles.

Several classes of models/ansätze for Yukawa couplings exist today, and we
list them below:

- Approximate flavor symmetries[1]-[5], in which the entries in the Yukawa
matrices are entered as small parameters by which the flavor symmetries are
broken.
- Fritzsch and/or GUT inspired models[6]-[17], in which some entries in the
Yukawa mass matrices are assumed to be zero (e.g. by discrete flavor symmetry), and others may be related by some GUT relation.

---

[1]This work was supported in part by the Director, Office of Energy Research, Office of
High Energy and Nuclear Physics, Division of High Energy Physics of the U.S. Department
of Energy under Contract DE-AC03-76SF00098.

[2]On leave of absence from the Ruder Bošković Institute, Zagreb, Croatia. Address after
September 1, 1994: Department of Physics, University of Maryland, College Park 20742.

- Flavor democracy models[18]-[19], in which all the entries in the Yukawa matrices are equal (i.e., no flavor symmetry), and hierarchy comes from diagonalization and RGE running.
- String inspired models, composite models, ...

This is in no way a complete list, of course. Also, most of the work done actually falls into several categories above, showing that there are common ideas, which will hopefully lead us to a certain trail beyond the Standard Model.

In this talk, we will concentrate on approximate flavor symmetries, since they are relatively simple and model independent.

2 Approximate flavor symmetries

In the Standard Model, the gauge interactions of the fermions

$$\mathcal{L}_0 = i\bar{Q}\not{D}Q + i\bar{U}\not{D}U + i\bar{D}\not{D}D + i\bar{L}\not{D}L + i\bar{E}\not{D}E \tag{2}$$

have global flavor symmetries and these are broken by Yukawa couplings (1). We can understand the Yukawa couplings as being naturally small[20], because in the limit that they become zero, the theory gets a larger (flavor) symmetry. This is certainly warranted by the smallness of fermion masses (except the top), which arise from Yukawa couplings.

One of the simplest assumptions one can make is that each of the chiral fermion fields carries a flavor symmetry which is broken by a small parameter, which we will call ϵ. For example, Froggatt and Nielsen[1] think of ϵ as $\epsilon \approx \frac{<\Phi_1>}{<\Phi_0>}$, where $< \Phi_1 >$ is the vev of the scalar that breaks the flavor symmetry, and $< \Phi_0 >$ is the vev of a superheavy field, therefore making ϵ small. Thus, for example

$$\lambda_{ij}^Q \approx \epsilon_{Q_i}\epsilon_{U_j} \tag{3}$$

where ϵ_{Q_i} is the breaking parameter of the flavor carried by Q_i, and ϵ_{U_j} is the breaking parameter of the flavor carried by U_j. Froggatt and Nielsen[1], as well as Leurer, Nir and Seiberg[5] use one or two ϵs, which enter the Yukawa matrices with different powers, thus explaining the hierarchy of masses. We rather keep different ϵs for different flavors, as it keeps the discussion more model independent. Notice that in (3) we assumed that the Higgs field does not carry any flavor symmetry. If it did, then the Yukawa couplings would be multiplied by another ϵ for the broken symmetry carried by the Higgs field.

What do we know about ϵs? Although their approximate value can be fixed by the known masses and mixings (at least in the quark sector[3]; lepton sector is much more speculative as the neutrino masses and mixings are not known[4]; we will talk more on this later), their exact values depend on the underlying theory, which we do not know. Therefore, relation like (3) is not meant to be an exact relation but rather an order of magnitude estimate, and we are interested here mainly in general features, rather than specific predictions. It was shown that flavor changing interactions, with couplings determined by approximate flavor symmetries, can involve new scalars at the scale as low as the weak scale, and still satisfy stringent experimental limits([2],[3]). This is opposed to the common view that, for example, $K_L - K_S$ mass difference, implies high bounds on the scale of new interactions, typically $\sim 1000\,TeV$. However, it is precisely because of the approximate flavor symmetries that the couplings of the new scalars are small, lowering the naive bound considerably.

3   Many Higgs doublet model

In this section we discuss the case of the minimal Standard Model extended only by the addition of an arbitrary number of Higgs doublets. In this case it is already known that, for the special case of Fritzsch-like Yukawa matrices, the additional scalars need not be heavier than a TeV [8]. However, our results are independent of the particular texture and depend only on the approximate flavor symmetry.

Let us look at a two Higgs doublet model (the generalization to many Higgs doublets is trivial). For example, the up quark Yukawa couplings are

$$
\left(\bar{Q}_1 \ \bar{Q}_2 \ \bar{Q}_3 \right)
\begin{pmatrix}
\sim \epsilon_{Q_1}\epsilon_{U_1} & \sim \epsilon_{Q_1}\epsilon_{U_2} & \sim \epsilon_{Q_1}\epsilon_{U_3} \\
\sim \epsilon_{Q_2}\epsilon_{U_1} & \sim \epsilon_{Q_2}\epsilon_{U_2} & \sim \epsilon_{Q_2}\epsilon_{U_3} \\
\sim \epsilon_{Q_3}\epsilon_{U_1} & \sim \epsilon_{Q_3}\epsilon_{U_2} & \sim \epsilon_{Q_3}\epsilon_{U_3}
\end{pmatrix}
\begin{pmatrix} U_1 \\ U_2 \\ U_3 \end{pmatrix} \frac{\tilde{H}_1}{\sqrt{2}}
$$

$$
+ \left(\bar{Q}_1 \ \bar{Q}_2 \ \bar{Q}_3 \right)
\begin{pmatrix}
\sim \epsilon_{Q_1}\epsilon_{U_1} & \sim \epsilon_{Q_1}\epsilon_{U_2} & \sim \epsilon_{Q_1}\epsilon_{U_3} \\
\sim \epsilon_{Q_2}\epsilon_{U_1} & \sim \epsilon_{Q_2}\epsilon_{U_2} & \sim \epsilon_{Q_2}\epsilon_{U_3} \\
\sim \epsilon_{Q_3}\epsilon_{U_1} & \sim \epsilon_{Q_3}\epsilon_{U_2} & \sim \epsilon_{Q_3}\epsilon_{U_3}
\end{pmatrix}
\begin{pmatrix} U_1 \\ U_2 \\ U_3 \end{pmatrix} \frac{\tilde{H}_2}{\sqrt{2}}
$$

and similarly for down type quark matrices, keeping in mind that each entry in the matrices is uncertain by a factor of 2 or 3 (denoted by $\sim$).

Notice that because of the numerical factors in front of the ϵs, the matrices for H_1 and H_2 are not equal in general. That means that if we diagonalize the matrix of H_1, the matrix of H_2 will not be diagonalized and will keep the same general form as above. In particular, we can always choose H_1 to be the only doublet that acquires a vev (rotating the doublets will not change the above form of matrices). Therefore we see that in the quark mass eigenstate basis, the new Higgs H_2 couplings are not diagonal: we have flavor changing couplings. The nice thing now is that since the flavor changing couplings are small the stringent experimental limits on flavor changing neutral currents (FCNC) actually translate only into lower limits on the mass of new scalars, about 1 TeV, as discussed above.

To avoid problems with large flavor-changing neutral currents, Glashow and Weinberg [21] argued that only one Higgs doublet could couple to up-type quarks and only one Higgs to down-type quarks. However, this naturality constraint, known as the Glashow-Weinberg criterion (or natural flavor conservation (NFC)), was based on an unusual definition of what is "natural." For them the avoidance of flavor-changing neutral currents was natural in a model only if it occured for all values of the coupling constants of that model. For us a model will be natural provided the smallness of any coupling is guaranteed by approximate symmetries [20], and we find that this implies the Glashow-Weinberg criterion is not necessary (however, see caveat below).

One potential problem arises from the smallness of the observed CP violation, as noted by Hall and Weinberg[3]. The CP violating parameter $\epsilon_{CP} = Im(\Delta M_K)/\sqrt{2}|\Delta M_K|$ would naively be expected to be of order unity in our case (we have no reason to assume a priori that the Yukawa couplings a real), contrary to the observed value of 10^{-3}. To avoid this problem one might go

back to NFC, or, more in the philosophy of naturalness of small couplings, just
say that CP is another approximately conserved quantity broken by ϵ_{CP}.

Here we would like to mention two different limits of the Yukawa matrices
which obey approximate flavor symmetries (as in (3)), one of which gives NFC,
and the other one which gives democratic matrices.

Notice that if the matrix for H_2 is nearly equal to the matrix for H_1, then
diagonalization of the H_1 matrix will almost diagonalize the second matrix. In
this case the flavor changing couplings become even smaller. This is of course
no surprise, because if the two matrices were exactly equal then only one linear
combination of Higgses ($H_1 + H_2$) couples to the quarks: we have NFC! This is
actually the starting point of Leurer, Nir and Seiberg [5]; i.e., use broken flavor
symmetries in combination with weakly broken NFC.

Also notice that if the Yukawa matrix elements are *exactly* a product of ϵs
by which the symmetries are broken, then the matrices have one large eigenvalue
and two eigenvalues equal to zero: we have flavor democracy([18],[19])! This
is easily understood, since this limit means that only one linear combination of
the left handed fields couples to one linear combination of right handed fields
(while the other two combinations remain massles).

None of these limits follow from the idea of approximate flavor symmetries
alone. Unless they are motivated by a specific model, we must stay with the
general relation of type (3).

4 Lepton sector

By adding the right-handed neutrinos N_i, $i = 1, 2, 3$, to the particle content of
the Standard Model we can allow for Dirac type masses. Under the action of
approximate flavor symmetries, whenever an N_i enters a Yukawa interaction,
the corresponding coupling must contain the symmetry breaking parameter ϵ_{N_i}.

A natural way to justify the smallness of neutrino masses is to use the *see-
saw mechanism[22]*, in which the smallness of the left-handed neutrino masses is
explained by the new scale of heavy right-handed neutrinos. The mass matrices
will have the structure*[4]*

$$m_{N_{D_{ij}}} \approx \epsilon_{L_i} \epsilon_{N_j} v_{SM} , \tag{4}$$

$$m_{N_{M_{ij}}} \approx \epsilon_{N_i} \epsilon_{N_j} v_{Big} , \tag{5}$$

$$m_{E_{ij}} \approx \epsilon_{L_i} \epsilon_{E_j} v_{SM} , \tag{6}$$

where m_{N_D} and m_E are the neutrino and charged lepton Dirac mass matrices,
m_{N_M} is the right-handed neutrino Majorana mass matrix, $v_{SM} = 174\,GeV$ and
v_{Big} is the new large mass scale. The generation indices i and j run from 1 to
3. In the following we assume a hierarchy in the ϵs (i.e. $\epsilon_{L_1} \ll \epsilon_{L_2} \ll \epsilon_{L_3}$,
etc.) as suggested by the hierarchy of quark and charged lepton masses. Then
the diagonalization of the neutrino mass matrix will give a heavy sector with

masses $m_{N_{Hi}} \approx \epsilon_{N_i}^2 v_{Big}$ and a very light sector with mass matrix

$$m_{N_{Lij}} \approx (m_{N_D} m_{N_M}^{-1} m_{N_D}^T)_{ij} \approx \epsilon_{L_i} \epsilon_{L_j} \frac{v_{SM}^2}{v_{Big}} , \qquad (7)$$

where the number $Tr(\epsilon_N(\epsilon_N\epsilon_N)^{-1}\epsilon_N)$ is assumed to be of order unity. We have the expected result: the heavy right-handed neutrino decouples from the theory leaving behind a very light left-handed neutrino. The masses and mixing angles are independent of the right-handed symmetry breaking parameters ϵ_{N_i}:

$$m_i^N \approx \epsilon_{L_i}^2 \frac{v_{SM}^2}{v_{Big}} ,$$
$$m_i^E \approx \epsilon_{L_i}\epsilon_{E_i} v_{SM} \quad (no\ sum\ on\ i) ,$$
$$V_{ij} \approx \frac{\epsilon_{L_i}}{\epsilon_{L_j}} \quad (i < j) . \qquad (8)$$

Therefore, besides the unknown scale v_{Big}, only two sets of ϵs are needed: ϵ_{L_i} and ϵ_{E_i}. In fact, the neutrino masses and mixings depend only on ϵ_{L_i} and they are approximately related through

$$V_{ij} \approx \sqrt{\frac{m_i^N}{m_j^N}} . \qquad (9)$$

Equation (9) reduces the number of parameters needed to describe neutrino masses and mixings by three; for example, given two mixing angles and one neutrino mass, we can predict the third mixing angle and the other two neutrino masses. These results are extremely general. They follow simply from the approximate factorization of the Dirac masses, regardless of the specific form of $m_{N_M}^{-1}$, which only contributes to set the scale.

To get further relations one needs some additional information about the ϵ_Ls and ϵ_Es. We tried several plausible ansätze[4] and found that in all of them the solar neutrino problem (SNP) can easily be accomodated with MSW[23] $\nu_e - \nu_\mu$ mixing solution. However, if we now fix the mass scale from the SNP solution (requiring m_{ν_μ} to be about $10^{-3} eV$), then all neutrino masses are too small to close the Universe. This comes about because the approximate flavor symmetries tell us that the ratios of neutrino masses are likely to be of the order of ratios of charged lepton masses (and therefore, the heaviest neutrino, ν_τ, is not likely to be heavier than $1 eV$), as opposed to some proposed quadratic relations.

5   Acknowledgement

This work was supported in part by the Director, Office of Energy Research, Office of High Energy and Nuclear Physics, Division of High Energy Physics of the U.S. Department of Energy under Contract DE-AC03-76SF00098.

References

[1]   C. D. Froggat and H. B. Nielsen, Nucl. Phys. **B147**, 277 (1979); Nucl. Phys. **B164**, 114 (1980).

[2] A. Antaramian, L. J. Hall and A. Rašin, Phys. Rev. Lett. **69**, 1871 (1992).

[3] L. Hall and S. Weinberg, Phys. Rev. **D48**, 979 (1993).

[4] A. Rašin and J. P. Silva, Phys. Rev. **D49**, 20 (1994).

[5] M. Leurer, Y. Nir and N. Seiberg, Nucl. Phys. **B398**, 319 (1993); Rutgers University preprint No. RU(93-43) (1993) (bulletin board No. 9310320); P. Pouliot and N. Seiberg, Phys. Lett. **318B**, 169 (1993).

[6] H. Fritzsch, Phys. Lett. **73B**, 317 (1978); Nucl. Phys. **B155**, 189 (1979).

[7] S. Dimopoulos, Phys. Lett. **129B**, 417 (1983).

[8] T. P. Cheng and M. Sher, Phys. Rev. **D35**, 3484 (1987).

[9] H. Georgi and C. Jarlskog, Phys. Lett. **86B**, 297 (1979).

[10] J. A. Harvey, P. Ramond and D.B. Reiss, Phys. Lett. **92B**, 309 (1989); Nucl. Phys. **B199**, 223 (1982).

[11] G. Giudice, Mod. Phys. Lett. **A7**, 2429 (1992).

[12] S. Dimopoulos, L. J. Hall and S. Raby, Phys. Rev. Lett. **68**, 1984 (1992); Phys. Rev. **D45**, 4192 (1992).

[13] P. Ramond, UFIFT-92-4 (1992); H. Arason *et al.*, Phys. Rev. **D47**, 232 (1993).

[14] G. Anderson *et al.*, Phys. Rev. **D47**, 3702 (1993); V. Barger *et al.*, Phys. Rev. Lett. **68**, 3394 (1992); V. Barger, M. S. Berger and P. Ohmann Phys. Rev. **D47**, 1093 (1993).

[15] K. S. Babu and Q. Shafi, Phys. Lett. **311B**, 172 (1993); A. Kusenko and R. Shrock, Institute for Theoretical Physics, U.C. Santa Barbara preprint No. ITP-SB-93-37 (1993) (bulletin board No. 9307344).

[16] G. Anderson *et al.*, Phys. Rev. **D49**, 3660 (1994).

[17] D. B. Kaplan and M. Schmaltz, Phys. Rev. **D49**, 3741 (1994).

[18] H. Harari, H. Haut and J. Weyers, Phys. Lett. **78B**, 459 (1978); H. Fritzsch and J. Plankl, Phys. Lett. **237B**, 451 (1990); M. Tanimoto, Mod. Phys. Lett. **A8**, 1387 (1993); P. Kaus and S. Meshkov, Phys. Rev. **D42**, 1863 (1990); P. M. Fishbane and P. Q. Hung, Phys. Rev. **D45**, 293 (1992); G. Cvetič and C. S. Kim, Nucl. Phys. **B407**, 290 (1993).

[19] See also talks by H. Fritzsch and P. Kaus, this workshop.

[20] G. 't Hooft, in *Recent Developments in Gauge Theories*, Cargese Summer Institute Lectures, 1979, edited by G. 't Hooft *et al.* (Plenum, New York, 1980).

[21] S. L. Glashow and S. Weinberg, Phys. Rev. **D15**, 1958 (1977).

[22] M. Gell-Mann, P. Ramond and R. Slansky, in *Supergravity*, edited by P. van Nieuwenhuizen and D. Freedman (North-Holland, Amsterdam, 1979); T. Yanagida, in *Proceedings of the Workshop on Unified Theories and Baryon Number in the Universe*, edited by A. Sawada and A. Sugamoto (KEK, Tsukuba, 1979); R. Mohapatra and G. Senjanović, Phys. Rev. Lett. **44**, 912 (1980); Phys. Rev. **D23**, 165 (1981).

[23] S. P. Mikheyev and A. Y. Smirnov, Nuovo Cimento **9C**, 17 (1986); L. Wolfenstein, Phys. Rev. **D17**, 2369 (1978).

Nonperturbative QCD Effects in Inclusive B Decays

SOO-JONG REY

PHYSICS DEPARTMENT & CENTER FOR THEORETICAL PHYSICS
SEOUL NATIONAL UNIVERSITY, SEOUL 151-742 KOREA
SJREY@PHYB.SNU.AC.KR

Abstract: In this talk I report calculable nonperturbative QCD effects to the inclusive semileptonic and radiative B-meson decays. Small instantons contribute to the Wilson coefficient function below m_b scale. It is found that the instantons give rise to a potentially large correction to the decay rate near the boundaries of the phase space for the semileptonic decay case, but totally negligible for the radiative $B \to X_s \gamma$ case. I explain how the difference is a simple consequence of different kinematics of the two cases.

1   Introduction

The decay of heavy flavors is an important probe to the fundamental parameters in the Standard Model and to new physics beyond the electroweak scale. Of particular interests are inclusive semileptonic and radiative decays of hadrons containing a single b quark. From the energy spectrum of charged lepton in $B \to X_q \ell \overline{\nu}_\ell$ decay one may extract the heavy quark masses m_b, m_c and the Cabbibo-Kobayashi-Maskawa matrix elements V_{qb} where $q = c, u$. From the $B \to X_s \gamma$ decay one may extract the indirect contribution of heavy particles to the decay rate through the penguin and Higgs exchange diagrams. Traditional theoretical description of these inclusive decays was based on models such as ACCMM model [1] or ISGW model [2].

Recently Chay, Georgi and Grinstein [3] have made an important theoretical progress to the model-independent approach of the inclusive B decays (See also Ref. [4].). Using the heavy quark effective field theoy(HQEFT) and the operator product expansion(OPE) Chay et.al. have shown that the weak decay of B mesons can be described by a systematic expansion in powers of Λ/m_b, in which the leading contribution reproduces the parton-model result.

Using the OPE, Chay et.al. have shown that next-order corrections are of order $(\Lambda/m_b)^2$, where Λ is the QCD scale, hence expected to be small. Mannel [5], Manohar and Wise [6] have analyzed these corrections and concluded that these corrections are singular at the boundaries of the Dalitz plot, which necessitates some smearing prescription before comparing with experimental data.

Experimentally it is necessary to make kinematical cuts to suppress the background. In order to extract V_{ub} from the shape of the electron energy spectrum, one has to concentrate to the endpoint region beyond the charm threshold. The available endpoint energy range is $\mathcal{O}(300 MeV)$, thus the final hadron is close to the resonance region of light hadrons. To extract a signal of new physics from the inclusive radiatve B meson decay, one has to concentrate to energetic photons near the kinematical endpoint beyond the threshold of

$K^* \to K\gamma$ secondary decay. Again this is the resonance region of the final light hadrons. In both cases the endpoint region is precisely the region at which theoretical prediction is least understood due to various low-energy QCD effects.

In this talk, I report theoretical results on the calculable nonperturbative QCD effects to the inclusive semileptonic and radiative B meson decays. In particular we calculate the effect of small QCD instantons to these decay rates [7, 8]. In section 2, theoretical framework of calculating the decay rate is summarized using the heavy quark effective field theory and the operator product expansion. In section 3, the instanton corrections are calculated. Size of the correction is different for each decay processes. In section 4, the origin of this difference is explained in terms of the kinematics and the analytic structure of the forward Compton scattering amplitude. In section 5, brief discussion of measuring the hadron energy spectra is made.

2 Operator Product Expansion for the Decay Rate

The basic idea of Chay et.al. is that the suitably averaged physical quantities may be calculated reliably using the perturbation theory. inclusive decay rate can be calsets the first scale in the problem which is far above the QCD scale. The other is the momentum transfer to the final hadron.

The effective Hamiltonian of weak B meson decay is obtained from integrating out the massive W-boson, Higgs and heavy fermions at the weak scale and scale down to $\mu = m_b$ using the renormalization group equation

$$H_{eff} = \sum_i c_i(m_b)\mathcal{O}_i(m_b) \tag{1}$$

The relevant local operators for $B \to X_u l\nu_l$ and $B \to X_s\gamma$ are [9, 10]

$$\mathcal{O} = \bar{u}_L\gamma^\mu b_L \bar{\nu}_L\gamma_\mu \ell_L$$
$$\mathcal{O} = \frac{e}{16\pi^2}F_{\mu\nu}\bar{s}\sigma^{\mu\nu}(m_b P_+ + m_s P_-)b \tag{2}$$

respectively. The decay rate of $B \to X + f$ (f denotes the non-nonhadronic final particles) is

$$d\Gamma = \frac{1}{2m_b}\sum_X (2\pi)^4\delta^{(4)}(p_B - p_X - q)|\langle X, f|c(\mu)\,\mathcal{O}(\mu)|B\rangle|^2 d(PS), \tag{3}$$

where $q = \sum_f p_f$ and

$$d(PS) = \prod_f \frac{d^3\vec{k}_f}{(2\pi)^3 2E_f}$$

$$= \frac{m_b^4}{2^7\pi^4}dv \cdot \hat{q}\,dy\,d\hat{q}^2 \tag{4}$$

$$= \frac{m_b^2}{4\pi^2}v \cdot \hat{q}\,d(v \cdot \hat{q}) \tag{5}$$

for semi-leptonic and radiative decays respectively. We have introduced dimensionless kinematic variables $\hat{q} = q/m_b$, $y = 2E_e/m_b$ and $\hat{q}^2 = q^2/m_b^2$.

In the matrix element-squared in Eq.(3) the non-hadronic part L is trivially factorized. The remaining hadronic matrix elements-squared W may be expressed in terms of the forward Compton scattering amplitude using the optical theorem. Thus

$$W(B \to X) = 2 Im T(B \to X \to B) \tag{6}$$

where the forward Compton scattering amplitude is

$$T = -i \int d^4 x e^{-iq \cdot x} \langle B | T\{J^\dagger(x) J(0)\} | B \rangle \tag{7}$$

To evaluate the decay rate using the optical theorem it is necessasry to examine the analytic structure of $T(q, v)$ [3]. In the complex $v \cdot \hat{q}$ plane, the analytic structure for both semi-leptonic and radiative decays are similar. For $B \to X \ell \bar{\nu}_\ell$ the "physical cut" relevant to the decay is located on the real axis $v \cdot \hat{q} \leq (1 + \hat{q}^2 - \rho)/2$ where $\rho \equiv m_u^2 / m_b^2$ is the mass ratio-squared of the u quark to the b quark. For $B \to X_s \gamma$ the "physical cut" is located on the real axis $v \cdot \hat{q} \leq 1 - r$ where $r = m_s^2 / m_b^2$. There are also another cut $v \cdot \hat{q} \geq (3 - \hat{q}^2)/2$ or $(2 + \sqrt{r})^2 - 1$ corresponding to other physical processes. The analytic structures are shown in Fig. 1 for both cases.

The double differential decay rate after integrating over $v \cdot \hat{q}$ is given by

$$\begin{aligned}
\frac{d^2\Gamma}{d\hat{q}^2 \, dy} &= \frac{G_F^2 m_b^5}{2^9 \pi^4} |V_{ub}|^2 \int_{\frac{y}{2} + \frac{\hat{q}^2}{2y}}^{\frac{1 + \hat{q}^2}{2}} d\hat{v} \cdot q \, \frac{W \cdot L}{m_b^2} \\
&= -\frac{G_F^2 m_b^5}{2^8 \pi^4} |V_{ub}|^2 \int_C dv \cdot \hat{q} \, \frac{T \cdot L}{m_b^2}
\end{aligned} \tag{8}$$

In the first equation the differential decay rate is evaluated from the discontinuity of $T \cdot L$ along the contour C' in Fig.1. In the second equation this is related to the contour integral along the contour C in the complex $v \cdot \hat{q}$ plane using the Cauchy theorem. The contour C is shown in Fig.1.

Near the boundaries of the Dalitz plot $\hat{q}^2 \to y$ or $y \to 1$ for semi-leptonic decay and $y_c = 1 - r$ for radiative decay, z shrinks to zero and we are in the resonance region. There we expect significant nonperturbative QCD effects. This limits the extent of our theoretical prediction of the decay rate to which the experimental data can be compared. Therefore it is necessary to introduce a smearing in order to use perturbation theory reliably [12]. The size of the smearing is determined by requiring that the nonperturbative effects be smaller than the parton-model result. Chay et.al. have observed that one can calculate the decay rate even in this limit perturbatively if it is possible to choose the integration contour away from the resonance region. The integral along C is reliably evaluated using perturbative QCD as long as $(m_b v - q)^2 \gg \Lambda^2$.

We have chosen the contour C in Eq. (8) as a circle of radius $z = (y - \hat{q}^2)(1 - y)/y$ centered at $v \cdot \hat{q} = (1 + \hat{q}^2)/2$ for $B \to X_u \ell \bar{\nu}_\ell$ and a circle of radius

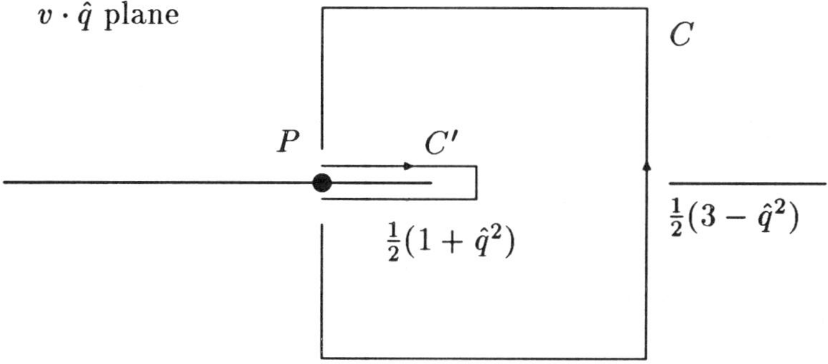

Figure 2.1. Analytic structure of T relevant to the semileptonic B decay.

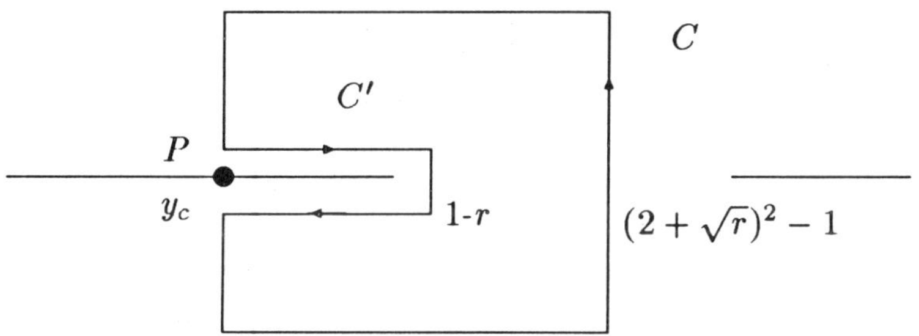

Figure 2.2. Analytic structure of T relevant to the radiative B decay

$z = 2(1-r-y_c)$ centered at $v\cdot\hat{q} = 2(1-r)$ for $B \to X_s\gamma$. The soft photon cutoff $2y_c$ is necessary experimentally to reduce the background. The contribution of the contour integral near the point P is negligibly small as long as $z \gg \Lambda/m_b$.

For the moment we consider $q^2 \ll m_b^2$ but $(m_b v - q)^2 \gg \Lambda^2$ in order to calculate the decay distribution reliably using perturbative QCD. In the end we are interested in how far the result may be extended to the endpoint region. For processes involving b-quark decay it is appropriate to use the HQEFT. In the HQEFT, the full QCD b field is expressed in terms of b_v for b quark velocity v. The b_v field is defined by

$$b_v = \frac{1+\not{v}}{2}e^{im_b v\cdot x}b + \cdots,\tag{9}$$

where the ellipses denote terms suppressed by powers of $1/m_b$. The equation of motion for b_v is $\not{v}b_v = b_v$. We can apply the techniques of the OPE to expand T in terms of matrix elements of local operators involving b_v fields

$$T = -i \int d^4x\, e^{iq\cdot x} \langle B|T\{J(x)^\dagger J(0)\}|B\rangle$$

$$= \sum_{n,v} C_n^{\mu\nu}(v,q)\, \langle B|\mathcal{O}_v^{(n)}|B\rangle. \tag{10}$$

Here $\mathcal{O}_v^{(n)}$ are local operators involving $\bar{b}_v b_v$ bilinears. The coefficient functions $C_n^{\mu\nu}$ depend explicitly on q and v because of the v dependence of $\mathcal{O}_v^{(n)}$. All large momenta are contained in the coefficient functions while the matrix elements describe low-energy physics.

In this scheme T is expanded as a double series in powers of α_s and $1/m_b$. To leading order in α_s and $1/m_b$ the forward Compton scattering amplitude for semi-leptonic and radiative decays are

$$T = -i \int d^4x\, e^{i\mathcal{Q}\cdot x}\, \langle B|T\{\bar{b}_v(x)\gamma^\mu(1+\gamma_5)S_0(x)\gamma^\nu(1+\gamma_5)b_v(0)\}|B\rangle$$

$$= -i \int d^4x\, e^{i\mathcal{Q}\cdot x}\, \langle B|\bar{b}_v(x)[\not{k},\gamma_\mu](1+\gamma_5)S_0(x)(1-\gamma_5)[\not{k},\gamma^\mu]b_v(0)|B\rangle \tag{11}$$

where $\mathcal{Q} = m_b v - k$ and $S_0(x)$ are the momentum and free propagator of the final quark.

When $\Lambda^2 \ll \mathcal{Q}^2 \ll m_b^2$, it is sufficient to keep the terms at leading order in $1/m_b$ only and expand Eq. (11) in powers of $k/|\mathcal{Q}|$. This generates coefficient functions and local operators in which k is replaced by the derivatives acting on the b_v fields. The leading term independent of k gives the parton-model result. Contracting the leading term with L we find

$$T_0 L = 128 m_b \frac{k_e \cdot \mathcal{Q} k_{\bar{\nu}} \cdot v}{\mathcal{Q}^2}$$

$$= 16 m_b^3 (1+r) \frac{k \cdot \mathcal{Q} k \cdot v}{\mathcal{Q}^2}, \tag{12}$$

where T_0 is T at leading order in $k/|\mathcal{Q}|$.

The next-order correction to Eq. (12) is suppressed by at least $(\Lambda/\mathcal{Q})^2$ [3] Corrections to T from higher dimensional operators involving $\bar{b}_v b_v$ bilinears have been studied systematically [4, 5, 6, 13, 14]. Furthermore Neubert [14] has resummed these corrections to get a "shape function". He has observed that the shape function is universal independent of the final quark flavor.

Usually the coefficient functions are calculated perturbatively. In addition there could be nonperturbative correction to the coefficient functions as well [15]. In the next section we calculate how instantons affect the coefficient function of the leading term, Eq. (12), in the OPE. The instantons contribute not only to the matrix elements but also to the coefficient functions. If there is an infrared divergence due to large instantons we attribute it to the contribution to the matrix elements. For infrared-finite parts we interpret them as the correction to the coefficient functions.

3 Calculable Instanton Correction

We now compute the contribution of instantons to the coefficient functions. In estimating the contribution we start from the Euclidean region where $\mathcal{Q}^2$ is large enough to use the OPE reliably. We expect that the main contribution comes from small instantons of size $\rho \lesssim 1/|Q|$, hence we use the dilute gas approximation in what follows. More specifically we calculate the instanton correction to the decay rate at leading order in both α_s and k in the OPE. This is the instanton correction to the parton model result Eq. (12).

In the background of an instanton ($+$ anti-instanton) of size ρ and instanton orientation U located at the origin, the Euclidean fermion propagator may be expanded in small fermion masses as [16]

$$S_{\pm}(x, y; \rho_{\pm}; U_{\pm}) = -\frac{1}{m}\psi_0(x)\psi_0^{\dagger}(y) + S_{\pm}^{(1)}(x, y; \rho_{\pm}; U_{\pm})$$

$$+ m \int d^4 w S_{\pm}^{(1)}(x, w; \rho_{\pm}; U_{\pm}) S_{\pm}^{(1)}(w, y; \rho_{\pm}; U_{\pm})$$

$$+ \mathcal{O}(m^2), \tag{13}$$

where $\pm$ denotes instanton, anti-instanton. ψ_0 is the fermion zero mode eigenfunction and $S_{\pm}^{(1)} = \sum_{E>0} \frac{1}{E} \Psi_{E\pm}(x) \Psi_{E\pm}^{\dagger}(y)$ is the Green's function of fermion nonzero modes. In evaluating the forward Compton scattering amplitude T, we should use the propagator in Eq. (13) instead of S_0. $T^{\mu\nu}$ can be written as

$$T = T_0 + T_{inst.}, \tag{14}$$

where the first term is the parton-model amplitude. The second term is the amplitude due to instantons of all orientation U, position z and size ρ.

After averaging over instanton orientations $T_{inst}^{\mu\nu}$ is given by

$$T_{inst.} = \int d^4\Delta\, e^{iQ\cdot\Delta} \sum_{a=\pm} \int d^4 z_a\, d\rho_a D(\rho_a)$$

$$\times \langle B|\bar{b}_v(x)\gamma^{\mu}(1+\gamma_5)\{S_a(X, Y; \rho_a) - S_0(\Delta)\}\gamma^{\nu}(1+\gamma_5)b_v(y)|B\rangle, \tag{15}$$

where $D(\rho)$ is the instanton density, $\Delta = x - y$, $X = x - z$ and $Y = y - z$. In Eq. (15), $\mathcal{S}_{\pm}(X, Y; \rho_{\pm})$ is the fermion propagator averaged over instanton (anti-instanton) orientations centered at z. Using the $\overline{MS}$ scheme with n_f flavors of light fermions, $D(\rho)$ is given by [17]

$$D(\rho) = K\,\Lambda^5\,(\rho\Lambda)^{6+\frac{n_f}{3}} \left(\ln\frac{1}{\rho^2\Lambda^2}\right)^{\frac{45-5n_f}{33-2n_f}}, \tag{16}$$

where

$$K = \left(\prod_i \frac{\hat{m}_i}{\Lambda}\right) 2^{\frac{12n_f}{33-2n_f}} \left(\frac{33-2n_f}{12}\right)^6 \frac{2}{\pi^2} \exp\left[\frac{1}{2} - \alpha(1) + 2(n_f - 1)\alpha(\frac{1}{2})\right] \tag{17}$$

in which the β function at two loops and the running mass at one loop are used and $\hat{m}_i$ are the renormalization-invariant quark masses. In Eq. (17) $\alpha(1) = 0.443307$ and $\alpha(1/2) = 0.145873$. From now on we replace the logarithmic term in $D(\rho)$ by its value for $\rho = 1/|Q|$. Corrections to this replacement are negligible since they are logarithmically suppressed.

Inserting Eq. (13) to Eq. (15), the leading contribution comes from the mass-independent part $\mathcal{S}_{\pm}^{(1)}$ due to the chiral structure of the left-handed weak currents. Keeping the chirality-conserving part of $\mathcal{S}_{\pm}^{(1)}$ in singular gauge [16] T_{inst} is evaluated straightforwardly. The z integral is convergent as $|z| \to \infty$. The integration over ρ is convergent for small instantons. On the other hand large instanton part is divergent. However since the integrand is analytic for large ρ, there are only a finite number of divergent terms when $T_{inst}^{\mu\nu}$ is expanded in $1/\rho$. We interpret these infrared divergent terms as the instanton contribution to the matrix elements of operators in the OPE [18]. The remaining terms are infrared convergent and are interpreted that they contribute to the coefficient functions. In order to calculate finite terms any convenient regularization prescription will do. To this end we analytically continue the exponent of ρ in the instanton density so that $D(\rho) \propto \rho^{M-4}$ [19] and the spacetime dimensions to $d = 4 + 2\epsilon$. In the end we let $M \to 11$ for $n_f = 3$ and $\epsilon \to 0$ to get the final answer.

It is now straightforward to evaluate the instanton contribution to the differential decay rate. So far we have worked in Euclidean spacetime. We now make a naive analytic continuation to Minkowski spacetime with timelike $\mathcal{Q}^2 = (m_b v - q)^2$ to evaluate the differential decay rate. For the semileptonic decay, the result of [7] for the double differential decay rate is

$$\frac{1}{\Gamma_0}\frac{d^2\Gamma_{inst}}{d\hat{q}^2 dy} = A\, y^5 \frac{5\hat{q}^2 - (1-y)(y - \hat{q}^2)}{(1-y)^6 (y - \hat{q}^2)^5}. \tag{18}$$

where

$$A = \frac{2^{11}\pi^2}{175}\Gamma^2(6)K\left(\frac{\lambda}{m_b}\right)^{12}. \tag{19}$$

The instanton effect is suppressed at $y \sim 0$. However it become singular near the boundaries of the phase space at $y \sim 1$ and $y \sim \hat{q}^2$. At these boundaries the final quark approaches the resonance region. As there are large instanton contributions near the resonance point, it is necessary to introduce a smearing to define sensible decay rate in perturbative QCD. We discuss this in detail in the next section.

We have calculated the instanton effect on the inclusive semileptonic b decay. As we have observed in the previous section the effect is singular at the boundaries of the phase space. In order to use perturbative expansions reliably a smearing prescription has to be introduced.

As a simple prescription we first consider the smeared single differential decay rate

$$\left\langle \frac{d\Gamma_{inst}}{dy} \right\rangle_\delta = \int_0^y d\hat{q}^2\, \theta(y - \hat{q}^2 - \delta)\frac{d\Gamma_{inst}}{dy} \tag{20}$$

in which we restrict the phase space so that the singular region $y = \hat{q}^2$ is avoided by δ. The size of smearing δ is determined by the requirement that the smeared instanton contribution to the single differential decay rate be smaller than that

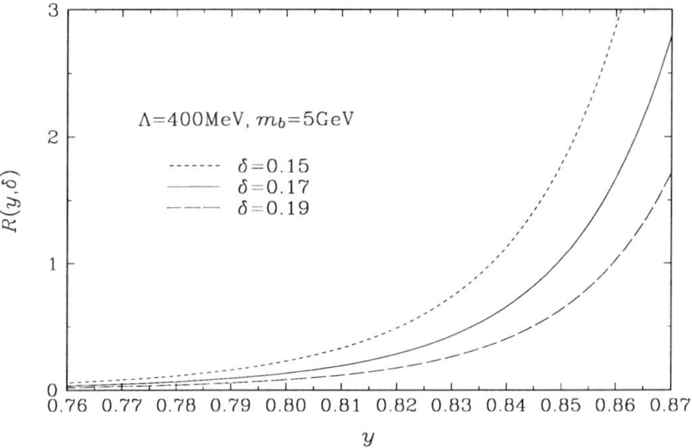

Figure 3.3. $R(y,\delta)$ for $\delta = 0.15$, 0.17, 0.19 with $\Lambda = 400$MeV and $m_b = 5$ GeV.

in the parton model $d\Gamma_0/dy = 2y^2(3 - 2y)\Gamma_0$. Eq.(20) is given by

$$\left\langle \frac{d\Gamma_{inst}}{dy} \right\rangle_\delta = \left(\frac{d\Gamma_0}{dy} \right) \frac{A}{24} \frac{1}{(1 - y)^6(3 - 2y)}$$

$$\times \left[9 - 4y - 24(\frac{y}{\delta})^3 + 15(\frac{y}{\delta})^4 + 4(\frac{y^4}{\delta^3}) \right]. \tag{21}$$

The ratio $R(y,\delta) = \langle d\Gamma_{inst}/dy \rangle_\delta/(d\Gamma_0/dy)$ is plotted in Fig.2 for three different values of $\delta = 0.15$, 0.17, 0.19 with $\Lambda = 400$ MeV and $m_b = 5$ GeV.

The numerical value of A is given by

$$A = \left(\frac{19.2GeV}{m_b} \right)^3 \left(\frac{\Lambda}{m_b} \right)^9, \tag{22}$$

in which we set $n_f = 3$ and the renormalization-invariant quark masses as [20] $\hat{m}_u = 8.2 \pm 1.5 MeV, \hat{m}_d = 14.4 \pm 1.5 MeV, \hat{m}_s = 288 \pm 48 MeV$. We see that for these choices of δ, $R(y,\delta)$ grows rapidly for $y \gtrsim 0.84$. From Fig.2 we see that a smearing of $\delta \approx 0.16 \sim 0.20$ at the boundary $y = \hat{q}^2$ of the phas space is needed to extract a reliable electron spectrum from the single differential decay rate.

We have also examined the behavior of $R(y,\delta)$ as we increase δ. The position at which $R(y,\delta)$ grows like a brick wall does not shift much from $y \sim 0.84$. Because of this singular behavior we need another smearing near $y = 1$. For simplicity we cut off the region near $y = 1$ by the same smearing width δ as in Eq. (20).

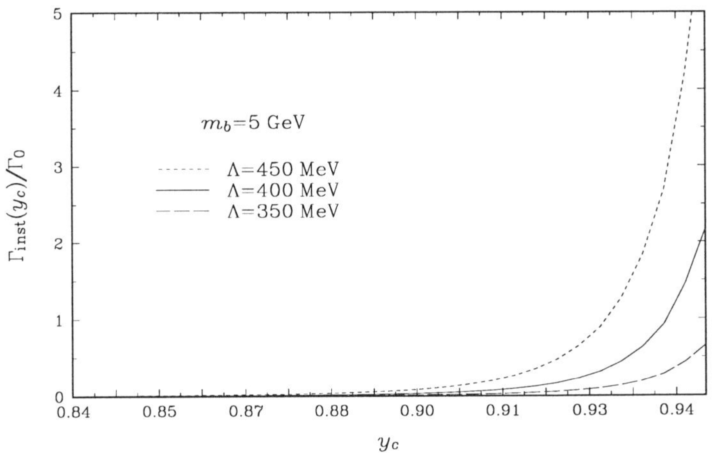

Figure 3.4. $R(\delta)$ for $\Lambda = 350, 400, 450$ MeV with $m_b = 5$ GeV.

Instanton contribution to the total decay rate after such a smearing prescription is given by

$$\langle \Gamma_{inst} \rangle_\delta = \int_0^1 dy \int_0^y d\hat{q}^2 \theta(y - \hat{q}^2 - \delta)\theta(1 - y - \delta)\frac{d^2\Gamma_{inst}}{d\hat{q}^2 dy}$$

$$= \Gamma_0 \frac{A}{24} \frac{1}{\delta^9} (6 - 53\delta + 198\delta^2 - 420\delta^3 + 612\delta^4$$

$$- 354\delta^5 + 16\delta^6 - 4\delta^7 - \delta^9 + 180\delta^5 \ln\delta). \tag{23}$$

The ratio $R(\delta) = \langle\Gamma\rangle_\delta/\Gamma_0$ is plotted in Fig.3 for three different values of $\Lambda = 350, 400, 450$ MeV and for $m_b = 5$ GeV.

We see that the ratio $R(\delta)$ also rises sharply like a brick wall as δ decreases. When δ is large, say $\delta \gtrsim 0.15$, $R(\delta)$ is insensitive to the choice of Λ/m_b. However, for small δ, say $\delta \approx 0.12$, $R(\delta)$ is sensitive to the value of Λ/m_b. If we require $R(\delta)$ be less than 20%, the size of the smearing is roughly at least $0.12 \sim 0.15$ for our choices of Λ.

Our results Eqs. (21), (23) are sensitive to the precise value of Λ. This is what we expect from the instanton effect. Nevertheless, as emphasized in Introduction, our results may still represent a typical size of nonperturbative effects.

We have also calculated the instanton effect for the inclusive radiative decay rate. Instanton effects on the $B \to X_s\gamma$ decay depend on the momentum cutoff of the final-state photon. The cutoff to remove soft photons should be introduced in experiments since it is difficult to isolate a soft photon from the background coming from the subsequent decay of, say, $K^* \to K + \gamma$. Experimentally the photon spectrum from $B \to X_s\gamma$ is concentrated in the region $2.2\ GeV \leq E_\gamma \leq 2.5\ GeV$. In CLEO the lower cut is taken at $E_\gamma \sim 2.2\mathrm{GeV}$ [21]. Thus we restrict the region of contour integral to $y_c \leq v \cdot \hat{q} \leq 1$ where y_c is the experimental cutoff for the scaled photon energy. Then the instanton decay rate is

$$\frac{\Gamma_{inst}(y_c)}{\Gamma_0} = \left(\frac{6.67\mathrm{GeV}}{m_b}\right)^3 \left(\frac{\Lambda}{m_b}\right)^9$$
$$\left[\frac{10}{(1-y_c)^6} - \frac{36}{(1-y_c)^5} + \frac{45}{(1-y_c)^4} - \frac{20}{(1-y_c)^3}\right]. \quad (24)$$

The ratio $\Gamma_{inst}(y_c)/\Gamma_0$ is shown in Fig. 4 as a function of the cutoff y_c for different values of $\Lambda = 350, 400, 450$ MeV with $m_b = 5$ GeV.

For $y_c \approx 0.82$ that CLEO has chosen, the instanton correction 13.1) $\times\ 10^{-4}$ as Λ varies from 350 MeV to 450 MeV. Therefore the instanton correction is negligibly small compared to other corrections. Note that the instanton contribution is much smaller as the cutoff y_c decreases. On the other hand, as Fig.4 shows, the contribution is appreciable at $y_c \geq 0.92$. In this region we need a smearing to make the perturbation theory valid. However this cutoff is too large to be significant experimentally since there is a very small fraction of the rate in this energy window.

The α_s-correction to the decay rate also becomes large as y_c approaches 1. Ali and Grueb [22] have examined the correction in detail and have concluded that the α_s-correction has to be exponentiated near the endpoint of the photon

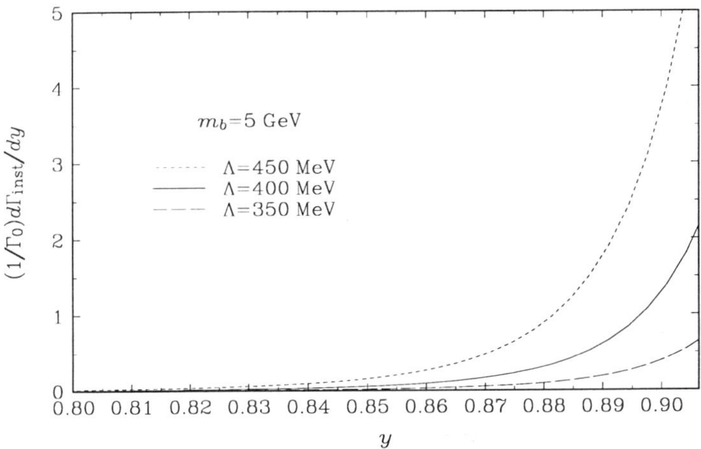

Figure 3.5. $\Gamma_{inst}(y_c)/\Gamma_0$ for $\Lambda = 350, 400, 450$ MeV with $m_b = 5$ GeV.

spectrum. Numerically they have found that the exponentiation is necessary for $y_c \geq 0.85$. Combined with their result our analysis indicates that there is a large theoretical uncertainty for the cutoff $y_c \geq 0.88$. Therefore in order to compare experiments with theory in a model-independent way the cutoff y_c has to be chosen below 0.88.

We also obtain the instanton contribution to the differential decay rate as

$$\frac{d\Gamma_{inst}}{dy} = \Gamma_0 \frac{N_{inst}}{m_b^{12}} \frac{y^3}{(1-y)^7}.\tag{25}$$

Numerically we find

$$\frac{1}{\Gamma_0} \frac{d\Gamma_{inst}}{dy} = \left(\frac{26.1 \text{ GeV}}{m_b}\right)^3 \left(\frac{\Lambda}{m_b}\right)^9 \frac{y^3}{(1-y)^7}.\tag{26}$$

The ratio in Eq. (26) is shown in Fig.5 for different values of $\Lambda = 350, 400, 450$ GeV with $m_b = 5$ GeV.

The ratio also increases sharply near the endpoint of the photon spectrum. Above $y \approx 0.87$ the instanton correction is appreciable and necessitates a smearing in this region.

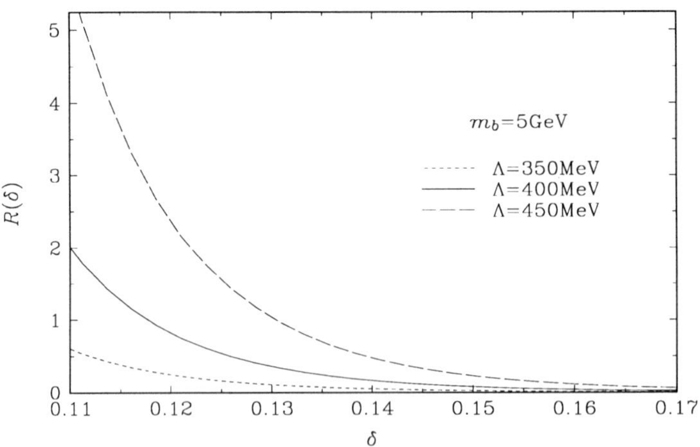

Figure 3.6. $(1/\Gamma_0)d\Gamma_{inst}/dy$ for $\Lambda = 350,\ 400,\ 450$ MeV with $m_b = 5$ GeV.

4   Kinematics and Analyticity

We have found that the instanton effect is quite different for different inclusive decay modes. For $B \to X_s \gamma$, the effect to the total decay rate is negligible as long as y_c is small enough. This is in contrast to the case of $B \to X_u e\bar{\nu}_e$ decay in which the differential decay rate receives a large instanton correction. This apparently big difference can be understood from kinematics.

As explained in Section 2 we have obtained the averaged total decay rate by deforming the integration contour (from C' to C in Fig.1). If the deformed contour is chosen sufficiently far away from the resonance region we can calculate the decay rate reliably using the perturbation theory. It depends on the details of the kinematics whether such contour deformation is possible.

In $B \to X_s\gamma$ decay the radius of the deformed contour C is fixed at $1 - y_c$. This radius represents the off-shell invariant mass-squared of the final hadrons. Note that this radius is independent of kinematic variables. As long as $1 - y_c \gg \Lambda^2/m_b^2$ the averaged decay rate can be calculated reliably at the scale $m_b^2(1-y_c)$. The instanton contribution is evaluated at this scale. As calculated in Section 3, the contribution is $\mathcal{O}(10^{-4})$.

In $B \to X_u e\bar{\nu}_e$ decay the radius of the deformed contour is given by $z = (y - \hat{q}^2)(1 - y)/y$ where $y = 2E_e/m_b$ and $\hat{q}^2 = (k_e + k_\nu)^2/m_b^2$ [7]. The off-shell invariant mass-squared of the final hadrons is equal to $m_b^2 z$. As long as $z \gg \Lambda^2/m_b^2$ we can calculate the averaged decay rate using perturbation theory. However unlike the $B \to X_s\gamma$ case the radius z depends on how the momentum transferred to the leptonic system is distributed between the electron and the anti-neutrino. In particular z vanishes at the boundaries of the phase space. Therefore the instanton contribution evaluated at the scale $m_b^2 z$ grows rapidly near the boundaries $y = 1$ and $y = \hat{q}^2$. This means that it is impossible to avoid the resonance region unless we introduce a model-dependent cut near the boundaries of the phase space.

Note that the dependence of the off-shell invariant mass-squared of the final hadrons on the kinematic variables arises only when there are more than one "nonhadronic" particle in the final state. In this respect it is interesting to compare our result with the inclusive hadronic τ decay $\tau \to \nu_\tau + X$ [23]. In this case the neutrino plays the same role as the photon in the $B \to X_s\gamma$ decay as far as the kinematics is concerned. The maximum invariant mass-squared $s = q^2$ of the final hadrons is fixed at m_τ^2 independent of the neutrino energy. Since this is much larger than Λ^2 the total inclusive decay rate can be calculated reliably using the deformed contour. While the specific kinematic variables under consideration are different, the analytic structure and the fact that the maximal invariant mass-squared of the final hadrons is independent of other kinematic variables are similar in both $B \to X_s\gamma$ and inclusive hadronic τ decays.

Indeed the instanton contributions are small in both cases. Nason and Porrati [19] have obtained the instanton contribution to the ratio of the hadronic to the leptonic width R_τ which is given by

$$\frac{R_\tau^{inst}}{R_\tau^0} = \left(\frac{3.64\Lambda}{m_\tau}\right)^9 \frac{\hat{m}_u \hat{m}_d \hat{m}_s}{m_\tau^3}. \tag{27}$$

Our result for the instanton contribution to the decay rate of $B \to X_s\gamma$ is written as

$$\frac{\Gamma_{inst}}{\Gamma_0} = \left(\frac{5.91\Lambda}{m_b}\right)^9 \frac{\hat{m}_u \hat{m}_d \hat{m}_s}{m_b^3} \qquad (28)$$

for $y_c = 0$. The fact that both have the same mass dependence and similar numerical factors confirms our expectation.

By the same argument we expect the instanton contribution to the $B \to X_s e^+ e^-$ decay rate is similar to that of the $B \to X_u e \bar{\nu}_e$ because kinematics and the analytic structure of the forward Compton scattering amplitudes are the same.

5 Discussions

In this talk I have reported on nonperturbative QCD effects to the inclusive semileptonic and radiative B-meson decays. The final hadrons are on the mass-shell at the boundaries of the phase space. If the final hadron has an invariant mass below the QCD scale, then the nonperturbative QCD effects affect not only the hadron matrix elements but also the coefficient functions. Our calculation is consistent with what we expect from the consideration of the analyticity and the kinematics of each decay channels. The instanton effect is quite sizable for semileptonic decays but negligibly small for the radiative decay.

I thank J.G. Chay for discussions and collaboration. This work was supported in part by Grants from KOSEF-SRC program, Ministry of Education BSRI 94-2108 and Korea Research Foundation Nondirected Research Grant '93.

References

[1] G. Altarelli, N. Cabibbo, G. Corbo, L. Maiani and G. Martinelli, Nucl. Phys. **B208**, 365 (1982).

[2] B. Grinstein, N. Isgur and M.B. Wise, Phys. Rev. Lett. **56**, 258 (1986); N. Isgur, D. Scora, B. Grinstein and M.B. Wise, Phys. Rev. **D39**, 799 (1989).

[3] J. Chay, H. Georgi and B. Grinstein, Phys. Lett. **247B**, 399 (1990).

[4] I.I. Bigi, M.A. Shifman, N.G. Uraltsev and A.I. Vainshtein, Phys. Rev. Lett. **71**, 496 (1993); B. Blok, L. Koyrakh, M. Shifman and A.I. Vainshtein, Phys. Rev. **D49** (1994) 3356.

[5] T. Mannel, Nucl. Phys. **B413**, 396 (1994).

[6] A.V. Manohar and M.B. Wise, Phys. Rev. **D49** (1994) 1310.

[7] J. Chay and S.-J. Rey, SNUTP 94-08 preprint, 1994 (unpublished).

[8] J. Chay and S.-J. Rey, SNUTP 94-54 preprint, 1994 (unpublished).

[9] T. Inami and C.S. Lim, Prog. Theor. Phys. **65** (1981) 297.

[10] B. Grinstein, R. Springer and M.B. Wise, Phys. Lett. **202B** (1988) 138; B. Grinstein and M.B. Wise, Phys. Lett. **B201** (1988) 274.

[11] I.I. Bigi, M.A. Shifman, N.G. Uraltsev and A.I. Vainshtein, Int. J. Mod. Phys. **A9** (1994) 2467.

[12] E.C. Poggio, H.R. Quinn and S. Weinberg, Phys. Rev. **D13** 1958 (1976).

[13] A. Falk, M. Luke and M.J. Savage, UCSD/PTH 93-23 preprint, 1993 (unpublished).

[14]   M. Neubert, CERN-TH.7087/93 preprint, 1993 (unpublished).

[15]   V.A. Novikov, M.A. Shifman, A.I. Vainshtein and V.I. Zakharov, Nucl. Phys. **B174**, 378 (1980).

[16]   N. Andrei and D. Gross, Phys. Rev. **D18**, 468 (1978).

[17]   G. 't Hooft, Phys. Rev. **D14** , 3432 (1976); G. 't Hooft, Phys. Rep. **142**, 357 (1986); C. Bernard, Phys. Rev. **D19**, 3013 (1979).

[18]   L. Balieu, J. Ellis, M.K. Gaillard and W.J. Zakrzewski, Phys. Lett. **77B**, 290 (1978).

[19]   I.I. Balitskii, M. Beneke and V.M. Braun, Phys. Lett. **318B**, 371 (1993); P. Nason and M. Porrati, Nucl. Phys. **B421**, 518 (1994).

[20]   C.A. Dominguez and E. de Rafael, Ann. Phys. **174**, 372 (1987).

[21]   For a comprehensive review, see T.E. Browder, K. Honscheid and S. Playfer, CLSN 93/1261, HEPSY 93-10 preprint, 1993 (unpublished).

[22]   A. Ali and C. Grueb, Phys. Lett. **B287** (1992) 191.

[23]   E. Braaten, S. Narison and A. Pich, Nucl. Phys. **B373** (1992) 581.

Upper Bounds
in
Low-Energy SUSY

G.L. KANE, CHRIS KOLDA,
LESZEK ROSZKOWSKI,[1] AND JAMES D. WELLS
RANDALL PHYSICS LABORATORY
UNIVERSITY OF MICHIGAN,
ANN ARBOR, MI 48190, USA

Abstract: In the constrained MSSM one is typically able to restrict the supersymmetric mass spectra below roughly 1-2 TeV *without* resorting to the ambiguous fine-tuning constraint.

1   Constrained MSSM (CMSSM)

Low-energy SUSY has been considered an attractive extension of the Standard Model (SM) ever since it was introduced over a decade ago. In the early days many expected supersymmetric masses to lie rather low, "just around the corner", often well within the reach of LEP and the Tevatron. Not finding SUSY signals there was consequently rather disappointing and one could hear from sceptics sarcastic comments that a SUSY discovery will always remain to be expected for the *next* round of accelerators, in a time-invariant manner. Theoretical arguments based on no fine-tuning limiting SUSY masses roughly below 1 TeV were greeted with even less trust. After all, theorists are known to be both creative and, at the same time, rather unwilling to give up their most beloved toys, as the continuing activity in alternatives to SUSY clearly shows.

In this talk I am going to show that, in the Minimal Supersymmetric Standard Model (MSSM) with a few sensible and relatively general assumptions, one is often able to limit the SUSY particle masses below about 1-2 TeV *by physical constraints alone [1], [2]*, without having to resort to an ill-defined fine-tuning constraint. Furthermore, the assumptions that we make are actually typically also made in most phenomenological studies of the MSSM and are well-motivated by GUTs. I will call this framework the constrained MSSM (CMSSM) *[1]*.

First, it is worth remembering that SUSY alone has been applied to particle physics in order to provide a sensible framework for GUTs, which are otherwise plagued by the (in)famous problems of naturalness and scale hierarchy. Without GUTs, or related attempts (like strings) to not only unify all interactions but also to close the gap between to Fermi scale and the only fundamental scale in high energy physics, the Planck scale, there is indeed little motivation to consider low-energy SUSY. Furthermore, precision measurements at LEP have provided us with a remarkable argument for gauge coupling unification within (even minimal) SUSY, while showing more than clearly that within the SM alone such unification does not take place *[3]*. We will thus require that gauge

---

[1]Speaker

couplings unify which will, for our purpose, fix the unification scale M_X. By doing so we actually are not forced to assume the existence of any specific GUT. We will only assume that $\sin^2 \theta_w(M_X) = \frac{3}{8}$ which also holds in many phenomenologically viable superstring-derived models.

Second, if the idea of unification is to be taken seriously, then one should expect not only the gauge couplings to emerge from a common source but also the same to be true for the various mass parameters of low-energy SUSY. In particular, one typically assumes that all the mass terms of the scalars in the model, like the squarks, sleptons and the Higgs bosons, originate from one "common" source m_0. Similarly, the masses of the gauginos (the gluino, winos and bino) should be equal to the "common" gaugino mass $m_{1/2}$ at M_X. These two assumptions are certainly not irrefutable but are at least sensible. In addition, they result from the simplest minimal supergravity framework and the simplest choice of the kinetic potential. Furthermore, there is at least some partial motivation for assuming the common scalar mass m_0 coming from experiment. The near mass degeneracy in the $K^0 - \bar{K}^0$ system implies a near mass degeneracy between $\tilde{s}_L$ and $\tilde{d}_L$ [4]. Similarly, some slepton masses have to be strongly degenerate from stringent bounds on $\mu \to e\gamma$ [4]. Needless to say, most phenomenological studies of SUSY rely on at least one of these two assumptions, at least for the sake of reducing the otherwise huge number of unrelated SUSY mass parameters. We also assume that the trilinear soft SUSY-breaking terms are equal to A_0 at M_X, although this assumption has actually almost no bearing here.

Furthermore, it has been long known that in SUSY there exists a remarkable "built-in" mechanism of radiative electroweak symmetry breaking (EWSB). When the Higgs mass-square parameters are run from the high scale down, at some point the Higgs fields develop vevs. We thus require that the conditions for EWSB be satisfied.

Having made these sensible and well-motivated assumptions, we can next derive complete mass spectra of all the Higgs and supersymmetric particles by running their 1-loop RGEs between M_X and m_Z. The spectra are parametrized in terms of just a few basic parameters which we conveniently choose to be: the top mass M_t, $\tan\beta$, $m_{1/2}$, m_0, as well as A_0. The parameters $|\mu|$ and B are determined through the conditions for EWSB, but the sign of μ remains undetermined. We thus consider both $sgn\,\mu = \pm 1$. We also employ the full 1-loop effective Higgs potential.

Besides requiring that EWSB occur, we demand that all physical mass-squares remain positive. We impose mass limits from current direct experimental searches and include the requirement that the solutions provide a BR($b \to s\gamma$) consistent with CLEO data. Furthermore, we calculate the relic density of the lightest SUSY particle (LSP), demanding only that the LSP be neutral, and, from limits on the age of the Universe of 10 billion years, we demand that $\Omega_{LSP}h_0^2 < 1$. Those solutions which finally remain after all these cuts comprise the allowed parameter space of the CMSSM.

2 Upper Limits

We have explored wide ranges of parameters, as described in detail in Refs. *[1]*, *[2]*. Clearly, in general one expects the emerging patterns of the SUSY mass spectra and properties (mixings, *etc.*) to vary strongly with the input parameters. While this is indeed true to some extent, nevertheless certain universal features emerge.

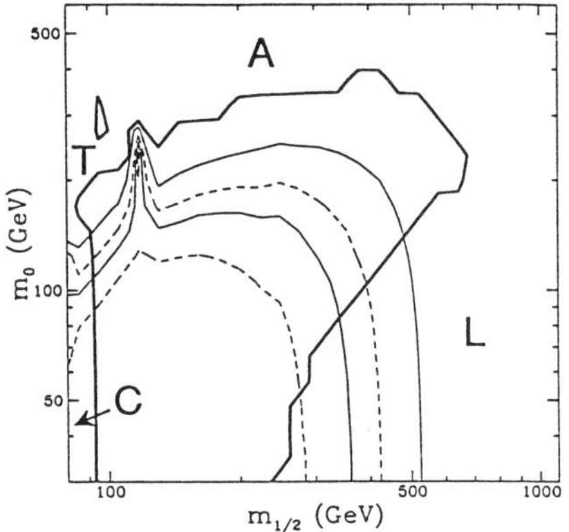

Figure 1. The regions of the $(m_{1/2}, m_0)$ plane consistent with low $\tan\beta$ b–τ mass unification, given all the constraints of the CMSSM, for $M_t = 170\,GeV$, $A_0/m_0 = 0$ and $\mu < 0$. Solutions outside the thick solid lines are excluded: on the left (small $m_{1/2}$) by the chargino mass bound (**C**) $m_{\chi^\pm} > 47\,GeV$ and by tachyonic $\widetilde{t}$'s (**T**); on the right (large $m_{1/2} \gg m_0$) by charged LSP (**L**); and from above by the age of the Universe, *i.e.* $\Omega_\chi h_0^2 \leq 1$ (**A**). We also indicate the sub-regions selected by either the hypothesis of cold dark matter ($0.25 \lesssim \Omega_\chi h_0^2 \lesssim 0.5$), between thin solid lines or the one of mixed dark matter ($0.16 \lesssim \Omega_\chi h_0^2 \lesssim 0.33$), between thin dashed lines.

These features are illustrated in Fig. 1 for $M_t = 170\,GeV$. The region of small $m_{1/2}$ is always excluded by either direct experimental searches for SUSY at LEP (typically the strongest bounds come from chargino or Higgs mass limits) or at the Tevatron (gluino). In some cases, in particular for $|A_0|/m_0$ significantly above zero, the lighter stop becomes too light, and even tachyonic, for $m_0 \gg m_{1/2} \lesssim 100\,GeV$. Also, for some but rare combinations of parameters, for either $m_0 \gg m_{1/2}$ or $m_{1/2} \gg m_0$ the conditions for EWSB fail to be satisfied.

Furthermore, since we assume unbroken R-parity here, the LSP is stable and should be present as a relic in the Universe. There are strong arguments against charged exotic relics. We thus require the LSP to be neutral. This rules out a significant region of the $(m_{1/2}, m_0)$ parameter space corresponding to $m_{1/2} \gg m_0$ where the LSP is the lighter stau $\widetilde{\tau}_1$. (Also $\widetilde{e}_R$ and $\widetilde{\mu}_R$ are not much heavier there.)

In the remaining region allowed by all the conditions listed above it is the lightest neutralino χ that is the LSP. (The sneutrino, another neutral sparticle, is the LSP in the region of small $m_{1/2}$ which is now completely excluded experimentally.) This is quite remarkable given the fact that the neutralino is a very attractive candidate for the dark matter (DM) in the Universe for which there seems to be inescapable need among astrophysicists [5]. Equally remarkable and non-trivial is the fact that χ comes out mostly bino-like which is essentially a necessary condition if one expects the neutralino to be a significant component of DM in the Universe. (The neutralino with a significant higgsino admixture has invariably very small relic abundance [5].) The only exceptions to this general rule can be found in some relatively rare case in very tiny regions of the $(m_{1/2}, m_0)$ on the border of the region where the conditions for the EWSB cannot be satisfied.

The fact that the lightest neutralino of bino-type comes out in the CMSSM as the unique neutral candidate for the LSP is not only interesting in itself. It also leads to a very remarkable upper bound on both $m_{1/2}$ and m_0. This comes about as follows. The neutralino relic density ρ_χ depends on how many neutralinos have pair-annihilated in the early Universe. Their number effectively froze when the expansion rate exceeded the annihilation rate. In order to calculate the neutralino relic density one thus needs to include all the annihilation channels of the neutralinos into ordinary particles, which we do. Since all the masses and mixings are determined in the CMSSM in terms of the basic independent parameters listed above, one can also express in terms of them the neutralino relic abundance $\Omega_\chi h_0^2$ (which is the neutralino relic density in units of the critical density times the squared reduced Hubble constant). The key point is that any significant contribution to the total mass-energy density of the Universe would have affected its evolution. In particular, the greater the total density the faster the Universe expands and the more quickly it reaches its present size. The age of the Universe, which is known to be *at least* 10 billion years, then puts an upper limit $\Omega_\chi h_0^2 < 1$. This is shown in Fig. 1. We see that *the whole* plane $(m_{1/2}, m_0)$ becomes limited within a few hundred GeV.

The dominant effect is played here by the annihilation of the neutralinos into light fermion-antifermion pairs via the t-channel exchange of the lightest sfermion(s); roughly $\Omega_\chi h_0^2 \propto m_{\tilde{f}}^4 / m_\chi^2$ [5, 6], although including other final states affects the exact location of the bound.

It is interesting to explore how these bounds vary with different choices of the input parameters. We find that at least for small $\tan\beta \lesssim 2$ one is able to close almost the whole plane $(m_{1/2}, m_0)$, and therefore the whole SUSY spectrum, from *above* for any combinations of other parameters, except for the region of large Z-pole enhancement ($m_0 \gg m_{1/2} \simeq 100\,GeV$). For larger values of $\tan\beta$ sometimes the conditions of EWSB cannot be satisfied in the regions of extreme $m_{1/2}$ or m_0 and very close to such regions the LSP is of higgsino-type. In such, relatively rare, cases one cannot close the plane $(m_{1/2}, m_0)$ from above completely. It is remarkable that small $\tan\beta$ is strongly favoured by the simple unification of the b- and τ-Yukawa coupling unification, like in $SU(5)$ [7]. It was recognized several years ago that, unlike in the SM, in the MSSM the b- and

τ-Yukawa running couplings meet at roughly the same mass scale at which the unification of the gauge couplings takes place *[8]*. In Ref. *[2]* we have studied in detail the various consequences of adding this sensible, but rather specific to $SU(5)$-type GUTs, assumption.

Can these upper bounds be improved? It is worth stressing that the assumption that the age of the Universe is at least 10 billion years is actually a rather conservative one. Many expect it to be no less than some 15 billion years which translates to $\Omega_\chi h_0^2 \lesssim 0.25$ and much tighter bounds on the parameters $m_{1/2}$ and m_0. Another attractive hypothesis is the one of cosmic inflation which predicts $\Omega = 1$ in which case most of the matter in the Universe must most likely hide in the form of DM. Two scenarios have attracted a lot of attention. In the purely cold DM (CDM) scenario the neutralino would constitute most of DM in the (flat) Universe in which case the range $0.25 \lesssim \Omega_\chi h_0^2 \lesssim 0.5$ would be favored. More recently (after COBE), a mixed CDM+HDM picture (MDM) became more popular as it apparently fits the astrophysical data better than the pure CDM model. In the mixed scenario one assumes about 30% HDM (like light neutrinos with $m_\nu \simeq 6\,eV$) and about 65% CDM (bino-like neutralino), with baryons contributing the remaining 5% of the DM. In this case the favored range for $\Omega_\chi h_0^2$ is approximately given by $0.16 \lesssim \Omega_\chi h_0^2 \lesssim 0.33$. Both ranges are plotted in Fig. 1. It is clear that their effect is to significantly reduce the allowed parameter space from *both above and below*. Consequently, the allowed mass ranges of the various SUSY (and Higgs) particles become much more restricted and, unfortunately, typically beyond the reach of LEP II and the upgraded Tevatron. More details can be found in Refs. *[1]* and *[2]*.

3   Conclusions

I have shown that, in the framework of Constrained MSSM (which is the MSSM with a few well-motivated assumptions stemming from grand unifications), one can often limit the SUSY particle masses below about 1-2 TeV *by physical constraints alone*. I have not used the ill-defined fine-tuning constraint *at all*. It certainly still makes sense to take it into account in expressing our expectations as to where SUSY might be realized. But relatively general physical constraints now do not allow us to push SUSY into a multi- TeV region even if we wanted. Especially with the improving knowledge of the top mass and the age of the Universe we soon will be able to make a definite statement, based purely on physics criteria, that (minimal) low-energy SUSY is either realized roughly below 1 TeV or is not realized in Nature at all.

4   Acknowledgements

I would like to thank Pierre Ramond for organizing a very stimulating meeting. His warm reception especially contrasted with the freezing weather back in Michigan.

References

[1]   G.L. Kane, C. Kolda, L. Roszkowski, and J. Wells, Michigan preprint UM-TH-93-24 (October 1993), Phys. Rev. **D**, *in press*.

[2]   C. Kolda, L. Roszkowski, J. Wells, and G.L. Kane, Michigan preprint UM-TH-94-03 (February 1994).

[3]   P. Langacker, in the *Proceedings of the PASCOS-90 Symposium*, eds. P. Nath and S. Reucroft (World Scientific, Singapore, 1990); P. Langacker and M.-X. Luo, Phys. Rev. **D44** (1991) 817; J. Ellis, S. Kelley, and D.V. Nanopoulos, Phys. Lett. **B260** (1991) 131; U. Amaldi, W. de Boer, and H. Fürstenau, Phys. Lett. **B260** (1991) 447; F. Anselmo, L. Cifarelli, A. Peterman, and A. Zichichi, Nuovo Cim. **104A** (1991) 1817, and Nuovo Cim. **105A** (1992) 581.

[4]   See, *e.g.*, M. Dine, A. Kagan, and S. Samuel, Phys. Lett. **B243** (1990) 250.

[5]   L. Roszkowski, "Supersymmetric Dark Matter - A Review", UM-TH-93-06, hep-ph/9302259 (February 1993), in the Proceedings of the XXIII Workshop *Properties of SUSY Particles*, Erice, Italy, September 28 - October 4, 1992, eds. L. Cifarelli and V.A. Khoze.

[6]   L. Roszkowski, Phys. Lett. **B262** (1991) 59.

[7]   V. Barger, M.S. Berger, P. Ohmann, and R.J.N. Phillips, Phys. Lett. **B314** (1993) 351.

[8]   H. Arason, D. Castaño, B. Keszthelyi, S. Mikaelian, E. Piard, P. Ramond, and B. Wright, Phys. Rev. Lett. **67** (1991) 2933.

Yukawa Unification: The Good, The Bad and The Ugly

URI SARID[1]
PHYSICS DEPARTMENT, STANFORD UNIVERSITY
STANFORD, CALIFORNIA 94305
IN COLLABORATION WITH RICCARDO RATTAZZI AND LAWRENCE J. HALL

Abstract: We analyze some consequences of grand unification of the third-generation Yukawa couplings, in the context of the minimal supersymmetric standard model. We address two issues: the prediction of the top quark mass, and the generation of the top-bottom mass hierarchy through a hierarchy of Higgs vacuum expectation values. The top mass is strongly dependent on a certain ratio of superpartner masses. And the VEV hierarchy always entails some tuning of the GUT-scale parameters. We study the RG equations and their semi-analytic solutions, which exhibit several interesting features, such as a focusing effect for a large Yukawa coupling in the limit of certain symmetries and a correlation between the A terms (which contribute to $b \to s\gamma$) and the gaugino masses. This study shows that non-universal soft-SUSY-breaking masses are favored (in particular for splitting the Higgs doublets via D-terms and for allowing more natural scenarios of symmetry breaking), and hints at features desired in Yukawa-unified models. Several phenomenological implications are also revealed.

1 Introduction

There is strong evidence to suggest that the three gauge couplings of the strong, electromagnetic and weak interactions are unified at a high energy scale in a single gauge interaction based on a simple group, such as SU(5) or SO(10), as long as the desert below the unification scale is described by the minimal supersymmetric extension of the standard model (the MSSM). Furthermore, the combination of supersymmetry (SUSY) and grand unification yields models with numerous attractive features: the embedding of the standard-model matter multiplets into a few irreducible representations of the GUT group, the technically natural preservation of a hierarchy between the weak and GUT scales, a longer proton lifetime to allow agreement with current experimental lower bounds, the correct prediction of the ratio of b quark to τ lepton masses, simple ansätze for the remaining fermion masses, and a picturesque scenario for radiative electroweak symmetry breaking. We have chosen, therefore, to look beyond the gauge unification prediction of the weak mixing angle and examine the unification of third-family Yukawa couplings [1], for the most part within its natural context of SO(10) unification. By Yukawa couplings we mean the couplings of the top, bottom and tau to the Higgs doublets which generate their masses when electroweak symmetry is broken. The third family is singled out because of its relatively large Yukawa couplings: it seems reasonable to suppose that they arise at tree-level from the simplest interactions, while the

---

[1] E-mail: sarid@squirrel.stanford.edu

masses and mixings of the other generations require more complex, perhaps also higher-order and certainly very model-dependent structures. Our study also applies more generally to scenarios in which these Yukawa couplings are unified but not in the context of an SO(10) GUT, or even models in which the Yukawas are only comparable near the GUT or Planck scales. We begin by considering the most immediate prediction of this Yukawa unification, namely the top quark mass [2]. However, we are quickly led to consider in some depth the more general question of how the top-bottom mass hierarchy could be generated in the MSSM, and how this hierarchy depends on the initial conditions of the renormalization group (RG) evolution at the GUT scale [3]. We will conclude with a discussion of how natural (or unnatural!) such a hierarchy seems in this context, what its other phenomenological predictions might be, and how one could hope to improve the theoretical picture or obtain experimental corroboration.

One consequence of Yukawa unification is immediate, and independent of other assumptions except for the qualitative nature of the RG evolution equations of the MSSM. Since the Yukawa couplings of the top and bottom quarks (and the τ lepton) are always comparable, the large ratio of the top mass versus the bottom (or tau) mass must be due to a large ratio of the Higgs vacuum expectation values (VEVs) which give rise to their masses. Namely, since the up-type and down-type matter fermion masses arise from couplings to the up-type and down-type Higgs multiplets ($H_{U,D}$), respectively, the large ratio $m_t/m_{b,\tau} = (\lambda_t v_U)/(\lambda_{b,\tau} v_D)$ is not a consequence of a large ratio of Yukawas $\lambda_t/\lambda_{b,\tau}$ but rather of large $v_U/v_D \equiv \tan\beta$. Thus Yukawa unification generically implies $\tan\beta \sim \mathcal{O}(50)$. Further assumptions are necessary to make any precise predictions. We will assume the following three throughout most of this work, although we will point out those conclusions which are more general:

(I)   The masses of the third generation, m_t, m_b and m_τ, originate from renormalizable Yukawa couplings of the form $\underline{16}_3 \, \mathcal{O} \, \underline{16}_3$ in a supersymmetric GUT with a gauge group containing (the conventional) SO(10); $\underline{16}_3$ denotes the 16-dimensional spinor representation of SO(10) containing the third-generation standard-model fermions (plus the right-handed neutrino which we assume to be superheavy) and their superpartners.

(II)   The evolution of the gauge and Yukawa couplings in the effective theory beneath the SO(10) breaking scale is described by the RG equations of the MSSM.

(III)   The two Higgs doublets lie predominantly in a single irreducible multiplet of SO(10).

The first and third assumptions serve to define what we mean by Yukawa unification, while the second allows us to relate this GUT-scale unification to weak-scale observables. From (I) and (III) it follows that the third-generation Yukawas must arise from either a $\underline{16}_3 \, \underline{10}_H \, \underline{16}_3$ or a $\underline{16}_3 \, \underline{126}_H \, \underline{16}_3$ interaction with an SO(10) Higgs multiplet. The latter leads to the boundary conditions $3\lambda_t^G = 3\lambda_b^G = \lambda_\tau^G \equiv \lambda_G$ at the GUT scale, but the resulting ratio of m_b/m_τ at low energies is far too low to be consistent with experiment (at least within the perturbative regime, and unless very large threshold corrections to the b mass

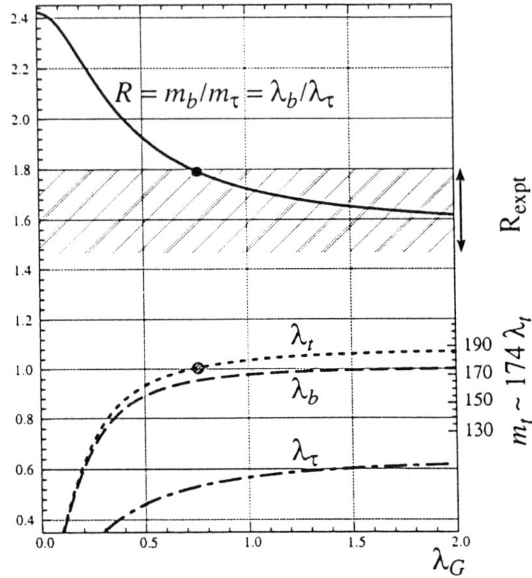

Figure 1. The dependence of the *low-energy* values $\lambda_{t,b,\tau}$ and of the ratio $R = \lambda_b/\lambda_\tau$ on the initial condition $\lambda_{t,b,\tau}^G = \lambda_G$, without any threshold corrections. The allowed range of R_{expt} is shown shaded, and the minimal value of λ_G allowed by this range in the absence of any corrections is indicated by the solid dot; lower values of λ_G require finite, negative δm_b (see the text). The corresponding value of λ_t is marked by the shaded dot. The vertical scale on the right indicates the approximate tree-level top mass $\sim 174\lambda_t\, GeV$ which would result from the values of λ_t on the left vertical scale; for example, the shaded dot predicts a heavy top, above 170 GeV or so.

arise at low energies *[2]*). Thus we are restricted to using the $\underline{10}_H$, and hence the boundary condition

$$\lambda_t^G = \lambda_b^G = \lambda_\tau^G \equiv \lambda_G \,. \tag{1}$$

With this boundary condition, and using the unification of gauge couplings to fix the unification scale and the gauge coupling at that scale, we can now evolve the Yukawa couplings down to the weak scale for any given value of λ_G. The idea is that the three observable masses m_t, m_b and m_τ are functions of the four parameters λ_t, λ_b, λ_τ and $\tan\beta$, and $\lambda_{t,b,\tau}$ are in turn determined by the unification in terms of λ_G and the GUT scale M_G. Since the latter is already known from gauge unification, we are left with three observable masses as functions of only two parameters, λ_G and $\tan\beta$. Thus we use two observables, m_b and m_τ to fix λ_G and $\tan\beta$, and thereby predict the third observable m_t. A detailed analysis of the RG evolution, and the consequent predictions, has already been presented *[2]*. The results, namely the values of $\lambda_{t,b,\tau}$ and the ratio $R \equiv m_b/m_\tau = \lambda_b/\lambda_\tau$ all at the weak scale, are plotted in Fig. 1 as functions of λ_G. (These curves actually use 2-loop RG evolution, but at this point the difference between 1- and 2-loop equations is not important. For the final predictions of m_t we use the full 2-loop evolution and 1-loop matching conditions.) Evidently, larger λ_G values correspond to a heavy top and to a smaller R ratio. The experimental value R_{expt} is found *[2]* from the QCD sum rules value and is evolved to the weak scale using 2-loop QCD running. We find, allowing for α_s to vary between roughly 0.11 and 0.12, the shaded range shown in Fig. 1. Thus, in the absence of any large corrections in the matching between the R evolved down from the GUT scale and the R_{expt} in the standard model, we find $\lambda_G > 0.75$, which implies a heavy top.

2 The top and bottom masses

For a precise prediction, 2-loop RG equations must be used along with 1-loop matching functions. These matching functions include logarithmic corrections from infinite counterterms as well as nonlogarithmic contributions from finite graphs. The former are given elsewhere *[2]*; they are generally quite small, and invariably increase the top mass as the superpartner masses increase. The latter are more interesting, since they can be very large *[2, 4]*. The dominant corrections arise from the graphs of Fig. 2, which match the value of the b mass as evolved down from the GUT scale to the value in the low-energy theory.

Typically the gluino graph dominates, yielding a corrected value $m_b = \lambda_b v_D + \delta m_b$ where $v_D = 174\,GeV$,

$$\frac{\delta m_b}{m_b} = \frac{8}{3} g_3^2 \frac{\tan\beta}{16\pi^2} \frac{m_{\tilde{g}}\mu}{m_{eff}^2},\tag{2}$$

and $m_{\tilde{g}}$ is the gluino mass while m_{eff} is the mass of the heaviest superpartner in the loop (more exact expressions may be found in our previous work). The important observation here is that the bottom mass gets a large contribution

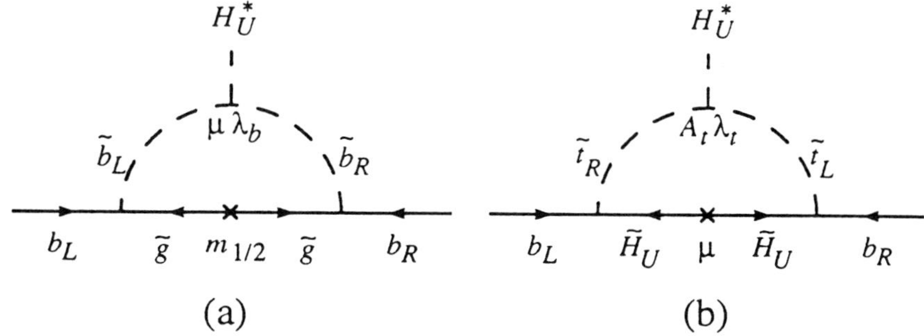

Figure 2. The leading (finite) 1-loop MSSM corrections to the bottom quark mass, namely δm_b.

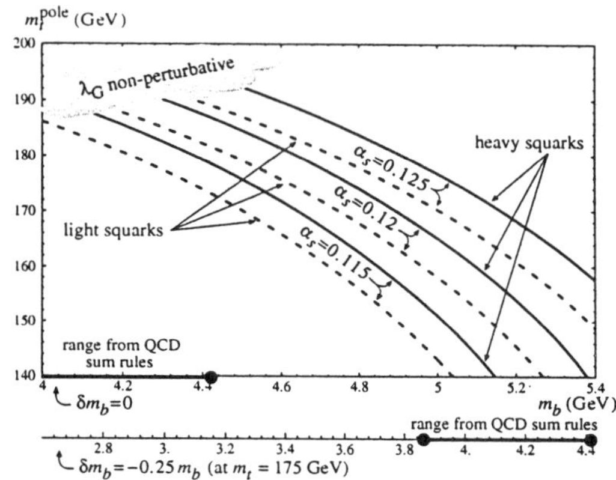

Figure 3. Our predictions *[2]* for the pole mass of the top quark, without superheavy corrections and using two qualitatively-different superpartner spectra, specifically $m_{higgsino} \sim \mu = 100\,GeV$, $m_{gluino} = 300\,GeV$, $m_{wino} = 100\,GeV$, $m_{squark} = m_{slepton} = 1000\,GeV$ and $m_A = 1000\,GeV$ for which the δm_b corrections are small, and $m_{higgsino} \sim \mu = 250\,GeV$, $m_{gluino} = 300\,GeV$, $m_{wino} = 100\,GeV$, $m_{squark} = m_{slepton} = 400\,GeV$ and $m_A = 400\,GeV$, for which $|\delta m_b/m_b| \sim 0.25$. The upper or lower horizontal axes should be used for these two spectra, respectively. The "cloud" indicates the region where the theory becomes nonperturbative at the GUT scale. Also shown are the estimated allowed mass ranges for the running parameter m_b as extracted in our previous work *[2]*.

from the up-type Higgs at 1-loop order, whereas its tree-level mass was small due to the small VEV of the down-type Higgs. In the usual scenario with small $\tan\beta$, the bottom was light because its Yukawa coupling was small, or in other words because it was protected by an approximate chiral symmetry. Thus any higher-order corrections would also be suppressed by this approximate symmetry. In the large $\tan\beta$ scenario these corrections are not suppressed: there is an enhancement of $v_U/v_D = \tan\beta$, which overcomes the usual $g_3^2/16\pi^2$ loop factor to give a correction of order 1 to the b mass—at least if $m_{\tilde{g}}\mu \sim m_{eff}^2$. Phenomenologically, the result is that m_b cannot be predicted with any certainty unless we know something about the superspectrum, and if m_b is uncertain then so is the top mass prediction.

In Fig. 3 we show the top pole mass prediction, now to full 2-loop order, as a function of the $\overline{MS}$ running parameter $m_b(m_b)$ in two cases. The top curves and the higher horizontal axis correspond to a hierarchical spectrum, in which the squarks are heavy whereas the μ parameter and gaugino masses are light. Then the corrections δm_b are small and the top mass is predicted to be above 180 GeV or so. The bottom curves and the lower horizontal axis correspond to a roughly degenerate spectrum in which the corrections to the b mass are $\mathcal{O}(25\%)$ and negative (i.e. R should be lowered by 25% in Fig. 1 before matching to the experimental value). Now the top can be significantly lighter. In fact, this last argument can be turned around: if the threshold corrections are too large in magnitude and negative then the top mass prediction will be below the experimental lower bound, while if they are large and positive then no value of λ_G will allow agreement with R_{expt}. We thus find the following bounds on the superspectrum: if λ_G is allowed to vary, then

$$-0.37 < \frac{m_{\tilde{g}}\mu}{m_{eff}^2} < 0.08, \tag{3}$$

whereas for fixed λ_G the appropriate limits can be read from Fig. 1.

3 Radiative bottom decay (I)

Do we have any experimental information about these corrections? Of course we have no direct evidence of any of the superpartners, but we can appeal to their indirect appearance in loop diagrams. Consider again the diagrams of Fig. 2, but with one of the b quarks replaced by a strange quark using a flavor-changing vertex, and with a photon attached in all possible ways. We see that the same processes will lead to a contribution to the rare decay $b \to s\gamma$, and with a similar enhancement of $\mathcal{O}(\tan\beta)$ over the usual MSSM scenario *[2, 5]*. The dependence of these diagrams on the superpartner masses is somewhat similar to that of δm_b, except that (a) now the operator is of higher dimension and so is suppressed by the mass of the heaviest superpartner, and (b) typically the higgsino-mediated diagram dominates. If the parameters appearing in this Higgsino diagram, namely μ, A (the trilinear soft SUSY-breaking parameter) and the squark masses, are all comparable and of order the Z mass, then this diagram gives a contribution to the amplitude for $b \to s\gamma$ many times bigger than the standard model or the usual MSSM amplitudes, and is clearly ruled out by the CLEO limit *[6]*. To restore agreement with experiment, either the overall superpartner mass scale must be raised far above the electroweak scale, or else μ or A or both must be suppressed relative to the squark mass in the loop. More quantitatively, we find that either the masses must all be raised to at least $\mathcal{O}(TeV)$, or else if both μ and A are near the Z mass then the squarks must be above either ~ 400 GeV or ~ 700 GeV, depending on whether these diagrams interfere destructively or constructively with the 2-Higgs standard model amplitudes. In any case, however, these restrictions do not yet tell us anything about the combination $m_{\tilde{g}}\mu/m_{eff}^2$ which appears in δm_b; the link between these two will be forged below, when we study the evolution of the entire set of MSSM parameters.

4 Electroweak symmetry breaking

We have assumed in the above that the top-bottom mass hierarchy would arise from a hierarchy of VEVs in the Higgs spectrum, namely $v_U/v_D \equiv \tan\beta \sim \mathcal{O}(50)$. We now consider how such a hierarchy could be generated in the MSSM. Clearly this question is of interest for any model in which the Yukawas themselves do not supply a sufficient hierarchy to explain the large ratio of third-generation quark masses, not just for the equal-Yukawas case to which we have specialized. In studying this question, however, we will pay attention to which spectra are favored by large $\tan\beta$ scenarios, and whether with such spectra and our unification assumptions we can pin down the top mass prediction.

The scalar potential of the neutral Higgs bosons which leads to electroweak symmetry breaking is given by

$$V_0 = m_U^2 |H_U|^2 + m_D^2 |H_D|^2 + \mu B(H_U H_D + h.c.) + (quartic\ terms) \qquad (4)$$

where $m_{U,D}^2 = \mu_{U,D}^2 + \mu^2$ contain the soft-breaking masses and the μ parameter in the superpotential, B is a soft-breaking mass parameter, and the quartic terms arise from D-terms and so are given by gauge couplings. We will evolve these parameters from the GUT to the weak scale using the 1-loop RG equations of the MSSM, stopping the evolution at some typical scale (of order the squark masses) which minimizes the effects of higher-order corrections. The subscript 0 indicates that we will restrict our attention to this (RG-improved) tree-level potential rather than calculate the full 1-loop effective potential or, better yet, explicitly integrate out massive particles and consider full 1-loop matching conditions. We expect [3] that our qualitative discussion of the radiative symmetry breaking will not be jeopardized by this simplification. That is, a more complete calculation will change the numerical values of the GUT parameters needed for correctly breaking the symmetry, but will not significantly alter the size of the domains in parameter space where such breaking is achieved. The conditions for this breaking are well-known:

$$m_U^2 + m_D^2 \geq 2|\mu B| \qquad (5)$$

ensures that the potential is bounded from below, and

$$m_U^2 m_D^2 < \mu^2 B^2 \qquad (6)$$

guarantees the existence of a minimum away from the origin and so breaks the symmetry. In practice, since $|\mu B|$ will always be much less than or at most comparable to $|m_U^2|$ and $|m_D^2|$, we can reduce these requirements to $m_A^2 = m_U^2 + m_D^2 > 0$ (using the expression for the pseudoscalar Higgs mass) and $m_U^2 < 0$ (noting that large $\tan\beta$ means that the up-type Higgs gets the large VEV).

In the usual—and very attractive—scenario of radiative breaking, the two mass parameters start out at the GUT scale with a universal positive value: $m_U^2 = m_D^2 = M_H^2 + \mu^2$, where M_H is the soft-breaking mass of the $\underline{10}_H$ of Higgs in SO(10), or of the $\underline{5}_H$ and $\underline{\bar{5}}_H$ in SU(5). Thus the symmetry is not broken at that scale. However, in the RG evolution to the electroweak scale, the large Yukawa coupling of the top quark to H_U, which gives the top its mass, also drives the mass-squared parameter μ_U^2 of H_U negative (with the help of the QCD coupling), while the absence of a large Yukawa in the down sector keeps the mass-squared of H_D positive. In fact, conditions (5) and (6) are easily satisfied for a large range of initial conditions if $\lambda_G \sim \mathcal{O}(1)$, resulting in a very natural picture of radiative symmetry breaking. This picture is essentially lost in the large $\tan\beta$ scenario, for the following two reasons:

1. Since both Yukawas are comparable [and in fact initially equal in the SO(10) case], the two Higgs doublets tend to run in the same way, so either both stay positive at the electroweak scale and the symmetry does

not break, or both become negative and the potential becomes unbounded from below (a situation which breaks the symmetry but in a Coleman-Weinberg fashion, yielding an "electroweak scale" orders of magnitude higher than the SUSY-breaking scale). The effects which differentiate the evolution of the two Higgs doublets, namely hypercharge and the absence of a right-handed neutrino, are small and a poor replacement for the usual $\lambda_t \gg \lambda_b$ splitting. Interestingly, an $\mathcal{O}(1)$ splitting between λ_t and $\lambda_{b,\tau}$ is still of little use since it is quickly diminished by the fixed-point behavior of these couplings. (Some of these observation have been previously made by T. Banks [4].)

2. Even when electroweak symmetry is broken, a large hierarchy of VEVs must be generated between the two similarly-evolving Higgs doublets. Minimizing the potential V_0 when $\tan\beta \gg 1$ yields

$$-2m_U^2 = m_Z^2 \tag{7}$$

and

$$\frac{1}{\tan\beta} = -\frac{\mu B}{m_u^2 + m_D^2} = -\frac{\mu B}{m_A^2}. \tag{8}$$

The first equation sets the scale, but from the second equation we see that a large hierarchy in VEVs requires the large hierarchy $\mu B \ll m_U^2 + m_D^2$. This, as we show below, implies a degree of fine-tuning between some parameters in the Lagrangian.

5 Solutions of the RG equations (I)

Before analyzing the implications of these two criticisms, we present the 1-loop solutions [3] of the RG equations for the MSSM mass parameters, integrated between $M_G = 3 \times 10^{16}\,GeV$ and a typical squark mass of 300 GeV. (None of our results is sensitive to the exact values of these starting and stopping scales.) The solutions depend on the dimensionless initial values of the gauge and Yukawa couplings α_G and λ_G and on the dimensionful GUT-scale parameters M_{sq}, M_H, μ_G, $M_{1/2}$, A_G, B_G and M_X. M_{sq} and M_H are the soft-breaking masses of the $\underline{10}_H$ and the $\underline{16}_3$ respectively, and M_X will be explained below; we note for now that it vanishes for universal soft-breaking masses. Whenever possible, capital letters denote values at the GUT scale. The RG equations themselves are well-known and will not be presented here. These equations dictate that the low-energy values of the various mass parameters depend very simply on the dimensionful initial values, with coefficients that depend only on the dimensionless ones. Since α_G is known from gauge unification, these coefficients depend only on λ_G. For the representative value $\lambda_G = 1$, the solutions are:

$$2\,m_U^2 = -5.1\,M_{1/2}^{G}{}^2 + 1.2\,M_H^{G2} - 1.6\,M_{sq}^{G2} + 2\mu^2 - 3.8\,M_X^{G2} \tag{9}$$

$$m_A^2 = -4.9\,M_{1/2}^{G}{}^2 + 1.1\,M_H^{G2} - 1.7\,M_{sq}^{G2} + 2\mu^2 + .01\,M_X^{G2} \tag{10}$$

$$m_Q^2 = +4.6\,M_{1/2}^{G}{}^2 - .25\,M_H^{G2} + .51\,M_{sq}^{G2} \qquad + 1.0\,M_X^{G2} \tag{11}$$

$$m_t^2 = +4.1\, M_{1/2}^{G}{}^2 - .27\, M_H^{G\,2} + .46\, M_{sq}^{G\,2} \qquad + .85\, M_X^{G\,2} \qquad (12)$$

$$m_b^2 = +4.2\, M_{1/2}^{G}{}^2 - .23\, M_H^{G\,2} + .55\, M_{sq}^{G\,2} \qquad - 2.9\, M_X^{G\,2} \qquad (13)$$

$$m_L^2 = +.53\, M_{1/2}^{G}{}^2 - .12\, M_H^{G\,2} + .77\, M_{sq}^{G\,2} \qquad - 3.1\, M_X^{G\,2} \qquad (14)$$

$$m_\tau^2 = +.15\, M_{1/2}^{G}{}^2 - .23\, M_H^{G\,2} + .55\, M_{sq}^{G\,2} \qquad + 1.2\, M_X^{G\,2} \qquad (15)$$

$$A_t = +.09\, A_G + 1.8\, M_{1/2}^{G} \qquad\qquad\qquad (16)$$

$$A_b = +.07\, A_G + 1.9\, M_{1/2}^{G} \qquad\qquad\qquad (17)$$

$$A_\tau = +.20\, A_G - .17\, M_{1/2}^{G} \qquad\qquad\qquad (18)$$

$$B = -..86\, A_G - 1.1\, M_{1/2}^{G} + 1.0\, B_G \qquad\qquad (19)$$

$$\mu = .44\, \mu_G \qquad\qquad\qquad\qquad (20)$$

Here Q and L are the squark and slepton doublets, respectively, and t, b and τ are the SU(2)-singlet squarks and sleptons. For clarity of presentation, we have dropped from the first 7 expressions above the terms proportional to A_G^2 and to $A_G M_{1/2}$, since their coefficients are small enough [$\mathcal{O}(0.01 - 0.1)$] and exhibit sufficiently small custodial-SU(2) breaking to be negligible for purposes of symmetry-breaking, at least if A_G is not very much larger than the other mass parameters. We have also used the low-energy value of μ in Eqs. (9-10). These solutions are useful references for the discussions below. These solutions can also be combined [3] with the more analytic approach briefly described in Eqs. (25) to give a complete, semi-analytic solution to the RG equations when custodial SU(2) is an approximate symmetry.

6   Obtaining a hierarchy of VEVs

To understand the second criticism above, let us examine in more detail how Eq. (8) may be satisfied. We concentrate for the moment on six relevant electroweak-scale parameters: the up- and down-type Higgs masses m_U^2 and m_D^2, a typical squark mass m_0^2, the B parameter in the scalar potential, a gaugino mass (specifically the wino mass) $m_{1/2}$, and the μ parameter. This last one may be set to zero by imposing a Peccei-Quinn symmetry on the Lagrangian, so the size of μ measures the breaking of this $\mathcal{PQ}$ symmetry; consequently, μ is multiplicatively renormalized. The previous two, B and $m_{1/2}$, along with the A parameter, transform in the same way under a continuous $\mathcal{R}$ symmetry (so they enter into each other's RG equations) and may be made arbitrarily small by imposing this $\mathcal{R}$ symmetry. With these two symmetries in mind [2], we consider three possible spectra having splittings which lead to a large $\tan\beta$ according to Eq. (8):

mass:	scenario A:		scenario B:		scenario C:	
$7\,m_Z$					$m_D\ m_0$	
m_Z	$m_U\ m_D\ m_0$		$m_U\ m_D\ m_0$	$m_{1/2}\ \mu$	m_U	$B\ m_{1/2}\ \mu$
$\frac{1}{7}\,m_Z$		$B\ m_{1/2}\ \mu$				
$\frac{1}{50}\,m_Z$			B			

Of course there are many other ways to split the parameters and obtain the correct hierarchy, but these will suffice to demonstrate the fine-tuning involved in the splittings. The value of $\tan\beta$ is determined directly only by m_U, m_D, μ and B; we include m_0 and $m_{1/2}$ to illustrate the symmetries. Scenario A involves no tuning at all (at this stage of the analysis): the only hierarchies present are those enforced by the two symmetries. However, as also pointed out by Nelson and Randall [7], such a scenario is ruled out for large $\tan\beta$ since it would imply a light chargino, in disagreement with bounds from LEP [8]. In fact, both μ and $m_{1/2}$ must be comparable to or above the Z mass to satisfy this bound. Therefore we *must* widely split some parameters without a symmetry justification, and this will entail fine-tuning. For example, in scenario B all parameters are kept at the Z mass but the B has been tuned to be light (namely, its initial value at the GUT scale is chosen to almost completely cancel the contributions induced by A and $m_{1/2}$ through the RG evolution). Alternatively, in scenario C, m_U is adjusted to end up much below the other scalar masses (yielding $m_A^2 \simeq 50 m_Z^2$) while the other parameters are kept at the Z mass using the approximate symmetries. In these two scenarios, and in fact *generically* whenever $\mu > m_Z$ and $m_{1/2} > m_Z$, the initial conditions at the GUT scale must be adjusted to at least a relative accuracy of $1/\tan\beta$ to obtain the necessary hierarchy of VEVs. We should point out, however, that such a tuning is no worse than the one which would be needed in the small $\tan\beta$ case if the squarks were experimentally determined to be above 700 GeV or so.

7 Splitting the Higgs doublets

We return now to the first criticism above, and address the splitting between the two Higgs doublets. Recall that after running we need $2m_U^2 < 0$ while $m_U^2 + m_D^2 > 0$. However, the two masses evolve almost in parallel, since custodial symmetry breaking effects, namely hypercharge and the absence of ν_R, are small. Thus, if at the GUT scale the mass parameters are custodial-SU(2) symmetric, the splitting of the two Higgs masses at the weak scale is small relative to a typical SUSY mass M_S at the GUT scale: $m_D^2 - m_U^2 \equiv \epsilon_c M_S^2$ ("c" for custodial). Putting these together, we learn that only within a window of size $\sim \epsilon_c$ in the GUT-scale parameter space can we simultaneously satisfy $m_U^2 < 0$ and $m_A^2 > 0$; if they are satisfied, then $m_Z^2 = -2m_U^2 < \epsilon_c M_S^2$ and $m_A^2 = m_U^2 + m_D^2 < \epsilon_c M_S^2$. In practice, this is usually accomplished [1] using the gaugino mass as the largest mass parameter, so $M_S = M_{1/2} \geq M_{sq,H}$: this is because, according to Eqs. (9,10), custodial breaking effects proportional to $M_{1/2}^2$ lower m_U^2 with respect to m_D^2, while those from the scalar masses $M_{sq,H}^2$ act in the opposite way. Furthermore μ must also typically be $\mathcal{O}(M_{1/2})$ in order to keep m_A^2 positive. Then, in addition to the $\mathcal{O}(\epsilon_c)$ fine-tuning of the Z mass, the large values of $m_{1/2}$ and μ mean that the B parameter must be adjusted beyond the $\mathcal{O}(1/\tan\beta)$ accuracy of the previous paragraph. To see this, we rewrite Eq. (8) in the form

$$\frac{B}{m_{1/2}} = \frac{1}{\tan\beta} \frac{m_U^2 + m_D^2}{\mu\, m_{1/2}} \qquad (21)$$

which quantifies the needed suppression of the electroweak-scale value of B (achieved by fine-tuning its GUT-scale value) relative to the minimum value it would naturally have, namely the value $\sim M_{1/2}$ induced through the RG evolution. In the present case, using $\mu \sim M_{1/2} \sim M_S$ we obtain $B/m_{1/2} \sim (1/\tan\beta)\,\epsilon_c$.

8 D-terms

This highly unnatural state of affairs arises partly because of the degeneracy of the Higgs doublets and their subsequent parallel evolution. A possible remedy is actually generic in SO(10) unification, due to the rank of this group which exceeds by one the rank of SU(5) or the standard model. Thus we write $SO(10) \supset SU(5) \otimes U(1)_X$, where $U(1)_X$ is proportional to $3(B - L) + 4T_{3R}$ [the generator of baryon- minus lepton-number symmetry and a generator of $SU(2)_R$] and couples to the scalar fields according to the following table:

field:	H_U	H_D	Q	t	b	L	τ	$\langle \underline{16}_H \rangle$	$\langle \underline{\overline{16}}_H \rangle$
$U(1)_X$ charge:	-2	2	1	1	-3	-3	1	5	-5

The $\underline{16}_H$ and $\underline{\overline{16}}_H$ are examples of extra superheavy Higgs representations which are typically added in order to break this $U(1)_X$ (in this case by acquiring VEVs in the "ν_R" direction) and reduce the rank of the group. As usual, the spontaneous breakdown of a U(1) leads to a VEV for its D-term, which can induce masses for the various fields which appear in this D-term. In particular, *if we do not assume universal soft-breaking masses* for all scalars, then the soft-breaking masses of the $\underline{16}_H$ and $\underline{\overline{16}}_H$ need not be equal, and therefore their VEVs are also split, in proportion to their mass splitting. This splitting in turn generates a mass splitting in the low-energy MSSM Lagrangian through the cross-term:

$$\mathcal{L} \supset \frac{1}{2}D_X^2 = \frac{1}{2}\left(\langle|\underline{16}_H|^2\rangle - \langle|\underline{\overline{16}}_H|^2\rangle + 2|H_U|^2 - 2|H_D|^2 + \ldots\right)^2 . \tag{22}$$

In fact, this mechanism splits any fields which have different charges under $U(1)_X$. Thus the boundary conditions for the scalar masses at the GUT scale become

$$\begin{aligned}
M_U^2 &= M_H^2 + \mu^2 - 2M_X^2 \\
M_D^2 &= M_H^2 + \mu^2 + 2M_X^2 \\
M_{Q,t,\tau}^2 &= M_{sq}^2 \quad\; + \; M_X^2 \\
M_{b,L}^2 &= M_{sq}^2 \quad\; - \; 3M_X^2
\end{aligned} \tag{23}$$

where the capital letters on the left-hand side serve as reminders that these are the values at the GUT scale, and

$$M_X^2 = \frac{1}{10}(M_{16}^2 - M_{\overline{16}}^2) \tag{24}$$

is a new soft-breaking mass parameter in the low-energy theory. With this mass we no longer need rely on large gaugino masses to split the Higgs doublets:

they can start out being different, and thus even with parallel evolution the symmetry-breaking conditions (5–6) can apparently be satisfied.

One problem with this mechanism is evident from the initial conditions in Eq. (23): not just the Higgs doublets but also the squarks and sleptons are split, so an excessively large M_X^2 could lower M_b^2 or M_L^2 sufficiently to make m_b^2 or m_L^2 negative at the electroweak scale, thereby spontaneously breaking the strong or electromagnetic gauge symmetries. If RG effects were irrelevant, namely for small λ_G, then M_{sq} could always be raised enough to prevent this without affecting $m_{U,D}^2$. However, for $\lambda_G \sim \mathcal{O}(1)$ the squark masses strongly affect the evolution of the Higgs doublets [see Eqs. (9-10)], and only for very constrained ranges of the initial parameters can the electroweak symmetry, and only that symmetry, be spontaneously broken at a reasonable scale. In fact, as we show in brief below, there is a focusing effect that is inherent in the MSSM RG equations when $\lambda_b \sim \lambda_t$, and which inevitably requires an adjustment of the GUT-scale parameters beyond the $1/\tan\beta$ level derived above. We first show this behavior of the RG equations for completely general initial conditions, and then return to discuss the specific case of Eqs. (23).

9 Solutions of the RG equations (II)

Consider the RG equations of the MSSM in the limit of exact $\mathcal{PQ}$ and $\mathcal{R}$ symmetries, in which $\mu = M_{1/2} = A = B = 0$ at all scales. For future reference, we call this scenario *the maximally symmetric case*. This limit is interesting for two reasons: First, no large corrections arise to the b quark mass, and the $R = m_b/m_\tau$ prediction for all values of $\lambda_G \sim \mathcal{O}(1)$ falls nicely within the range allowed by experiment (see Fig. 1); in other words, a heavy top quark near its fixed-point mass favors small δm_b. Second, as we saw above, having a large μ and $m_{1/2}$ calls for fine-tuning B (or some equivalent adjustment), so we'd like to explore the opposite limit to see whether a more natural scenario can be achieved. Of course, eventually we must relax this limit to agree with LEP bounds, but the qualitative behavior we shall discover will persist. If we further approximate $\lambda_b \simeq \lambda_t \equiv \lambda$ and neglect the sleptonic contributions (thereby restoring custodial symmetry), the RG solutions simplify considerably. There are now five relevant parameters. In terms of their initial conditions at the GUT scale, M_U^2, M_D^2, M_Q^2, M_t^2 and M_b^2, the solutions at the electroweak scale are:

$$
\begin{aligned}
-2m_U^2 &= -\tfrac{3}{7}\epsilon_\lambda X - \tfrac{3}{5}\epsilon_\lambda' X' - I - I' \\
m_A^2 &= \tfrac{3}{7}\epsilon_\lambda X + I - I' \\
m_Q^2 &= \tfrac{1}{7}\epsilon_\lambda X - \tfrac{1}{4}I + \tfrac{1}{4}I'' \\
m_t^2 &= \tfrac{1}{7}\epsilon_\lambda X + \tfrac{1}{5}\epsilon_\lambda' X' - \tfrac{1}{4}I - \tfrac{1}{2}I' - \tfrac{1}{4}I'' \\
m_b^2 &= \tfrac{1}{7}\epsilon_\lambda X - \tfrac{1}{5}\epsilon_\lambda' X' - \tfrac{1}{4}I + \tfrac{1}{2}I' - \tfrac{1}{4}I''
\end{aligned}
\tag{25}
$$

where

$$
\epsilon_\lambda = \exp\left(-\frac{7}{8}\int_{\ln m_Z}^{\ln M_G} \frac{\lambda^2}{\pi^2}\, d\ln\mu\right)
\tag{26}
$$

$$\sim 0.085 \qquad (for\ \lambda_G \simeq 1),$$

$$\epsilon'_\lambda = \epsilon_\lambda^{5/7}, \tag{27}$$

and

$$
\begin{aligned}
X &= M_U^2 + M_D^2 + 2M_Q^2 + M_t^2 + M_b^2 \\
X' &= M_U^2 - M_D^2 \qquad\quad + M_t^2 - M_b^2 \\
I &= \tfrac{4}{7}(M_U^2 + M_D^2) - \tfrac{3}{7}(2M_Q^2 + M_t^2 + M_b^2) \\
I' &= \tfrac{2}{5}(M_U^2 - M_D^2) - \tfrac{3}{5}(M_t^2 - M_b^2) \\
I'' &= \qquad\qquad\qquad 2M_Q^2 - M_t^2 - M_b^2 .
\end{aligned}
\tag{28}
$$

(Note again the use of capital letters to denote GUT-scale initial parameters, and recall that U, D, Q, t and b refer to the up-type Higgs, the down-type Higgs, the SU(2)-doublet third-generation squarks, the SU(2)-singlet stop and the SU(2)-singlet sbottom, respectively.)

Evidently, two linear combinations of masses, labeled by X and X' at the GUT scale, renormalize multiplicatively and exponentially contract at low energies for $\lambda_G \sim \mathcal{O}(1)$. The three other linear combinations, I, I' and I'', are invariant. The important observation here is that in the first contraction— the sum rule $m_A^2 + 2m_Q^2 + m_t^2 + m_b^2 = \epsilon_\lambda X$—the coefficient of every term is positive, while we already know that each mass-squared itself must be positive for a proper electroweak-breaking scenario. Therefore each term by itself must be small, less than $\epsilon_\lambda X$. This can only happen if the various combinations of invariants (and possibly also $\epsilon'_\lambda X'$) in the expressions (25) for these terms are adjusted to be small relative to X. The exact constraints that follow from this requirement are given explicitly elsewhere [3]. They are of the form $I, I', I'' \lesssim \max(\epsilon_\lambda X, \epsilon'_\lambda X')$. We learn that, for $\lambda_G \sim \mathcal{O}(1)$ where this focusing effect is important, any given model for the GUT-scale soft-breaking masses must either provide an explanation of why each invariant should be small relative to the sum $X = M_U^2 + M_D^2 + 2M_Q^2 + M_t^2 + M_b^2$, or else that invariant must be tuned by hand to be small. We also learn that the conditions for successful symmetry-breaking are sensitive to any other small effects. One such effect is custodial symmetry violation, which is parametrized above by ϵ_c and results from hypercharge and λ_τ (or the absence of ν_R). In the running of the Yukawas, both of these cause λ_t to slightly exceed λ_b, and thus drive m_U^2 below m_D^2, as in the conventional scenario of electroweak symmetry breaking. In the running of the masses, the contributions of the τ Yukawa have an opposite and numerically more relevant impact. Therefore the custodial-breaking effects make it harder to break the symmetry correctly—significantly harder in the specialized case discussed below. In any realistic scenario there are also contributions from the gauginos and μ, so in the end the sum rule, and therefore the general limit which must be set on I, I' and I'', takes the form

$$
\{I, I', I'', m_A^2 + 2m_Q^2 + m_t^2 + m_b^2\} \lesssim \mathcal{O}\left[\max\left(\epsilon_\lambda, \epsilon_c, \frac{\mu^2}{M_S^2}, \frac{m_{1/2}^2}{M_S^2}\right)\right] M_S^2 \tag{29}
$$

where M_S is the largest mass parameter in the initial conditions at the GUT scale.

If we now return to the more specialized boundary conditions of Eq. (23), we find (after setting $\mu = 0$):

$$
\begin{aligned}
X &= 2M_H^2 + 4M_{sq}^2 \\
X' &= 0 \\
I &= \tfrac{4}{7}(2M_H^2 - 3M_{sq}^2) \\
I' &= -4M_X^2 \\
I'' &= +4M_X^2 \, .
\end{aligned}
\tag{30}
$$

For this choice of boundary conditions, keeping the invariants small imposes only two requirements:

$$
M_H^2 - \tfrac{3}{2}M_{sq}^2 = \epsilon_{\lambda c}M_S^2 \ll M_S^2
\tag{31}
$$

and small M_X. The upper bound on the invariants involves both ϵ_λ and ϵ_c, and as we hinted above they partially cancel in the combination $\epsilon_{\lambda c}$ which enters into the requirements: $|\epsilon_{\lambda c}| < |\epsilon_\lambda|, |\epsilon_c|$. (One sum rule which can be formed under these boundary conditions, $-2m_U^2 + \tfrac{4}{3}m_A^2 + \tfrac{4}{3}m_b^2$, is particularly sensitive to this cancellation[3].) The first requirement, Eq. (31), entails a definite tuning of parameters to a precision of $\epsilon_{\lambda c}$, which does not apparently follow from any symmetry. Note that without this requirement, either color breaks when $M_H^2 > \tfrac{3}{2}M_{sq}^2$ or a Coleman-Weinberg mechanism operates when $M_H^2 < \tfrac{3}{2}M_{sq}^2$. The requirement of small M_X may on the other hand be natural, since [see Eq. (24)] the value of M_X^2 is smaller by an order of magnitude than the soft-breaking masses whose splitting generates the D-terms, and those masses may be expected to be comparable to M_{sq} and M_H. In any case, we see that because of the focusing effect of the RG equations, the D-terms cannot be *allowed* to induce splittings bigger than those we already had through custodial SU(2)-breaking effects. Hence they do not eliminate the criticism that the electroweak symmetry is hard to break when the Yukawas are comparable. But there is still a significant advantage in using these D-terms, since they can now substitute for large values of $m_{1/2}$ and μ, and with light gauginos and μ it is much easier to obtain a large $\tan\beta$, according to Eqs. (8) and (21).

10   Radiative bottom decay (II)

Before putting the various observations to use in examining specific scenarios and their merits, we point out another feature of the solutions to the RG equations which will further constrain the scenarios. As is evident from Eqs. (16-18) (or directly from the RG equations), the initial value A_G hardly affects the low-energy values of $A_{t,b,\tau}$; they are instead largely determined in magnitude and sign by the gaugino mass $M_{1/2}^G$, which also fixes the low-energy gluino mass. (It is difficult, though perhaps not impossible for sufficiently small λ_G, to construct models in which $A_G \gg M_{1/2}$ and yet the electroweak symmetry but neither color nor charge breaks spontaneously and correctly, so we shall disregard this possibility in these proceedings. The implications of tuning A_G to cancel the gaugino mass at low energies in the expression for A_t will be considered elsewhere [3].) This observation, which was also emphasized by Carena et al. [9], directly relates the δm_b corrections of Eq. (2) to the large $b \to s\gamma$ graphs discussed above. (More precisely, the gluino- and higgsino-exchange diagrams for each process are directly related.) The sign of this correlation [9] is such that when $\delta m_b < 0$ (i.e. the predicted R is lowered, and therefore so is the top mass) then the large $b \to s\gamma$ graphs interfere *constructively* with the usual 2-Higgs standard model amplitude, and vice-versa. On one hand, we see from Eq. (3) or from Fig. 1, that the bounds on δm_b are more severe when $\delta m_b > 0$. On the other hand, as noted above, when $\delta m_b < 0$ the interference is constructive and the bounds on the large $b \to s\gamma$ graphs are stricter. Thus these two bounds are much stronger when taken together, and translate into the following statement: either (a) the gauginos or μ or both are significantly lighter than the squarks, or (b) the superpartners are much heavier than the Z.

11 Case studies

With these remarks in mind, we first examine the popular [1] case of universal soft-breaking masses. This scenario has also been recently studied in some detail by Carena et al. [9]. If all soft-breaking scalar masses are equal then the D-term contributions vanish ($M_X \equiv 0$), and we are left with the three parameters μ, $M_{1/2}$ and $M_0 \equiv M_{sq} = M_H$, in addition to B_G which is adjusted at the end to obtain the correct $\tan\beta$ [see Eq. (8)]. We have already mentioned that in this case we need μ comparable to $M_{1/2}$, and both at least as big as M_0, to break electroweak symmetry correctly. We have also seen that these three parameters must be tuned in order to obtain a positive m_Z^2 and m_A^2:

$$m_Z^2 \sim m_A^2 \sim \epsilon_c M_{1/2}^2 \,. \tag{32}$$

Next, to achieve a hierarchy of Higgs VEVs, B_G must be adjusted very precisely such that, at low energies,

$$\frac{B}{m_{1/2}} = \frac{1}{\tan\beta} \frac{m_A^2}{\mu\, m_{1/2}} \sim \frac{\epsilon_c}{\tan\beta} \,. \tag{33}$$

Finally, since μ and the gauginos are not lighter than the squarks, δm_b is rather large, and so must be negative, as can be seen from Eq. (3) or from Fig. 1. Hence the $b \to s\gamma$ constraint is strong, necessitating large superpartner masses of at least $\mathcal{O}(TeV)$ and therefore a *further* tuning (by roughly another order of magnitude) of the three parameters to achieve correct electroweak breaking. Of course, such a highly-tuned scenario is also highly predictive: for example, the spectrum is highly constrained, and the top mass is predicted by the large δm_b corrections to be light.

In search of a more natural scenario, we next relax the assumption of universal soft-breaking masses, which was perhaps arbitrary to begin with. We begin with the maximally symmetric scenario studied previously (see also scenario C above), in which $\mu \sim M_{1/2} \sim A_G \sim B_G\, (\sim m_Z) \ll m_A, \ldots, m_b\, (\ll M_{sq} \sim M_H)$

so that the $\mathcal{PQ}$ and $\mathcal{R}$ symmetries are approximately obeyed. Since this hierarchy directly implies a small δm_b and therefore (from Fig. 1) a relatively large λ_G, the focusing effect of Eq. (26) takes its toll, and once again the initial parameters—this time M_{sq}, M_H, and M_X—must be adjusted to at least $\mathcal{O}\left[\max\left(\epsilon_\lambda, \epsilon_c\right)\right]$. Then, to truly get a maximally symmetric scenario, we choose to obtain a large $\tan\beta$ not by tuning to get a small B but rather by (equivalently) tuning to get a small $m_Z^2 \sim m_A^2/\tan\beta$, which then allows small μ and $m_{1/2}$ relative to the typical low-energy soft-breaking masses and therefore establishes approximate $\mathcal{PQ}$ and $\mathcal{R}$ symmetries even at the electroweak scale:

$$-2m_U^2 = m_Z^2 \sim \frac{\max\left(\epsilon_\lambda, \epsilon_c\right)}{\tan\beta} M_S^2 \tag{34}$$

where $M_S \equiv M_H \sim M_{sq}$. After this adjustment, large $\tan\beta$ is automatic:

$$\frac{B}{m_{1/2}} = \frac{1}{\tan\beta} \frac{m_A^2}{\mu\, m_{1/2}} \sim \mathcal{O}(1)\,. \tag{35}$$

And no further adjustment is necessary to suppress $b \to s\gamma$ since μ and $M_{1/2}$ are small. So this scenario requires much less adjusting than the universal case, although more than just the inevitable $1/\tan\beta$ tuning. (The actual tuning needed is in reality slightly more than indicated, due to the squark- and slepton-splittings induced by the D-terms and due to the effects of custodial symmetry violation, as mentioned above; qualitatively, though, the picture we described remains.) It is also predictive: the superspectrum is hierarchical with light charginos and neutralinos but heavy squarks, and since λ_G is large so is the top mass.

We can continuously retreat from this maximally symmetric case by increasing μ or $M_{1/2}$ (or the related parameters), thereby losing $\mathcal{PQ}$ or $\mathcal{R}$, first at low energies when μ or $m_{1/2}$ become comparable to $\epsilon_\lambda^{1/2} M_S$, and then at all energies when μ or $m_{1/2}$ become comparable to M_S itself. Both may be interesting for model-building (for example, if μ is radiatively generated by A terms at the GUT scale, then $\mathcal{PQ}$ rather than $\mathcal{R}$ symmetry should be evident) and for comparison with experiments once the superspectrum is measured. In either case, the tuning is comparable to the maximally symmetric case, though less tuning is needed in Eq. (34) and more in Eq. (35). We will however defer their discussion to our more complete study *[3]*, and instead consider the case where *both* $\mathcal{PQ}$ and $\mathcal{R}$ symmetries are abandoned in favor of a smaller λ_G. We will call this scenario, unimaginatively, the asymmetric case. It alleviates the focusing effect of Eq. (26) since now $\epsilon_\lambda \sim \mathcal{O}(1)$, which in turn allows the initial conditions (specifically the D-terms) to split the Higgs doublets and the other multiplets by a large amount, so now $M_X \sim M_S$. Small λ_G also entails larger (negative) δm_b corrections to correctly predict m_b/m_τ, and to this end we take $\mu \sim M_{1/2} \sim M_{sq} \sim M_H \equiv M_S$. We then see, however, that what we gain by eliminating the focusing effect we lose by restoring the $b \to s\gamma$ problem: since μ and $m_{1/2}$ are no longer small, we are forced to raise the SUSY scale to $\sim \mathcal{O}(TeV)$, and therefore again to tune the initial parameters to make the Z light:

$$- 2m_U^2 = m_Z^2 \sim \left(\frac{1}{10}M_S\right)^2 . \tag{36}$$

Since the D-term splitting of the Higgs is now large, $m_A^2 \sim M_S^2$, we find that B_G requires only the typical tuning

$$\frac{B}{m_{1/2}} = \frac{1}{\tan\beta}\frac{m_A^2}{\mu\, m_{1/2}} \sim \frac{1}{\tan\beta} . \tag{37}$$

This scenario is comparable in its naturalness (or lack thereof) to the symmetric case. The superspectrum is uniformly heavy rather than hierarchical, and the top is light since λ_G is small.

12 Conclusions

These last two scenarios are qualitatively the best one can hope for from an SO(10) model with Yukawa unification, for two reasons. First, obtaining a large $\tan\beta$ and thereby the top-bottom hierarchy is never natural in the MSSM, due to the LEP bounds on μ and $m_{1/2}$ which force either m_A^2 to be much heavier than the Z or B to be much lighter than the gauginos. Second, the last two scenarios illustrate how, for large λ_G, the *inherent* focusing property of the RG equations in the symmetric limit necessitates a further tuning of the initial parameters, while for small λ_G a similar tuning is mandated by bounds on the rate of $b \to s\gamma$. There are also intermediate scenarios, with only one of the symmetries, but they are apparently no more natural. These various possibilities are considered in more detail elsewhere [3]. The universal case is generically much more tuned than most of these scenarios, as we have shown. Hence departures from universality, although possibly dangerous for the flavor-changing neutral current interactions they can induce in some models, are strongly favored in achieving a large $\tan\beta$.

What is the status of the predictions? Perhaps surprisingly, the top mass is not an independent prediction of Yukawa unification, but rather depends strongly (i.e. non-logarithmically) on a certain ratio of superpartner masses appearing in δm_b. Since the large- and small-λ_G cases are equally fine-tuned, naturalness arguments do not single out any particular top mass within this SO(10) framework. Instead, information about the top mass can be combined with Fig. 1 and Eq. (2) to restrict the mass parameters which can be consistent with Yukawa unification, and to determine a favored superspectrum. It must be admitted that, with Yukawa unification, the attractive conventional picture of *radiative* electroweak symmetry breaking due to a large λ_t but small λ_b is largely lost: the symmetry is either broken radiatively using small custodial isospin-violating effects and extraordinarily fine-tuned initial conditions, or else the Higgs doublets are already split at the GUT scale. Furthermore, we have seen that all the large $\tan\beta$ scenarios are technically unnatural. On the other hand (and to some extent because of the necessary fine-tuning), they are certainly predictive and so will be tested in future accelerators: for example, if the charginos are light but the squarks are heavy, the top should also be heavy; if the $SU(2)_L$-singlet bottom squark or the doublet sleptons were lighter than

the other squarks and sleptons, this would be a sign that the D-term splittings were large *[10]*; and finally, $\tan\beta$ can itself eventually be measured to decisively confirm or dismiss the large $\tan\beta$ hypothesis.

Even before any further experimental input, there are a few theoretical avenues worth pursuing which would make Yukawa unification much more attractive. First, as discussed also by J. Lykken in these proceedings *[11]*, string models which lead to true grand-unified models as their low-energy effective Lagrangians can exhibit higher symmetries in certain sectors of the GUT than in other sectors. In particular, a string theory with an SO(10) gauge symmetry might break to an effective low-energy SU(5) GUT in a stringy way, leaving *all three* Yukawa couplings unified at the Planck scale à la SO(10) but splitting the $\underline{5}_H$ from the $\underline{\bar{5}}_H$ soft masses, namely M_U^2 from M_D^2. Note that unlike the D-term splittings we have considered before, M_U^2 and M_D^2 could now be split without necessarily decreasing some squark or slepton masses. This approach can thus provide more freedom in the choice of boundary conditions—although as we have shown, some features of the RG equations, in particular the focusing effect, are inherent in the equations themselves in certain limits and apply to any boundary conditions, so that freedom could either be the key to a more natural scenario (see the following remark) or could necessitate still more arbitrary fine-tunings at the GUT scale. Second, we have shown that in the symmetric case with large λ_G, the ratio of the soft-breaking masses for the $\underline{10}_H$ and the $\underline{16}_3$ needs to have a certain value if the electroweak symmetry is to break correctly. While such a value is arbitrary in the context of an SO(10) model and therefore apparently requires a fine-tuning, perhaps this value could be explained as a ratio of integral conformal weights in the context of a string theory into which the GUT is embedded. (Notice that this value is favored simply by the unification of Yukawas at some large scale, and does not depend on an SO(10) symmetry.) If such an explanation could be found, then the symmetric large-λ_G case would now be strongly favored: it would only require the minimal $1/\tan\beta$ tuning (and would predict a heavy top!). In fact, if the squarks were to be *experimentally determined* to be heavy while the gauginos were light, then this case would be no more fine-tuned than the conventional small $\tan\beta$ scenario, but would have the advantage of explaining the top-bottom mass hierarchy (through the $\mathcal{PQ}$ and $\mathcal{R}$ symmetries)—which, after all, was historically the motivation for studying Yukawa unification.

13   Acknowledgments

We would like to thank M. Carena and C. Wagner for enjoyable discussions, in particular regarding the sign of the correlation between the diagrams in $b \to s\gamma$ and those in δm_b. We have also benefited from discussions with J. Lykken and F. Zwirner regarding string models. U.S. would like to thank the organizers of the Second IFT Workshop on Yukawa Couplings and the Origins of Mass for a stimulating conference. This work was supported in part by NSF grants PHY-89-17438 (U.S.), PHY-91-21039 (R.R.) and PHY-90-21139 (L.J.H.) and in part by the United States Department of Energy under contract no. DE-AC03-76SF00098 (L.J.H.).

References

[1] G.F. Giudice and G. Ridolfi, Z. Phys. C **41**, 447 (1988); M. Olechowski and S. Pokorski, Phys. Lett. B **214**, 393 (1988); W. Majerotto and B. Mösslacher, Z. Phys. C **48**, 273 (1990); P.H. Chankowski, Phys. Rev. D **41**, 2877 (1990); B. Ananthanarayan, G. Lazarides, and Q. Shafi, Phys. Rev. D **44**, 1613 (1991); Q. Shafi and B. Ananthanarayan, *Proceedings of the 1991 Trieste Summer School*, edited by E. Gava *et al.* (World Scientific, Singapore, 1992), p. 233 [Bartol Research Institute Report No. BA-91-76 (1991)]; H. Arason, D.J. Castaño, B.E. Keszthelyi, S. Mikaelian, E.J. Piard, P. Ramond, and B.D. Wright, Phys. Rev. Lett. **67**, 2933 (1991); S. Kelley, J.L. Lopez, and D.V. Nanopoulos, Phys. Lett. B **274**, 387 (1992); M. Drees and M.M. Nojiri, Nucl. Phys. **B369**, 54 (1992); V. Barger, M.S. Berger, and P. Ohmann, Phys. Rev. D **47**, 1093 (1993); B. Ananthanarayan, G. Lazarides, and Q. Shafi, Phys. Lett. B **300**, 245 (1993); M. Carena, S. Pokorski and C.E.M. Wagner, Nucl. Phys. **B406** 59 (1993); B. Ananthanarayan and Q. Shafi, Bartol Research Institute Report No. BA-93-25 (1993); and Carena et al. *[9]*.

[2] L. J. Hall, R. Rattazzi and U. Sarid, Lawrence Berkeley Laboratory Report No. LBL-33997 (June 1993, revised March 1994), submitted to Phys. Rev. D.

[3] R. Rattazzi and U. Sarid, Stanford University Report No. SU-ITP-94-16, in preparation.

[4] T. Banks, Nucl. Phys. **B303**, 172 (1988).

[5] R. Garisto and J.N. Ng, Phys. Lett. **B315**, 372 (1993); M.A. Díaz, Phys. Lett. **B322**, 207 (1994); F.M. Borzumati, DESY Report No. DESY 93-090 (October 1993).

[6] E. Thorndike, CLEO Collab., talk given at the 1993 Meeting of the American Physical Society, Washington, D.C., April 1993.

[7] A.E. Nelson and L. Randall, Phys. Lett. **B316**, 516 (1993).

[8] F. Zwirner, in *Physics and experiments with linear colliders*, edited by R. Orava, P. Eerola and M. Nordberg (World Scientific, Singapore, 1992), Vol. I, p. 309 [CERN Theoretical Report No. CERN-Th.6357/91], and references therein.

[9] M. Carena, M. Olechowski, S. Pokorski and C.E.M. Wagner, CERN Theoretical Report No. CERN-Th.7163/94 (1994).

[10] Y. Kawamura, H. Murayama and M. Yamaguchi, Phys. Lett. **B324**, 52 (1994).

[11] J. Lykken, this volume.

General Analysis of Phases in Fermion Mass Matrices

ROBERT SHROCK
INSTITUTE FOR THEORETICAL PHYSICS
STATE UNIVERSITY OF NEW YORK AT STONY BROOK
STONY BROOK, NY 11794-3840

Abstract: We present a general method to determine (1) the number of unremovable, physically meaningful phases in fermion mass matrices and (2) which elements of these matrices can be made real by rephasings of fermion fields. The results are applied to several models.

1 Introduction

Understanding fermion masses and quark mixing remains one of the most important outstanding problems in particle physics. In an effort to gain insight into this problem, many studies of models of fermion mass matrices have been carried out. A given model is characterized by the number of parameters (amplitudes and phases) which specify the quark and lepton mass matrices. Thus, a very important problem is to determine, for any model, how many unremovable, and hence physically meaningful, phases occur in the fermion mass matrices and which elements of these matrices can be made real by rephasings of fermion fields. In collaboration with A. Kusenko, we have recently presented general solutions to this problem for the quark case *[1, 2]* and lepton case *[3, 4]*. We discuss our methods for quarks first and then leptons.

2 Theorem on the Number of Physical Phases

The quark mass terms are taken to arise from interactions which are invariant under the standard-model gauge group $G_{SM} = SU(3) \times SU(2) \times U(1)$, via the spontaneous symmetry breaking of G_{SM}. In the standard model and its supersymmetric extensions, the resultant mass terms appear at the electroweak level via (dimension-4) Yukawa couplings (which in general have contributions from higher-dimension operators present at a high mass scale near to that of the string scale). These mass terms can thus be written in terms of the G_{SM} quark fields as

$$ -\mathcal{L}_m = \sum_{j,k=1}^{N_G} \left[(\bar{Q}_{jL})_1 M_{jk}^{(u)} u_{kR} + (\bar{Q}_{jL})_2 M_{jk}^{(d)} d_{kR} \right] + h.c. \tag{1} $$

where j and k are generation labels; $N_G = 3$ is the number of generations of standard-model fermions; Q_{jL} is an SU(2) doublet, with $Q_{1L} = \binom{u}{d}_L$, $Q_{2L} = \binom{c}{s}_L$, $Q_{3L} = \binom{t}{b}_L$; the subscript a on $(Q_{jL})_a$ is the SU(2) index; and the SU(2)-singlet right-handed quark fields are denoted as u_{kR} with $u_{1R} = u_R$, $d_{1R} = d_R$, $u_{2R} = c_R$, $d_{2R} = s_R$, etc. The diagonalization of the mass matrices $M^{(u)}$ and $M^{(d)}$ yields the mass eigenstates and the CKM matrix V. It is convenient to write these mass matrices in terms of dimensionless Yukawa matrices as

$$ Y^{(f)} = 2^{1/2} v_f^{-1} M^{(f)} , \quad f = u,d \tag{2} $$

where v_u and v_d are Higgs vacuum expectation values ($v_u = v \sin \beta$ and $v_d = v \cos \beta$ in the minimal supersymmetric standard model (MSSM)).

To count the number of unremovable, and hence physically meaningful, phases in the quark mass matrices, we rephase the fermion fields in (1) so as to remove all possible phases in these matrices. One can perform the rephasings defined by

$$Q_{jL} = e^{-i\alpha_j} Q'_{jL} \tag{3}$$

$$u_{jR} = e^{i\beta_j^{(u)}} u'_{jR} \tag{4}$$

$$d_{jR} = e^{i\beta_j^{(d)}} d'_{jR} \tag{5}$$

for $1 \leq j \leq 3$. In terms of the rephased fermion fields, the Yukawa matrices have elements

$$Y_{jk}^{(f)\prime} = e^{i(\alpha_j + \beta_k^{(f)})} Y_{jk}^{(f)} \tag{6}$$

for $f = u, d$. Thus, if $Y^{(f)}$ has N_f nonzero, and, in general, complex elements, then the N_f equations for making these elements real are

$$\alpha_j + \beta_k^{(f)} = -\arg(Y_{jk}^{(f)}) + \eta_{jk}^{(f)} \pi \tag{7}$$

for $f = u, d$, where the set $\{jk\}$ runs over each of these nonzero elements, and $\eta_{jk}^{(f)} = 0$ or 1. The η_{jk} term allows for the possibility of making the rephased element real and negative rather than positive. This will not affect the counting of unremovable phases, and henceforth we set $\eta_{jk}^{(f)} = 0$. We define the 9-dimensional vector

$$v = (\{\alpha_i\}, \{\beta_i^{(u)}\}, \{\beta_i^{(d)}\})^T \tag{8}$$

and the vector

$$w = (\{-\arg(Y_{jk}^{(u)}) + \eta_{jk}^{(u)}\pi\}, \ \{-\arg(Y_{mn}^{(d)}) + \eta_{mn}^{(d)}\pi\})^T \tag{9}$$

of dimension equal to the number of rephasing equations $N_{eq} = N_u + N_d$. We can then write (7) for $f = u, d$ as

$$Tv = w \tag{10}$$

which defines the N_{eq}-row by 9-column matrix T. Clearly, $rank(T) < \min\{N_{eq}, 9\}$. For the viable models which we have studied, $N_{eq} \geq 9$; in this case, one has the more restrictive bound $rank(T) \leq 8$.

Our first theorem is as follows: The number of unremovable phases N_p in $Y^{(u)}$ and $Y^{(d)}$ is

$$N_p = N_{eq} - rank(T) \tag{11}$$

The proof is given in Refs. [1, 2]. There is actually a one-to-one correspondence between each such unremovable phase and an independent phase of a certain product of elements of Yukawa matrices which is invariant under fermion field rephasings. We note that, in general, the result (11) does not depend on whether or not $Y_{jk}^{(f)} = Y_{kj}^{(f)}$ initially. Hence making $Y^{(f)}$ (complex) symmetric for either $f = u$ or $f = d$ does not, in general, result in any reduction in N_p.

3 Rephasing Invariants and Theorems on Locations of Phases

A fundamental question concerns which elements of $Y^{(u)}$ and $Y^{(d)}$ can be made real by fermion rephasings. Our general method to answer this is to construct all independent complex products of elements of the $Y^{(f)}$, $f = u, d$, having the property that these products are invariant under the rephasings (3)-(5). We let N_{ia} denote the number of linearly independent arguments among the N_{inv} independent complex invariants.

We thus construct first a set of rephasing-invariant products depending on the up and down quark sectors individually:

$$P^{(f)}_{2n;j_1 k_1,\ldots j_n k_n;\sigma_L} = \prod_{a=1}^{n} Y^{(f)}_{j_a k_a} Y^{(f)*}_{\sigma_L(j_a) k_a} \tag{12}$$

where $f = u, d$, and σ_L is an element of the permutation group S_n.

Secondly, we construct a set of invariants connecting the up and down quark sectors:

$$Q^{(s,t)}_{2n;\{j\},\{k\},\{m\};\sigma_L,\sigma_u;\sigma_d} = (\textstyle\prod_{a=1}^{s} Y^{(u)}_{j_a k_a})(\prod_{b=1}^{t} Y^{(d)}_{j_{s+b} m_b})(\prod_{c=1}^{s} Y^{(u)*}_{\sigma_L(j_c)\sigma_u(k_c)}) \times$$
$$(\textstyle\prod_{e=1}^{t} Y^{(d)*}_{\sigma_L(j_{s+e})\sigma_d(m_e)}) \tag{13}$$

where $s, t \geq 1$, $s + t = n$, $\sigma_L \in S_n$, $\sigma_u \in S_s$, and $\sigma_t \in S_t$.

At quartic order, $2n = 4$, $\sigma_L \in S_2$ in eq. (12) and we obtain the complex invariants

$$P^{(f)}_{j_1 k_1, j_2 k_2} \equiv P^{(f)}_{4;j_1 k_1, j_2 k_2;\sigma_L=\tau} = Y^{(f)}_{j_1 k_1} Y^{(f)}_{j_2 k_2} Y^{(f)*}_{j_2 k_1} Y^{(f)*}_{j_1 k_2} \tag{14}$$

for $f = u, d$, where τ is the transposition in S_2. At this quartic order, there is only one Q-type complex invariant,

$$Q_{j_1 k_1, j_2 m_1} \equiv Q^{(1,1)}_{4;\{j_1,j_2\},\{k_1\},\{m_1\};\sigma_L=\tau,\sigma_u=1} = Y^{(u)}_{j_1 k_1} Y^{(d)}_{j_2 m_1} Y^{(u)*}_{j_2 k_1} Y^{(d)*}_{j_1 m_1} \tag{15}$$

These satisfy the symmetries $P^{(f)}_{j_1 k_1, j_2 k_2} = P^{(f)}_{j_2 k_2, j_1 k_1}$, $P^{(f)}_{j_1 k_1 j_2 k_2} = P^{(f)*}_{j_1 k_2, j_2 k_1}$, and $Q_{j_1 k_1, j_2 m_1} = Q^{*}_{j_2 k_1, j_1 m_1}$.

At order $2n = 6$, we find one independent P-type complex invariant for each quark sector $f = u, d$, and two independent Q-type complex invariants:

$$P^{(f)}_{j_1 k_1, j_2 k_2, j_3 k_3} = Y^{(f)}_{j_1 k_1} Y^{(f)}_{j_2 k_2} Y^{(f)}_{j_3 k_3} Y^{(f)*}_{j_3 k_1} Y^{(f)*}_{j_1 k_2} Y^{(f)*}_{j_2 k_3} \tag{16}$$

and

$$Q^{(fff')}_{j_1 k_1, j_2 k_2, j_3 m_1} = Y^{(f)}_{j_1 k_1} Y^{(f)}_{j_2 k_2} Y^{(f')}_{j_3 m_1} Y^{(f)*}_{j_1 k_2} Y^{(f)*}_{j_3 k_1} Y^{(f')*}_{j_2 m_1} \tag{17}$$

where $(fff') = (uud)$ or (ddu). These satisfy the symmetries

$$P^{(f)}_{j_1 k_1, j_2 k_2, j_3 k_3} = P^{(f)*}_{j_3 k_2, j_2 k_1, j_1 k_3} \tag{18}$$

and

$$Q^{(fff')}_{j_1 k_1, j_2 k_2, j_3 m_1} = Q^{(fff')*}_{j_1 k_2, j_3 k_1, j_2 m_1} \tag{19}$$

For a given model, one constructs the maximal set of N_{ia} independent complex invariants of lowest order(s), whose arguments (phases) are linearly independent. Then we have the theorem that (a) each of these invariants implies a constraint that the elements contained within it cannot, in general, all be made simultaneously real; (b) this constitutes the complete set of constraints on which elements of $Y^{(u)}$ and $Y^{(d)}$ can be made simultaneously real; and hence (c) $N_p = N_{ia}$. Further, (d) for the physical case of $N_G = 3$ generations, the complex invariants of order 4 and 6 suffice to describe all of the phase constraints in an arbitrary model of quark mass matrices. The proof of (a) is obvious; the proofs of (b), (c) and (d) are given in Refs. [1, 2].

Note that if in a given model there are as as many independent quartic complex invariants with independent arguments as there are unremovable phases, then the arguments of complex invariants of not only 8'th and higher order, but also of 6'th order are expressible in terms of the arguments of the quartic invariants and hence yield no new phase constraints. In Ref. [2] we have presented a useful graphical representation which expedites the enumeration and construction of the complex invariants.

One may gain further insight into the physical $N_G = 3$ case by considering the generalization to arbitrary N_G. One finds [2], for example, that the maximal number of independent complex quartic P-type invariants of each type f $(= u$ or $d)$ is

$$(N_{P_4,f})_{max} = \binom{N_G}{2}^2 \tag{20}$$

while the maximal number of quartic Q-type invariants is

$$(N_{Q_4})_{max} = \frac{N_G^3(N_G - 1)}{2} \tag{21}$$

It follows that the maximal number of complex quartic invariants is

$$(N_{inv,4})_{max} = \frac{N_G^2(N_G - 1)(2N_G - 1)}{2} \tag{22}$$

Corresponding results were obtained for the numbers of higher-order invariants as a function of N_G.

It is important to stress the difference between the unremovable and hence physically meaningful phases in the quark mass matrices, and the unremovable phase(s) in the Cabibbo-Kobayashi-Maskawa (CKM) quark mixing matrix, V. Consider general N_G. The rephasing properties of V are quite different from those of the $Y^{(f)}$, as is obvious from the fact that the charged weak current is chirality-preserving, connecting left-handed with left-handed chiral components of individual physical mass eigenstates of quark fields, whereas the Yukawa terms connect left-handed with right-handed chiral components. Furthermore, the Yukawa matrices, being defined via $SU(3) \times SU(2) \times U(1)$-invariant Yukawa couplings, involve the actual $SU(3) \times SU(2) \times U(1)$ fields, i.e. the N_G left-handed $SU(2)$ doublets Q_{jL} and the $2N_G$ right-handed SU(2) singlets u_{jR} and d_{jR} (not mass eigenstates). The rephasing equations for the Yukawa matrices were given above in (3)-(5). In contrast, when one determines the unremovable phases in

V, since it is defined in terms of the mass eigenstates themselves, the rephasing equations which one uses are

$$u_{j,Lm} = e^{i\theta_j} u'_{j,Lm} \tag{23}$$

$$d_{j,Lm} = e^{i\phi_j} d'_{j,Lm} \tag{24}$$

for $j = 1, ..., N_G$. The number of these rephasing equations is $2N_G$, as opposed to the $3N_G$ rephasing equations (3)-(5) for the Yukawa matrices. Recall that there are $(N_G - 1)^2$ real parameters specifying V, of which $N_G(N_G + 1)/2$ are rotation angles and the remaining $N_{p,V} = (N_G - 1)(N_G - 2)/2$ are unremovable, physically meaningful, phases. Note that $N_{p,V}$ is only a function of N_G, whereas the number of unremovable phases in the Yukawa matrices, N_p, does not just depend on N_G but rather on the detailed structure of these matrices. An important inequality is that the number of unremovable phases in the quark Yukawa matrices is at least as large as the number of unremovable phases in the CKM matrix V:

$$N_p \geq N_{p,V} \tag{25}$$

4 Applications to Specific Models

We illustrate our general methods with some specific models. Although our methods do not depend on the detailed origin of a particular model, we note that in an appealing class of models one assumes the minimal supersymmetric standard model, which stabilizes the hierarchy and gives rise to gauge coupling unification at a mass $M_{unif.} \simeq 1 - 2 \times 10^{16}$ GeV. One hypothesizes some simple form for the Yukawa matrices at this mass scale, and then evolves the Yukawa matrices down to the electroweak scale, where they are diagonalized to yield the quark and lepton masses and the CKM matrix. For appropriate choices of parameters, these models give acceptable fits to the data on quark masses and mixing. Since our results do not depend on whether or not the Yukawa matrices are symmetric at some mass scale, we shall consider the general case of (complex) non-symmetric Yukawa matrices.

The first model is defined by the Yukawa matrices (at $M_{unif.}$)

$$Y^{(u)} = \begin{pmatrix} 0 & A_{12} & 0 \\ A_{21} & A_{22} & 0 \\ 0 & 0 & A_{33} \end{pmatrix} \tag{26}$$

$$Y^{(d)} = \begin{pmatrix} 0 & B_{12} & 0 \\ B_{21} & B_{22} & B_{23} \\ 0 & B_{32} & B_{33} \end{pmatrix} \tag{27}$$

Special cases of this model have been studied in Refs. *[5]* and *[6]*. This model has $N_{eq} = 10$. We obtain the 10-row by 9-column matrix

$$T = \begin{pmatrix} 1 & 0 & 0 & 0 & 1 & 0 & 0 & 0 & 0 \\ 0 & 1 & 0 & 1 & 0 & 0 & 0 & 0 & 0 \\ 0 & 1 & 0 & 0 & 1 & 0 & 0 & 0 & 0 \\ 0 & 0 & 1 & 0 & 0 & 1 & 0 & 0 & 0 \\ 1 & 0 & 0 & 0 & 0 & 0 & 0 & 1 & 0 \\ 0 & 1 & 0 & 0 & 0 & 0 & 1 & 0 & 0 \\ 0 & 1 & 0 & 0 & 0 & 0 & 0 & 1 & 0 \\ 0 & 1 & 0 & 0 & 0 & 0 & 0 & 0 & 1 \\ 0 & 0 & 1 & 0 & 0 & 0 & 0 & 1 & 0 \\ 0 & 0 & 1 & 0 & 0 & 0 & 0 & 0 & 1 \end{pmatrix} \tag{28}$$

and calculate that $rank(T) = 8$. Our theorem (11) then implies that this model has $N_p = N_{eq} - rank(T) = 2$ unremovable phases in the Yukawa matrices $Y^{(f)}$, $f = u, d$. According to our theorem on invariants, there are therefore $N_{ia} = N_p = 2$ complex invariants with independent arguments. We find two corresponding quartic complex invariants

$$P^{(d)}_{22,33} = B_{22} B_{33} B^*_{32} B^*_{23} \tag{29}$$

and

$$Q_{12,22} = A_{12} B_{22} A^*_{22} B^*_{12} \tag{30}$$

Since, in general, $\arg(P^{(d)}_{22,33}) \neq 0, \pi$, it follows that (i) at least one of the $N_p = 2$ unremovable phases must reside among the set

$$S_{P^{(d)}_{22,33}} = \{B_{22}, B_{23}, B_{32}, B_{33}\} \tag{31}$$

and (ii) the 2×2 submatrix in $Y^{(d)}$ formed by the set (31),

$$\begin{pmatrix} B_{22} & B_{23} \\ B_{32} & B_{33} \end{pmatrix} \tag{32}$$

and hence also $Y^{(d)}$ itself, cannot be made real or hermitian. Second, since, in general, $\arg(Q_{12,22}) \neq 0, \pi$, it follows that the elements in the set

$$S_{Q_{12,22}} = \{A_{12}, A_{22}, B_{12}, B_{22}\} \tag{33}$$

(iii) cannot in general be made simultaneously real and thus, (iv) if one chooses $Y^{(u)}$ real, then it is not possible to make B_{12} and B_{22} both real. These constitute the complete set of rephasing constraints on the $Y^{(f)}$, $f = u, d$. Our results show that it is not, in general, true that a (complex) symmetric Yukawa or mass matrix can be transformed to a hermitian matrix by the rephasings of fermion fields. As an explicit example, after rephasings, one could obtain

$$Y^{(u)'} = \begin{pmatrix} 0 & |A_{12}| & 0 \\ |A_{21}| & |A_{22}| & 0 \\ 0 & 0 & |A_{33}| \end{pmatrix} \tag{34}$$

and

$$Y^{(d)'} = \begin{pmatrix} 0 & |B_{12}|e^{-i\,\arg(Q_{12,22})} & 0 \\ |B_{21}| & |B_{22}|e^{i\,\arg(P^{(d)}_{22,33})} & |B_{23}| \\ 0 & |B_{32}| & |B_{33}| \end{pmatrix} \tag{35}$$

There is one 6'th order complex invariant for this model:

$$Q^{(ddu)}_{32,23,12} = B_{32}B_{23}A_{12}B^*_{33}B^*_{12}A^*_{22} \tag{36}$$

but our theorem shows that since all of the phase constraints are already accounted for by the quartic complex invariants, this 6'th order invariant cannot yield any new constraint. Indeed, an explicit calculation shows that the argument (phase) of the invariant (36) is a linear combination of the arguments of the two quartic invariants (29) and (30).

A second model is defined by $Y^{(d)}$ as in (26) and

$$Y^{(u)} = \begin{pmatrix} 0 & A_{12} & 0 \\ A_{21} & 0 & A_{23} \\ 0 & A_{32} & A_{33} \end{pmatrix} \tag{37}$$

For this model, $N_{eq} = 11$, and we calculate $rank(T) = 8$, so that there are $N_p = 3$ unremovable phases. We find four complex quartic invariants, $P^{(d)}_{22,33}$ as in model 1, together with

$$Q_{12,32} = A_{12}B_{32}A^*_{32}B^*_{12} \tag{38}$$

$$Q_{23,32} = A_{23}B_{32}A^*_{33}B^*_{22} \tag{39}$$

and

$$Q_{23,33} = A_{23}B_{33}A^*_{33}B^*_{23} \tag{40}$$

Of these, the arguments of the first, third, and fourth invariants are related according to

$$\arg(P^{(d)}_{22,33}) + \arg(Q_{23,32}) - \arg(Q_{23,33}) = 0 \tag{41}$$

so that these four quartic invariants have three independent phases. Each of the invariants yields the phase constraint that not all of the Yukawa elements comprising it can be made real. These constitute a complete set of phase constraints. An example of a rephasing of the Yukawa matrices allowed by these constraints is $Y^{(u)}$ real,

$$Y^{(u)'} = \begin{pmatrix} 0 & |A_{12}| & 0 \\ |A_{21}| & 0 & |A_{23}| \\ 0 & |A_{32}| & |A_{33}| \end{pmatrix} \tag{42}$$

and

$$Y^{(d)'} = \begin{pmatrix} 0 & |B_{12}|e^{-i\,\arg(Q_{12,32})} & 0 \\ |B_{21}| & |B_{22}|e^{-i\,\arg(Q_{23,32})} & |B_{23}| \\ 0 & |B_{32}| & |B_{33}|e^{i(\arg(P^{(d)}_{22,33})+\arg(Q_{23,32}))} \end{pmatrix} \tag{43}$$

In Refs. [1, 2], several other models were also analyzed. Although the quartic invariants usually yielded all of the phase constraints, an example was given of a model where a sixth order complex invariant provides an independent phase constraint.

5 Phases in Leptonic Mass Matrices

The analysis of phases for the lepton sector is more complicated than that for the quark sector because of the general presence, in the neutral lepton case, of three types of fermion bilinears, Dirac, left-handed Majorana, and right-handed Majorana, each with its own gauge and rephasing properties. The mass terms for the charged leptons can be written in terms of the G_{SM} lepton fields as

$$-\mathcal{L}_{mass,\ell} = \sum_{j,k=1}^{3} \left[(\bar{L}_{jL})_2 M_{jk}^{(\ell)} \ell_{kR} \right] + h.c. \tag{44}$$

where $j, k = 1, 2, 3$ denote generation indices, and $L_{jL} = \begin{pmatrix} \nu_{\ell_j} \\ \ell_j \end{pmatrix}_L$, are the left-handed SU(2) doublets of leptons (where $\ell_1 = e$, $\ell_2 = \mu$, $\ell_3 = \tau$). The index $a = 2$ in $(\bar{L}_{jL})_2 = \bar{\ell}_{jL}$ is the SU(2) index. Further, $\ell_{k,R}$ are the right-handed SU(2) singlets, and $M^{(\ell)}$ is the charged lepton mass matrix. We use the result here from LEP and SLC that there are three generations of standard model fermion with associated light neutrinos.

In contrast to the charged lepton sector, where one at least knows the relevant fields, in the neutral lepton sector, *a priori*, one does not; in addition to the three known left-handed $I = 1/2$, $I_3 = 1/2$ Weyl neutrino fields ν_{jL}, $j = 1 - 3$, there could be some number n_s of electroweak-singlet neutrino fields $\chi_{j,R}$, $j = 1, 2, ... n_s$. The general neutrino mass matrix is given by

$$-\mathcal{L}_{mass,N} = \frac{1}{2} (\bar{\nu}_L \ \bar{\chi}^c_L) \begin{pmatrix} M^{(L)} & M^{(D)} \\ (M^{(D)})^T & M^{(R)} \end{pmatrix} \begin{pmatrix} \nu^c_R \\ \chi_R \end{pmatrix} + h.c. \tag{45}$$

where $\nu_L = (\nu_e, \nu_\mu, \nu_\tau)^T_L$, $\chi_R = (\chi_1, ..., \chi_{n_s})^T_R$; $M^{(L)}$ and $M^{(R)}$ are 3×3 and $n_s \times n_s$ Majorana mass matrices, and $M^{(D)}$ is a 3-row by n_s-column Dirac mass matrix. (In the context of supersymmetric extensions of the standard model, we assume unbroken R parity so that the neutrinos do not mix with the neutralinos (higgsinos and neutral color-singlet gauginos).) We denote the $(3+n_s) \times (3+n_s)$ neutrino mass matrix in (45) as $M^{(N)}$, where N denotes "neutral", $Q = 0$. Recall that $M^{(f)} = M^{(f)T}$ for $f = L, R$, and hence $M^{(N)} = M^{(N)T}$.

As in the quark case, one writes down a set of $N_{eq,\ell}$ leptonic rephasing equations, which can be expressed compactly by a leptonic $T_\ell v_\ell = w_\ell$ equation analogous to (10). The number of unremovable phases in the leptonic mass matrices is then given by $N_{p,\ell} = N_{eq,\ell} - rank(T_\ell)$.

The constraints on the rephasings of individual elements are determined by constructing and analyzing the maximal set of complex invariant products with independent arguments. There is again a 1-1 correspondence between unremovable phases and complex invariant products with independent arguments. However, in contrast to the quark case, the leptonic invariants of 4'rth and 6'th order do not suffice to describe all phase constraints. As an illustration of this difference, in Ref. *[4]* we exhibited a model in which there is one unremovable phase, but in which there are no complex invariants of 4'rth or 6'th order; the phase is contained in an 8'th order complex invariant.

For the general construction of complex invariants, we note first that at the quartic order there are two P-type invariants each involving a single charge sector,

$$P^{(f)}_{j_1 k_1, j_2 k_2} = M^{(f)}_{j_1 k_1} M^{(f)}_{j_2 k_2} M^{(f)*}_{j_2 k_1} M^{(f)*}_{j_1 k_2} \tag{46}$$

for $f = \ell, N$. These may be denoted $\ell\ell\ell^*\ell^*$ and NNN^*N^* in a convenient shorthand. The P_4 complex invariants of the form NNN^*N^* may be further classified into six subtypes according to which submatrices in $M^{(N)}$ they involve, $M^{(L)}$, $M^{(R)}$, and/or $M^{(D)}$:

$$
\begin{array}{lll}
NNN^*N^* : & LLL^*L^* & LDL^*D^* \\
& RRR^*R^* & RDR^*D^* \\
& DDD^*D^* & LRD^*D^* \tag{47}
\end{array}
$$

There are two Q-type quartic invariants linking the different charge sectors:

$$Q^{(D\ell)}_{j_1 k_1, j_2 m_1} = M^{(D)}_{j_1 k_1} M^{(\ell)}_{j_2 m_1} M^{(D)*}_{j_2 k_1} M^{(\ell)*}_{j_1 m_1} \tag{48}$$

and

$$\Pi^{(L\ell)}_{j_1 j_2, j_3 m_1} = M^{(L)}_{j_1 j_2} M^{(\ell)}_{j_3 m_1} M^{(L)*}_{j_1 j_3} M^{(\ell)*}_{j_2 m_1} \tag{49}$$

In the short notation, these are denoted $D\ell D^*\ell^*$ and $L\ell L^*\ell^*$.

The higher order invariants are again more complicated than in the quark case; these are constructed in Refs. *[3, 4]*. There we also addressed the case of grand unified theories in which there are relations between the quark and lepton mass matrices. Just as these relations may reduce the total number of independent amplitudes in these matrices, they may also reduce the number of phases. Applications to several models of lepton mass matrices were given in Refs. *[3, 4]*.

6   Conclusions

The goal of understanding fermion masses and mixing remains a great challenge. In studying models of fermion mass matrices, it is necessary to determine how many unremovable phases there are in these matrices, and what allowed forms the matrices can take after rephasing of fermion fields. We have presented here a general methodology which answers both of these questions.

This research was done in collaboration with A. Kusenko and was supported in part by NSF grant PHY-93-09888. I would like to thank Pierre Ramond and his colleagues for a very enjoyable and stimulating workshop.

References

[1]   A. Kusenko and R. Shrock, Phys. Rev. **D50** (1994) R30 (hep-ph/9310307).

[2]   A. Kusenko and R. Shrock, (hep-ph/9401274).

[3]   A. Kusenko and R. Shrock, Phys. Lett. **B323** (1994) 18 (hep-ph/9311307).

[4]   A. Kusenko and R. Shrock, preprint ITP-SB-93-68 (hep-ph/9403315).

[5]   P. Ramond, R. Roberts, and G. G. Ross, Nucl. Phys. **B406** (1993) 19 (hep-ph/9303320).

[6]   A. Kusenko and R. Shrock, Phys. Rev. **D49** (1994) 4962 (hep-ph/9307344).

FERMION MASSES IN SOME MODELS WITH HORIZONTAL SYMMETRIES[1]

WILLIAM A. PONCE

DEPARTAMENTO DE FÍSICA,

CENTRO DE INVESTIGACIÓN Y DE ESTUDIOS AVANZADOS DEL I.P.N.

APARTADO POSTAL 14-740, 07000 MÉXICO D.F., MÉXICO

AND

DEPARTAMENTO DE FÍSICA,

UNIVERSIDAD DE ANTIOQUIA

A.A. 1226, MEDELLÍN, COLOMBIA

ARNULFO ZEPEDA

DEPARTAMENTO DE FÍSICA,

CENTRO DE INVESTIGACIÓN Y DE ESTUDIOS AVANZADOS DEL I.P.N.

APARTADO POSTAL 14-740,

07000 MÉXICO D.F., MÉXICO

Abstract: We briefly review the models $SU(6)_L \otimes U(1)_Y$, $[SU(6)]^3 \times Z_3$ and $[SU(6)]^4 \times Z_4$ which unify nongravitational forces with flavors, making emphasis in the implications of each one of those models in the mass spectrum of the known elementary fermions

1 $SU(6)_L \otimes U(1)_Y$

In Ref.*[1]* we presented a model based on the gauge group $SU(6)_L \otimes U(1)_Y$ which unifies the weak isospin $SU(2)_L$ of the Standard Model (SM) with a horizontal gauge group $SU(3)_{HL}$. With the usual definition for the electric charge operator $(Q = T_Z + Y/2)$, the known fermion fields are included in the following set if irreducible representations of $SU(6)_L \times U(1)_Y$ (one for each color in the case of quarks)

- $[6_{(1/3)}]_L = \psi_L^\alpha(1/3) = (u, d, c, s, t, b)_L$
- $[6_{(-1)}]_L = \psi_L^\alpha(-1) = (\nu_e, e, \nu_\mu, \mu, \nu_\tau, \tau)_L$
- $[1^f_{(-4/3)}]_L = q_L^{c,f}(-4/3) = u_L^c, c_L^c, t_L^c$, for $f = 1, 2, 3$.
- $[1^f_{(2/3)}]_L = q_L^{c,f}(2/3) = d_L^c, s_L^c, b_L^c$, for $f = 1, 2, 3$.
- $[1^f_{(2)}]_L = l_L^{c,f}(2) = e_L^+, \mu_L^+, \tau_L^+$, for $f = 1, 2, 3$.

where f is a flavor index, α is an $SU(6)_L$ tensor index and the upper c symbol indicates a charge-conjugated field (the model so defined is not renormalizable, so exotic leptons in the representation $2[\overline{6}_{(-1)}]_L + 2[\overline{6}_{(1)}]_L$ must be introduced in order to cancel the anomalies). The symmetry breaking chain $SU(6)_L \otimes U(1)_Y \to SU(2)_L \otimes U(1)_Y \xrightarrow{M_L} U(1)_Q$ is implemented with three different sets of Higgs fields, one for each step. In the last step a Higgs field

---

[1]Presented at the II IFT Workshop on Yukawa Couplings and the Origins of Mass by A. Zepeda

$\phi_\alpha(1) = [\bar{6}_{(1)}]$ with Vacuum Expectation Values (VEV's) in the most general direction available $\langle\phi_\alpha(1)\rangle = (v_1, 0, v_2, 0, v_3, 0)$ is used. This Higgs field produces a mass matrix for the Up quark sector of the form

$$\sum_f Y_f q_L^{c,f}(-4/3)C\phi_\alpha(1)\psi_L^\alpha(1/3) = \begin{pmatrix} Y_u v_1 & Y_u v_2 & Y_u v_3 \\ Y_c v_1 & Y_c v_2 & Y_c v_3 \\ Y_t v_1 & Y_t v_2 & Y_t v_3 \end{pmatrix}$$

where $Y_{u,c,t}$ are Yukawa couplings. This is a rank one matrix with a tree level mass term only for the t quark $m_t^2 = (v_1^2 + v_2^2 + v_3^2)(Y_u^2 + Y_c^2 + Y_t^2)$. The other known fermions (5 quarks and 6 leptons) get only radiative masses in the context of this model, by the introduction of appropriate multiplets of elementary Higgs scalars and discrete symmetries[1].

2 $[\mathbf{SU(6)}]^3 \times \mathbf{Z_3}$

$SU(6)_L \otimes U(1)_Y$ is made left-right symmetric with $SU(6)_L \otimes SU(6)_R \otimes U(1)_{Y(B-L)}$, and in Ref.[2] the gauge group is made simple by the introduction of $SU(6)_C$, a vector-like color group which contains three hadronic colors ($SU(6)_C \supset SU(3)_C$ the QCD gauge group) and three leptonic colors. The phenomenology of $G \equiv SU(6)_L \otimes SU(6)_C \otimes SU(6)_R \times Z_3$ is presented in Refs.[3].

$G \supset SU(4)_{PS} \supset SU(3)_C \otimes U(1)_{Y(B-L)}$, where $SU(4)_{PS}$ is the four color group of Pati-Salam and $Y(B-L) \sim Dg.(1/3, 1/3, 1/3, -1, 1, -1)$. The flavor group $SU(6)_L \otimes SU(6)_R \otimes U(1)_{Y(B-L)} \subset G$ contains $SU(2)_L \otimes SU(2)_R \otimes U(1)_{Y(B-L)}$, the left-right symmetric extension of the SM. $SU(3)_{HL} \otimes SU(3)_{HR}$ is the horizontal symmetry implicit in G.

The known fermions are included in $\psi(108) = Z_3\psi(6,1,\bar{6}) \equiv \psi(6,1,\bar{6}) + \psi(1,\bar{6},6) + \psi(\bar{6},6,1)$. The quantum numbers of $\psi(108)$ with respect to $[SU(3)_C, SU(2)_L, U(1)_Y]$ are:

$\psi(\bar{6},6,1) = 3[3,2,1/3] + 6[1,2,-1] + 3[1,2,1]$
$\psi(1,\bar{6},6) = 3[\bar{3},1,-4/3] + 3[\bar{3},1,2/3] + 6[1,1,2] + 3[1,1,-2] + 9[1,1,0]$
$\psi(6,1,\bar{6}) = 9[1,2,2] + 9[1,2,-2]$.

As can be seen the model does not contain exotic quarks, and all the exotic leptons in $\psi(108)$ have electric charges ± 1. Also all the exotic fields in $\psi(108)$ belong to vectorlike representations of the SM gauge group and the implementation of the survival hypothesis[4] is simple.

In Refs.[3], G is broken down to $SU(3)_c \otimes U(1)_Q \equiv G_U$ in the following way

$G \xrightarrow{M} SU(6)_L \otimes SU(4)_{PS} \otimes SU(2)_c \otimes SU(4)_R \otimes SU(2)_r \xrightarrow{M_H} SU(3)_c \otimes SU(2)_L \otimes U(1)_Y \xrightarrow{M_L} G_U$. The renormalization group equations imply that $M \sim 10^{12} > M_H \sim 10^9 \gg M_L \sim 10^2$ GeVs. An outstanding feature of this model is that the baryon number operator, which is defined in the linear sub-space of $SU(6)_C$ by the diagonal operator $B \equiv Dg.(1/3, 1/3, 1/3, 0, 0, 0)$, is conserved. So the proton is perturbatively stable in G.

With the VEV's of the Higgs Fields introduced in Ref.[3], which break the symmetry and implement at the same time the survival hypothesis, the following tree level mass spectrum for the particles in $\psi(108)$ is obtained:

- Two neutrinos get see-saw masses of order M_L^2/M_H.

- The t quark gets a mass $\sim M_L$ times a Yukawa coupling.

- One neutral exotic gets a mass M_H^2/M.

- All the other exotics get masses of order M_H and greater.

The other known particles (five quarks, three charged leptons and one neutrino) get masses via radiative corrections, with loops of gauge fields and loops of the elementary Higgs fields used to break the symmetry.

To be more specific, both the Up and Down quark matrices are dominated by flavor democratic mass matrices[5]

$$\mathcal{M} = \delta \begin{pmatrix} 1 & 1 & 1 \\ 1 & 1 & 1 \\ 1 & 1 & 1 \end{pmatrix}$$

where δ is proportional to M_L times a Yukawa coupling for the Up quarks and it is proportional to a constant in the Higgs potential times a loop with Higgses for the Down quarks. The texture of this flavor democratic mass matrices is given by radiative loops with gauge fields alone.

3 $[SU(6)]^4 \times Z_4$ without mirror fermions

G is extended to a chiral color symmetry in ReF.[6] by the structure $[G', \psi(144)]$, where $G' = SU(6)_L \otimes SU(6)_{CL} \otimes SU(6)_{CR} \otimes SU(6)_R \times Z_4$, and $\psi(144) = Z_4\psi(\overline{6}, 61, 1) = \psi(\overline{6}, 6, 1, 1) + \psi(6, 1, 1, \overline{6}) + \psi(1, 1, \overline{6}, 6) + \psi(1, \overline{6}, 6, 1)$.

To see that $\psi(144)$ does not contains mirror fermions we write its quantum numbers with respect to $[SU(3)_C, SU(2)_L, U(1)_Y]$

$\psi(\overline{6}, 6, 1, 1) = 3[3, 2, 1/3] + 6[1, 2, -1] + 3[1, 2, 1]$

$\psi(1, 1, \overline{6}, 6) = 3[\overline{3}, 1, -4/3] + 3[\overline{3}, 1, 2/3] + 6[1, 1, 2] + 3[1, 1, -2] + 9[1, 1, 0]$

$\psi(6, 1, 1, \overline{6}) = 9[1, 2, 2] + 9[1, 2, -2]$

$\psi(1, \overline{6}, 6, 1) = [8 + 1, 1, 0] + 2[3, 1, 4/3] + 2[\overline{3}, 1, -4/3] + [3, 1, -2/3] + [\overline{3}, 1, 2/3] + 5[1, 1, 0] + 2[1, 1, 2] + 2[1, 1, -2].$

As can be seen, not only there are not mirror fermion fields in $\psi(144)$, but all the exotic fields belong to vector-like representations with respect to the SM quantum numbers. The symmetry breaking pattern is now $G' \xrightarrow{M} G_I \xrightarrow{M_L} G_U$, where G_I is $SU(3)_C \otimes SU(2)_L \otimes U(1)_Y$, or $SU(3)_{CL} \otimes SU(3)_{CR} \otimes SU(2)_L \otimes U(1)_Y$. For the first case $M \sim 10^6$ GeV's, and for the second case $M \sim 10^4$ GeV's. The mass spectrum of the structure $[G', \psi(144)]$ depends upon the intermediate gauge group in the symmetry breaking chain, but the known fermion fields follow almost the same pattern of the mass spectrum for the structure $[G, \psi(108)]$, the main difference being now that the b quark can get a tree level mass via a flavor democratic mass matrix and a universal see-saw mechanism[7] produced by the exotic quark $[3, 1, -2/3] + [\overline{3}, 1, 2/3]$ in $\psi(1, \overline{6}, 6, 1)$. As it is shown in Ref.[6], the proton is also perturbatively stable in the context of this new model.

References

[1]   A.H.Galeana, W.A.Ponce and A.Zepeda, Z. Physik **C55**, 423 (1992)

[2]   W.A. Ponce, in *"Proceedings of the Third Mexican School of Particles and Fields"*, Oaxtepec, Mexico, 1988, edited by J.L. Lucio and A. Zepeda (World Scientific, Singapore, 1989), pp. 90-129

[3]   A.H. Galeana, R. Martinez, W.A. Ponce and A. Zepeda, Phys. Rev. **D44**, 2166 (1991); W.A. Ponce and A. Zepeda, Phys. Rev. **D48**, 240 (1993); W.A.Ponce, A.Zepeda and J.B. Flórez, Phys. Rev. **D49**, May (1994); W.A.Ponce, A.Zepeda and R.G. Lozano, Phys. Rev. **D49**, May (1994)

[4]   H.Georgi,Nucl.Phys.**B156**,126(1979); R.Barbieri and D.V. Nanopoulos, Phys. Lett. **91B**, 369 (1980)

[5]   H.Fritzsch, Nucl.Phys. Proc. Suppl. **1B**, 447 (1988); P.Kaus and S.Meshkov, Mod. Phys. Lett. **A3**,1251 (1988)

[6]   W.A.Ponce and A.Zepeda; "*A* $[SU(6)]^4 \times Z_4$ *model without mirror fermions*", to appear in Z. PhysiK **C61**

[7]   A.Davidson and K.C.Wali, Phys. Rev. Lett. **59**, 393 (1987); S.Rajpoot, Phys. Rev. **D36**, 1497 (1987)

Flavour Mixing and Fermion Mass Generation as a Result of Symmetry Breaking

HARALD FRITZSCH[1]

THEORY DIVISION, CERN, GENEVA
MAX–PLANCK–INSTITUT FÜR PHYSIK UND
ASTROPHYSIK WERNER– HEISENBERG–INSTITUT
FÜR PHYSIK, MUNICH

Invited talk given at the
Workshop on Yukawa Couplings and the Origins of Mass,
Gainesville, Florida (Feburary 1994)[2]

Abstract: It is shown that a simple breaking of the subnuclear democracy leads to a successful description of the mixing between the second and third family. In the lepton channel the $\nu_\mu - \nu_\tau$ oscillations are expected to be described by a mixing angle of $2.65°$ which might be observed soon in neutrino experiments.

In the standard electroweak model, both the masses of the quarks as well as the weak mixing angles enter as free parameters, and any further insight into the yet unknown dynamics of mass generation would imply a step beyond the physics of the electroweak standard model. At present it seems far too early to attempt an actual solution of the dynamics of mass generation, and one is invited to follow a strategy similar to the one which led eventually to the solution of the strong interaction dynamics by QCD, by looking for specific patterns and symmetries as well as specific symmetry violations.

The mass spectra of the quarks are dominated strongly by the masses of the members of the third family, i. e. by t and b. Furthermore, the masses of the first family are small compared to those of the second one. Thus a clear hierarchical pattern exists. Also, the CKM–mixing matrix exhibits a hierarchical pattern – the transitions between the second and third family as well as between the first and the third family are small compared to those between the first and the second family.

About 15 years ago it was emphasized[1] that the observed hierarchies signify that nature seems to be close to the so–called "rank–one" limit in which all mixing angles vanish and both the u– and d–type mass matrices are proportional to the rank-one matrix

$$M_0 = const. \cdot \begin{pmatrix} 0 & 0 & 0 \\ 0 & 0 & 0 \\ 0 & 0 & 1 \end{pmatrix}. \tag{1}$$

[1]Supported in part by DFG-contract 412/22–1
[2]On leave from Sektion Physik, Universität München

Whether the dynamics of the mass generation allows this limit to be achieved in a consistent way remains an unsolved issue. Encouraged by the observed hierarchical pattern of the masses and the mixing parameters, we shall assume that this is the case. In itself it is a non-trivial constraint and can be derived from imposing a chiral symmetry, as emphasized in ref. (2). This symmetry ensures that an electroweak doublet which is massless remains unmixed and is coupled to the W–boson with full strength. As soon as mass is introduced, at least for one member of the doublet, the symmetry is violated and mixing phenomena are expected to show up. That way a chiral evolution of the CKM matrix can be considered.[2] At the first stage only the t and b quark masses are introduced, due to their non-vanishing coupling to the scalar "Higgs" field. The CKM–matrix is unity in this limit. At the next stage the second generation acquires a mass also. Since the (u, d)–doublet is still massless, only the second and the third generations mix, and the CKM–matrix is given by a real 2×2 rotation matrix in the $(c, s) - (t, b)$ subsystem, describing e. g. the mixing between s and b. Only at the next step, at which the u and d masses are introduced, does the full CKM–matrix appear, described in general by three angles and one phase. It has been emphasized some time ago[3] that the rank-one mass matrix (see eq. (1)) can be expressed in terms of a "democratic mass matrix":

$$M_0 = c \begin{pmatrix} 1\ 1\ 1 \\ 1\ 1\ 1 \\ 1\ 1\ 1 \end{pmatrix}, \tag{2}$$

which exhibits an $S(3)_L \times S(3)_R$ symmetry. Writing down the mass eigenstates in terms of the eigenstates of the "democratic" symmetry, one finds e.g. for the lepton channel:

$$e^0 = \frac{1}{\sqrt{2}}(l_1 - l_2)$$
$$\mu^0 = \frac{1}{\sqrt{6}}(l_1 + l_2 - 2l_3) \tag{3}$$
$$\tau^0 = \frac{1}{\sqrt{3}}(l_1 + l_2 + l_3)$$

(l_i: symmetry eigenstates). Note that e^0 and μ^0 are massless in the limit considered here, and any linear combination of the first two state vectors given in eq. (3) would fulfil the same purpose, i. e. the decomposition is not unique, only the wave function of the coherent state τ^0 is uniquely defined. This ambiguity will disappear as soon as the symmetry is violated, and as a result, the masses for the members of the second family are introduced.

Introducing the symmetry eigenstates does not, on its own, imply introducing new physics concepts. However, a new basis might be very welcome if one is trying to find new patterns and symmetries in the quark and lepton spectrum. I like to quote from Feynman's Nobel talk: "Theories of the known, which are described by different physical ideas, may be equivalent in all their predictions and are hence scientifically indistinguishable. However, they are not psychologically identical when trying to move from that base into the unknown. For different views suggest different kinds of modifications which might be made. I, therefore, think that a good theoretical physicist today might find it useful to have a wide range of physical viewpoints and mathematical expressions of the same theory available to him[4]."

The wave functions given in eq. (3) are reminiscent of the wave functions of the neutral pseudoscalar mesons in QCD in the $SU(3)_L \times SU(3)_R$ limit:

$$\pi_0^0 = \frac{1}{\sqrt{2}}(\bar{u}u - \bar{d}d) \tag{4}$$

$$\eta_0 = \frac{1}{\sqrt{6}}(\bar{u}u + \bar{d}d - 2\bar{s}s)$$

$$\eta_0' = \frac{1}{\sqrt{3}}(\bar{u}u + \bar{d}d + \bar{s}s).$$

(Here the lower index denotes that we are considering the chiral limit.) Also the mass spectrum of these mesons is identical to the mass spectrum of the leptons and quarks in the "democratic" limit: two mesons (π_0^0, η_0) are massless and act as Nambu–Goldstone bosons, while the third coherent state η_0' is <u>not</u> massless due to the QCD anomaly.

In the chiral limit, the $(\text{mass})^2$–matrix of the neutral pseudoscalar mesons is also a "democratic" mass matrix when written in terms of the $(\bar{q}q)$– eigenstates $(\bar{u}u)$, $(\bar{d}d)$ and $(\bar{s}s)$[5]:

$$M^2(ps) = \lambda \begin{pmatrix} 1 & 1 & 1 \\ 1 & 1 & 1 \\ 1 & 1 & 1 \end{pmatrix} \tag{5}$$

where the strength parameter λ is given by $\lambda = M^2(\eta_0') / 3$. The mass matrix (5) describes the result of the QCD–anomaly which causes strong transitions between the quark eigenstates (due to gluonic annihilation effects enhanced by topological effects). Likewise, one may argue that analogous transitions are the reason for the lepton–quark mass hierarchy. Here we shall not speculate about a detailed mechanism of this type, but merely study the effect of symmetry breaking.

In the case of the pseudoscalar mesons, the breaking of the symmetry down to $SU(2)_L \times SU(2)_R$ is provided by a direct mass term $m_s\bar{s}s$ for the s–quark. This implies a modification of the (3,3) matrix element in eq. (5), where λ is replaced by $\lambda + M^2(\bar{s}s)$ where $M^2(\bar{s}s)$ is given by $2M_k^2$, which is proportional to $< \bar{s}s >_0$, the expectation value of $\bar{s}s$ in the QCD vacuum. This direct mass term causes the violation of the symmetry and generates at the same time a mixing between η_0 and η_0', a mass for the η_0, and a mass shift for the η_0'.

It would be interesting to see whether an analogue of the simplest violation of the "democratic" symmetry which describes successfully the mass and mixing pattern of the $\eta - \eta'$–system is also able to describe the observed mixing and mass pattern of the second and third family of leptons and quarks. Let us replace the (3,3) matrix element in eq. (2) by $1 + \varepsilon_i$; (i = l (lepton), u (u–quark), d (d–quark)) respectively. The small real parameter ε describes the departure from democratic symmetry and leads

a)   to a generation of mass for the second family and

b)   to flavour mixing between the third and the second family. Since ε is directly related (see below) to a fermion mass and the latter is <u>not</u> restricted to be positive, ε can be positive or negative. (Note that a negative Fermi–Dirac mass can always be turned into a positive one by a suitable γ_5–transformation of the spin $\frac{1}{2}$ field.) Since the original mass term is represented by a symmetric matrix, we take ε to be real.

First we study the mass and mixing pattern of the charged leptons. The mass operator (trace Θ^μ_μ of the energy–momentum tensor $\Theta_{\mu\nu}$) can be written as

$$\Theta^\mu_\mu = \Theta^{0\mu}_\mu + c_l \varepsilon_l \bar{l}_3 l_3 \tag{6}$$

where $\Theta^{0\mu}_\mu$ describes the mass term in the symmetry limit. The modification of the spectrum and the induced mixing can be obtained by considering the matrix elements:

$$< \mu^0 |c_l \varepsilon_l \bar{l}_3 l_3| \mu^0 > = +\frac{2}{3} c_l \varepsilon_l$$

$$< \tau^0 |c_l \varepsilon_l \bar{l}_3 l_3| \tau^0 > = +\frac{1}{3} c_l \varepsilon_l \tag{7}$$

$$< \mu^0 |c_l \varepsilon_l \bar{l}_3 l_3| \tau^0 > = -\frac{\sqrt{2}}{3} c_l \varepsilon_l \ .$$

One observes that

a)   the muon acquires a mass given by $c_l \cdot \varepsilon_l$ i. e. $m(\mu)/m(\tau) \cong \frac{2}{9}\varepsilon_l$;

b)   the τ–lepton mass is changed slightly $(m(\tau)/m(\tau^0) \cong 1 + \frac{1}{9}\varepsilon_l)$;

c)   the flavour mixing is induced by the fact that the perturbation proportional to $\bar{l}_3 l_3$ leads to a non–vanishing transition matrix element between μ^0 und τ^0.

This phenomenon is analogous to the chiral symmetry violation of QCD, where the s–quark mass term $m_s \bar{s} s$ leads to a mass for the η–meson, a mass shift for the η'–meson and a mixing between η and η'.

It is instructive to rewrite the mass matrix in the hierarchical basis, where one obtains, using the relations (7):

$$M = c_l \begin{pmatrix} 0 & 0 & 0 \\ 0 & +\frac{2}{3}\varepsilon_l & -\frac{\sqrt{2}}{3}\varepsilon_l \\ 0 & -\frac{\sqrt{2}}{3}\varepsilon_l & 3+\frac{1}{3}\varepsilon_l \end{pmatrix} . \tag{8}$$

In lowest order of ε one finds the mass eigenvalues $m_\mu = \frac{2}{9}\varepsilon_l \cdot m_\tau$, $m_\tau = m_{\tau^0}$, $\Theta_{\mu\tau} = |\sqrt{2}\cdot\varepsilon_l/9|$.

The exact mass eigenvalues and the mixing angle are given by:

$$m_1/c_l = \frac{3+\varepsilon_l}{2} - \frac{3}{2}\sqrt{1 - \frac{2}{9}\varepsilon_l + \frac{1}{9}\varepsilon_l^2}$$

$$m_2/c_l = \frac{3+\varepsilon_l}{2} + \frac{3}{2}\sqrt{1 - \frac{2}{9}\varepsilon_l + \frac{1}{9}\varepsilon_l^2} \tag{9}$$

$$\sin\Theta_l = \frac{1}{\sqrt{2}}\left(1 - \frac{1-\frac{1}{9}\varepsilon_l}{(1-\frac{2}{9}\varepsilon_l+\frac{1}{9}\varepsilon_l^2)^{1/2}}\right)^{1/2} .$$

The ratio m_μ/m_τ, observed to be 0.0595, gives $\varepsilon_l = 0.286$ and a $\mu-\tau$ mixing angle of 2.65°. Whether this mixing angle is directly relevant for neutrino oscillations or not depends on the neutrino sector. For massless neutrinos the mixing angle generated in the charged lepton channel by the introduction of the muon mass cannot be observed, i. e. it can be rotated away. If neutrinos have a mass, the neutrino mass matrix will in general induce further mixing angles. A general discussion will not be attempted here.

However, we should like to consider an interesting scenario which is being discussed in connection with cosmological aspects. Let us suppose that the $\tau-$neutrino mass is of the order of 10 eV in order to be relevant for the "missing matter problem" in cosmology, the muon neutrino is in the milli–eV range, i. e. $m(\nu_\mu) < 10^{-2}eV$, and the electron neutrino mass is neglected. The mass generation for the ν_μ–mass proceeds in an analogous way as discussed above for the muon mass. However, the ε–parameter for the neutrino sector turns out to be tiny $(< 5\cdot 10^{-2})$, and the mixing angle induced via the ν_μ–mass generation can safely be neglected. Thus the angle relevant for the $\nu_\mu - \nu_\tau$ oscillations remains 2.65°, i. e. $\sin^2 2\Theta = 0.0085$. This value is essentially the lowest limit given by the Charm II experiment[7], i. e. it is not ruled out for any value of $\Delta m^2 = m(\nu_\tau)^2 - m(\nu_\mu)^2$.

However, the E531 experiment[8] gives a limit of about $16eV^2$ for Δm^2, i. e. $m(\nu_\tau) < 4eV$. This limit seems to rule out a cosmological role with respect to the "missing matter" for the $\tau-$neutrino. However, one might caution this conclusion since our mixing angle of 2.65° is not far from the limit of $(\sin^2 2\Theta = 0.004)$, at which, according to the E531 experiment, all values of $m(\nu_\tau)$ are allowed. New experiments, e. g. the CHORUS and NOMAD experiments now or soon under way at CERN, will clarify this issue. If the mixing angle is 2.65°

as argued above and the ν_τ–mass above 10 eV, one should observe the $\nu_\mu - \nu_\tau$–oscillations within one year[9].

Replacing ε_l by ε_u, ε_d respectively, we can determine the symmetry breaking parameters for the quark sector. The ratio m_s/m_b is allowed to vary in the range $0.022\ldots0.044$ (see ref. (9)). According to eq. (9) one finds ε_u to vary from $\varepsilon_d = 0.11$ to 0.21. The associated $s - b$ mixing angle varies from $\Theta(s,b) = 1.0°$ ($\sin\Theta = 0.018$) and $\Theta(s,b) = 1.95°$ ($\sin\Theta = 0.034$). As an illustrative example we use the values $m_b(1GeV) = 5200MeV$, $m_s(1GeV) = 220MeV$. One obtains $\varepsilon_d = 0.20$ and $\sin\Theta(s,b) = 0.032$.

To determine the amount of mixing in the (c,t)–channel, a knowledge of the ratio m_c/m_t is required. As an illustrative example we take $m_c(1GeV) = 1.35GeV$, $m_t(1GeV) = 260GeV$ (i. e. $m_t(m_t) = 160GeV$), which gives $m_c/m_t \cong 0.005$. In this case one finds $\varepsilon_u = 0.023$ and $\Theta(c,t) = 0.21°$ ($\sin\Theta(c,t) = 0.004$) .

The actual weak mixing between the third and the second quark family is combined effect of the two family mixings described above. The symmetry breaking given by the ε–parameter can be interpreted, as done in eq. (7), as a direct mass term for the $l_3(u_3,d_3)$ fermion system. However, a direct fermion mass term need not be positive, since its sign can always be changed by a suitable γ_5–transformation. What counts for our analysis is the relative sign of the m_s–mass term in comparison to the m_c–term, discussed previously. Thus two possibilities must be considered:

a)   Both the m_s– and the m_c–term have the same relative sign with respect to each other, i. e. both ε_d and ε_u are positive, and the mixing angle between the second and third family is given by the difference $\Theta(sb) - \Theta(ct)$. This possibility seems to be ruled out by experiment, since it would lead to $V_{cb} < 0.03$.

b)   The relative signs of the breaking terms ε_d and ε_u are different, and the mixing angle between the (s,b) and (c,t) systems is given by the sum $\Theta(sb) + \Theta(ct)$. Thus we obtain $V_{cb} \cong \sin(\Theta(sb) + \Theta(ct))$.

According to the range of values for m_s discussed above, one finds $V_{cb} \cong 0.022...0.038$. For example, for $m_s(1GeV) = 220MeV$, $m_c(1GeV) = 1.35GeV$, $m_t(1GeV) = 260GeV$ one finds $V_{cb} \cong 0.036$.

Before discussing the experimental situation, we add a comment about the mass generation for the first family, which at the same time will also generate the other mixing elements, e.g. V_{us} and V_{ub}, of the CKM matrix. These masses can be generated by a further breakdown of the symmetry, e. g. in the matrix of eq. (5) by a small departure of a second diagonal matrix element from unity. (This would correspond to a direct mass term for that state.) Due to the small values of the masses of the first family in comparison to the λ–scale, given by the mass of the third generation fermion (e.g. $m_e/\lambda = 0.0009$), the strength of this symmetry breaking is much smaller than the primary symmetry breaking, which leads to the masses for the second family. (The situation is analogous to the one in hadronic physics, where the breaking of the chiral symmetry is given primarily by the mass of the s–quark, and the m_u/m_d mass terms can be neglected to a good approximation.) In general it is expected, both from the arguments considered here and more generally from the analysis on chiral symmetry given in ref. (2), that the matrix elements V_{cb} and V_{ts} will be affected only by small corrections of order 10^{-3} or less in absolute magnitude (of order $\frac{m_d}{m_b}$, $\frac{m_u}{m_t}$ respectively). Thus the primary breaking of the democratic symmetry leads solely to a mixing between the second and the third family, and the secondary breaking, responsible for the Cabibbo angle etc., will not affect the 2×2 submatrix of the CKM–matrix describing the $s - b$ mixing in a significant way.

The experiments give $V_{cb} = 0.032 \ldots 0.054^{10)}$. We conclude from the analysis given above that our ansatz for the symmetry breaking reproduces the lower part of the experimental range. According to a recent analysis the experimental data are reproduced best for $V_{cb} = 0.038 \pm 0.003^{11)}$, i. e. it seems that V_{cb} is lower than previously thought, consistent with our expectation. Nevertheless we obtain consistency with experiment only if the ration m_s/m_b is relatively large, implying $m_s(1GeV) \geq 180MeV$.

It is remarkable that the simplest ansatz for the breaking of the "democratic symmetry", one which nature follows in the case of the pseudoscalar mesons, is able to reproduce the experimental data on the mixing between the second and third family. We interpret this as a hint that the eigenstates of the symmetry l_i, q_i respectively, and not the mass eigenstates, play a special rôle in the physics of flavour dynamics, a rôle which needs to be investigated further.

References

[1] H. Fritzsch, Nucl. Phys. **B155** (1979) 189;
See also: H. Fritzsch, in: Proc. Europhysics Conf. on Flavor Mixing, Erice, Italy (1984).

[2] H. Fritzsch, Phys. Lett. **B184** (1987) 391.

[3] H. Harari, H. Haut and J. Weyers, Phys. Lett. **78B** (1978) 459;
Y. Chikashige, G. Gelmini, R.P. Peccei and M. Roncadelli, Phys. Lett. **94B** (1980) 499;
C. Jarlskog, in: Proc. of the Int. Symp. on Production and Decay of Heavy Flavors, Heidelberg, Germany (1986);
P. Kaus and S. Meshkov, Mod. Phys. Lett. **A3** (1988) 1251; **A4** (1989) 603 (E);
H. Fritzsch and J. Plankl, Phys. Lett. **B237** (1990) 451.
G.C. Branco, J. I. Silva–Marcos and M.N. Rebelo, Phys. Lett. **B237** (1990) 446.

[4] R. P. Feynman, *The Development of the Space Time View of Quantum Electrodynamics*, Nobel Lecture, reprinted in: Physics Today, Aug. 1966, 31.

[5] H. Fritzsch and P. Minkowski, Nuovo Cimento **30A** (1975) 393;
H. Fritzsch and D. Jackson, Phys. Lett. **66B** (1977) 365.

[6] H. Fritzsch and D. Holtmannsp"otter, CERN preprint CERN–TH 7236/94 (1994).

[7] M. Gruwé et al., Phys. Lett. **B309** (1993) 463.

[8]   N. Ushida et al., Phys. Rev. Lett. **57** (1986) 2897.

[9]   K. Winter, private communication.

[10]   J. Gasser and H. Leutwyler, Physics Reports **87** (1982) 77.

[11]   S. Stone, Syracuse preprint HEPSY 93–11 (1993)

CKM Matrix Elements: Magnitudes, Phases, Uncertainties

JONATHAN L. ROSNER

ENRICO FERMI INSTITUTE AND DEPARTMENT OF PHYSICS
UNIVERSITY OF CHICAGO, CHICAGO, IL 60637

Abstract: Schemes for fermion masses should predict elements of the Cabibbo-Kobayashi-Maskawa (CKM) matrix. We review the freedom allowed by present experiments and how the parameter space will shrink in the next few years. In addition to experiments which directly affect CKM parameters, we discuss constraints arising from precise electroweak tests.

1 Introduction

The present workshop is devoted to ways in which the bewildering pattern of fermion masses and mixings might be understood. The purpose of this talk is to describe the allowed parameter space of mixings and how it may be expected to shrink as a result of improved experiments and theory. A more detailed account may be found in Ref. [1]; we take the opportunity to update some of the numbers presented there. Some of the latest developments since this talk was presented will be mentioned, but are not included in the fits to data.

We set out the frameworks for the discussion in Section 2. The determination of CKM parameters is described in Section 3. A long digression on the top quark is contained in Section 4. Electroweak tests lead primarily to a correlation between the top quark and Higgs masses, as discussed in Section 5. Returning to the CKM matrix itself, we note in Section 6 several ways to obtain improved information on magnitudes and phases of its elements. Among these is the study of CP violation in the decays of neutral B mesons. Recent progress in identifying the flavor of neutral B mesons is reported in Section 7. We conclude in Section 8.

2 Frameworks

2.1 The CKM Matrix

The weak charge-changing interactions lead primarily to the transtions $u \leftrightarrow d$, $c \leftrightarrow s$, $t \leftrightarrow b$ between left-handed quarks $(u,\ c,\ t)$ of charge 2/3 and those $(d,\ s,\ b)$ of charge $-1/3$. However, as noted by Cabibbo [2] and Glashow-Iliopoulos-Maiani [3] for two families of quarks and by Kobayashi and Maskawa [4] for three, additional transitions of lesser strength can be incorporated into this framework in a universal manner. The charge-changing transitions then connect u, c, t not with d, s, b but with a rotated set $(d',\ s',\ b') = V(d,\ s,\ b)$, where V is a unitary 3×3 matrix now known as the Cabibbo-Kobayashi-Maskawa (CKM) matrix.

The elements of V are as mysterious as the quark masses, and are intimately connected with them since the matrix arises as a result of diagonalization of the quark mass matrices (see, e.g., Ref. [1]). Moreover, the phases in the matrix are candidates for the source of CP violation as observed in the decays of neutral kaons. We shall assume that to be the case in the present analysis.

2.2 *Precise electroweak tests*

The top quark plays an indirect role in the extraction of CKM parameters from data on $B - \bar{B}$ mixing and CP-violating $K\bar{K}$ mixing. Thus, it is important to know its mass. At this time this talk was given, the best source of information on the top quark was its indirect effects on the W and Z bosons' self-energies. Some updated information may be found at the end of Section 4.

In addition to diagrams involving top quarks, W and Z self-energies can be affected by Higgs bosons and by various new particles which can appear in loop diagrams. In conjunction with measurement of the top quark mass, precise electroweak tests then will be able to shed first light on these contributions.

3 Determination of CKM parameters

We turn now to a description of the CKM matrix elements. More details on the measurement of the elements V_{cb} and V_{ub} may be found in Refs. *[5, 6]*.

3.1 *Parametrization*

We adopt a convention in which quark phases are chosen *[7]* so that the diagonal elements and the elements just above the diagonal are real and positive. The parametrization we shall introduce and employ is one suggested by Wolfenstein *[8]*.

The diagonal elements of V are nearly 1, while the dominant off-diagonal elements are $V_{us} \simeq -V_{cd} \simeq \sin\theta \equiv \lambda \simeq 0.22$. Thus to order λ^2, the upper 2×2 submatrix of V is already known from the Cabibbo-GIM four-quark pattern:

$$V \simeq \begin{pmatrix} 1 - \frac{\lambda^2}{2} & \lambda & \cdot \\ -\lambda & 1 - \frac{\lambda^2}{2} & \cdot \\ \cdot & \cdot & 1 \end{pmatrix} \quad . \tag{1}$$

The empirical observation that $V_{cb} \simeq 0.04$ allows one to express it as $A\lambda^2$, where $A = \mathcal{O}(1)$. Unitarity then requires $V_{ts} \simeq -A\lambda^2$ as long as V_{td} and V_{ub} are small enough (which they are). Finally, V_{ub} appears to be of order $A\lambda^3 \times \mathcal{O}(1)$. Here one must allow for a phase, so one must write $V_{ub} = A\lambda^3(\rho - i\eta)$. Finally, unitarity specifies uniquely the form $V_{td} = A\lambda^3(1 - \rho - i\eta)$. To summarize, the CKM matrix may be written

$$V \approx \begin{bmatrix} 1 - \lambda^2/2 & \lambda & A\lambda^3(\rho - i\eta) \\ -\lambda & 1 - \lambda^2/2 & A\lambda^2 \\ A\lambda^3(1 - \rho - i\eta) & -A\lambda^2 & 1 \end{bmatrix} \quad . \tag{2}$$

We shall note below that $V_{cb} = 0.038 \pm 0.005$, so that $A = 0.79 \pm 0.09$. The measurement of semileptonic charmless B decays *[9, 10]* gives $|V_{ub}/V_{cb}|$ in the range from 0.05 to 0.11, where most of the uncertainty is associated with the spread in models *[11, 12]* for the lepton spectra. Taking 0.08 ± 0.03 for this ratio, we find that the corresponding constraint on ρ and η is $(\rho^2 + \eta^2)^{1/2} = 0.36 \pm 0.14$.

The form (2) is only correct to order λ^3 in the matrix elements. For certain purposes it may be necessary to exhibit corrections of higher order to the elements.

The unitarity of V implies that the scalar product of any row and the complex conjugate of any other row, or of any column and the complex conjugate of any other column, will be zero. In particular, taking account of the fact that V_{ud} and V_{tb} are close to 1, we have

$$V_{ub}^* + V_{td} \simeq A\lambda^3 \tag{3}$$

or, to the order of interest in small parameters,

$$\rho + i\eta + (1 - \rho - i\eta) = 1 \quad . \tag{4}$$

The point (ρ, η) forms the apex of a triangle in the complex plane, whose other vertices are the points $(0,0)$ and $(1,0)$. This "unitarity triangle" [13] and its angles are depicted in Fig. 1.

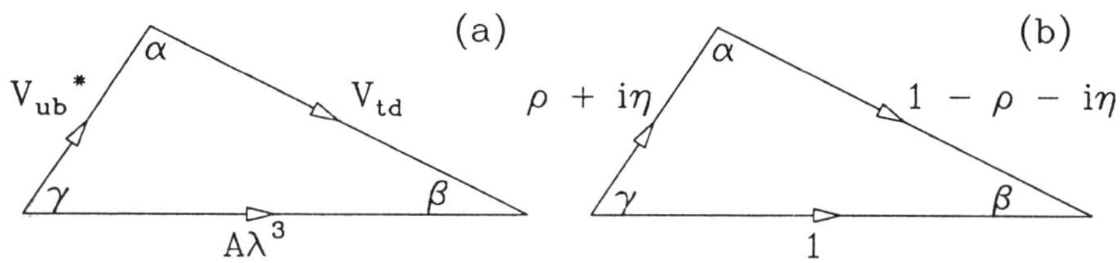

FIGURE 1. The unitarity triangle. (a) Relation obeyed by CKM elements; (b) relation obeyed by (CKM elements)$/A\lambda^3$

The main indeterminacy in the CKM matrix concerns the magnitude of V_{td}, for which only indirect evidence exists. Correspondingly, we are still quite uncertain about Arg $V_{ub}^* = \arctan(\eta/\rho)$. Most of our effort will be devoted to seeing how these quantities can be pinned down better.

3.2 Measuring the Cabibbo-GIM submatrix

The elements of the 2×2 submatrix connecting the quarks u, c of charge $2/3$ with those (d, s) of charge $-1/3$ are described satisfactorily by the single parameter λ in the form (1). The only lingering question is whether $|V_{ud}|^2 + |V_{us}|^2$ really is 1 (up to corrections of order $|V_{ub}|^2$, which are negligible), and present data appear to be consistent with this [1, 14].

3.3 Measuring V_{cb}

In this subsection and the next we give a cartoon version of a discussion which is set forth much more completely in Ref. [6].

The decay of a b quark to a charmed quark c and a lepton pair offers the best hope for determining V_{cb}. One would use the b quark lifetime and the branching ratio for the process $b \to c\ell\nu$ to estimate the rate for the process, which would then be proportional to a known kinematic factor times $|V_{cb}|^2$. Even if b and c were free, we would have to know their masses accurately in order to make a useful estimate.

Since the b quark and charmed quark are incorporated into hadrons such as a B meson and a D meson, the problem becomes one of estimating hadronic effects. There are several ways to do this.

Free quarks

A good deal of indeterminacy of the rate for $b \to c\ell\bar{\nu}_\ell$ is associated merely with uncertainty in quark masses. However, the uncertainty in the predicted decay rate can be reduced by constraints on the mass difference $m_b - m_c$ from hadron spectroscopy [15, 16]. Taking $m_b = 5.0 \pm 0.3$ GeV/c^2, $m_b - m_c$ ranging from 3.34 to 3.40 GeV/c^2, $B(B \to charm + \ell + \bar{\nu}_l) = 10.5\%$, and $\tau_B = 1.49$ ps, we obtained [1] $V_{cb} = 0.038 \pm 0.003$.

Free quarks and QCD

One can take account of the effects of the light quarks by means of Fermi momentum and can apply QCD corrections to the decay of the free b quark [11]. The result should be an average over the excitation of individual final states of the charmed quark and the spectator antiquark.

Models for final states

One can calculate B semileptonic decay rates to specific final states, such as $D\ell\bar{\nu}_\ell$, $D^*\ell\bar{\nu}_\ell$, and so on [12]. It is then necessary to include all relevant states, so an important question is what charmed states besides D and D^* play a role.

Use of the "zero-recoil" point

When the lepton pair has its maximum invariant mass, the b quark decays to a charmed quark without causing it to recoil [17]. Thus, hadronic effects are kept to a minimum. The limitation on this method is mainly one of statistics at present.

Averages

When the various methods are combined, one gets an idea of the spread in theoretical approaches. In Ref. [1] we quoted the value $V_{cb} = 0.038 \pm 0.005$ obtained in Ref. [5] on the basis of such averages. That is the value which we will use in the present analysis, corresponding to $A = 0.79 \pm 0.09$. More recently Stone [6] estimates $V_{cb} = 0.038 \pm 0.003$, in accord with our original free-quark value and corresponding to an error $\Delta A = 0.06$.

3.4   Measuring V_{ub}

In order to see the effects of the process $b \to u\ell\bar{\nu}_\ell$, one has to study leptons beyond the end point for charm production. As a result, one sees only a very small part of the total phase space for the process of interest. The question then becomes one of how the decay populates this small region of phase space. The final u quark can combine with the initial $\bar{u}$ or $\bar{d}$ in the decaying B meson to form a nonstrange hadron such as π, ρ, a_1, ... One can either describe this recombination in an average sense [11, 18] or employ models for excitation of individual resonances [12].

The range of theoretical approaches allows values of $|V_{ub}/V_{cb}|$ between 0.05 and about 0.11 when the more recent CLEO data are used [10]. Somewhat larger values (up to a factor of 2, in some models) are implied by earlier ARGUS data [9]. These results correspond to a fraction of b decays without charmed particles of between 1 and 2%. For present purposes we shall take $|V_{ub}/V_{cb}| = 0.08 \pm 0.03$. For comparison, Stone [6] quotes $|V_{ub}/V_{cb}| = 0.08 \pm 0.02$.

3.5   *Arg V_{ub}*

The phase of V_{ub} is one of the least well known parameters of the CKM matrix. For it, we must rely upon indirect information.

$B^0 - \overline{B}^0$ *mixing*

The original evidence for $B^0 - \bar{B}^0$ mixing came from the observation [19] of "wrong-sign" leptons in B meson semileptonic decays. The diagrams of Fig. 2 give rise to a splitting between mass eigenstates

$$\Delta m \sim f_B^2 m_t^2 |V_{td}|^2 \tag{5}$$

times a slowly varying function of m_t. (See, e.g., Ref. [1] for detailed expressions.) Here f_B is the "B meson decay constant," which expresses the overlap of a $b\bar{q}$ state at zero separation with the wave function of the B meson. Information on f_B is improving, but still is a major source of indeterminacy. The top quark mass m_t is becoming better known, as we shall see at the end of Section 4. The CKM element V_{td} has a magnitude proportional to $|1 - \rho - i\eta|$, which is what we would like to learn.

The average value of data used for the present analysis gives $\Delta m/\Gamma = 0.66 \pm 0.10$ [20]. The resulting constraint on (ρ, η) for fixed values of f_B and m_t is a circular band with radius approximately 1 and center at the point $(1,0)$, as bounded by the dashed arcs in Fig. 3. Uncertainty in f_B and, to a lesser extent, m_t, is a source of spread in this band.

CP-*violating* $K^0 - \overline{K}^0$ *mixing*

The box diagrams shown in Fig. 4 give rise to a CP-violating term in the matrix element between a K^0 and a $\bar{K}^0$,

$$Im \; \mathcal{M}_{12} \sim f_K^2 \; Im \; (V_{td}^2) \sim \eta(1 - \rho) \quad , \tag{6}$$

so that the constraint in the (ρ, η) plane is a band bounded by hyperbolae with focus at the point $(1,0)$, as shown by the solid lines in Fig. 3]. Here, again, the top quark mass enters. There are small corrections (not completely negligible) from charmed quarks in the loop. Neglecting these, however, one can take the quotient of the constraints (5) and (6) to find a constraint on Arg V_{td}. As we shall see, such a constraint is useful in predicting the expected asymmetry in certain CP-violating decays of B mesons.

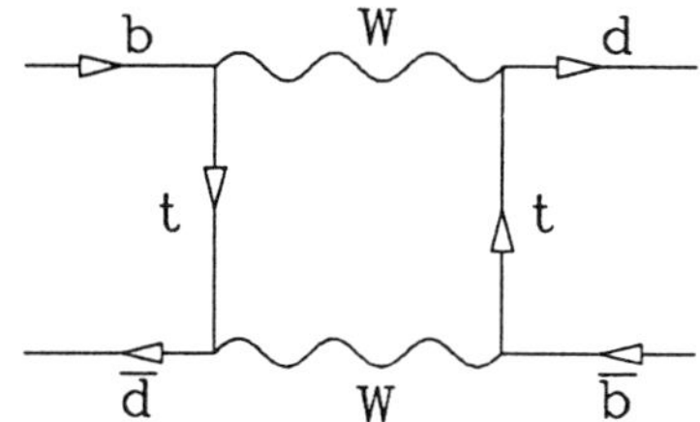

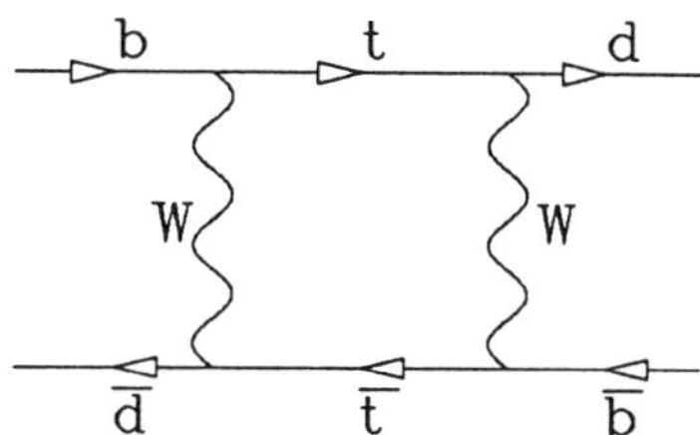

FIGURE 2. Dominant box diagrams for mixing of B^0 and $\bar{B}^0$

3.6 *Allowed region of parameters*

When the constraints of Eqs. (5) and (6) are combined with that on $|V_{ub}/V_{cb}|$ [shown as the circular band bounded by the dotted arcs in Fig. 3], one gets the allowed region of parameters shown in Fig. 5 and described by the first line in Table 1.

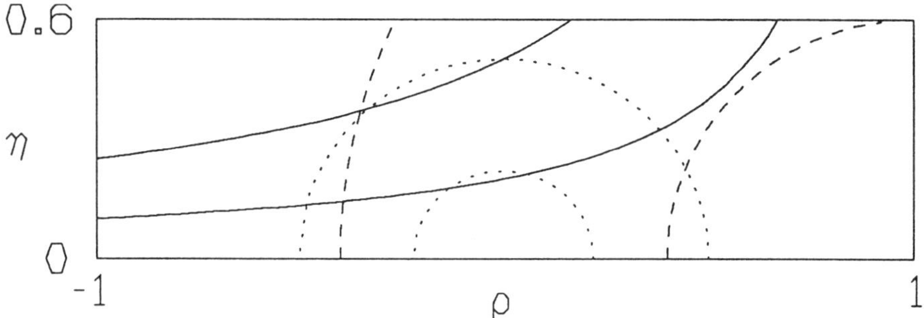

FIGURE 3. Constraints in (ρ, η) plane from $B - \bar{B}$ mixing (dashes), CP-violating $K - \bar{K}$ mixing (solid), and $|V_{ub}/V_{cb}|$ (dots)

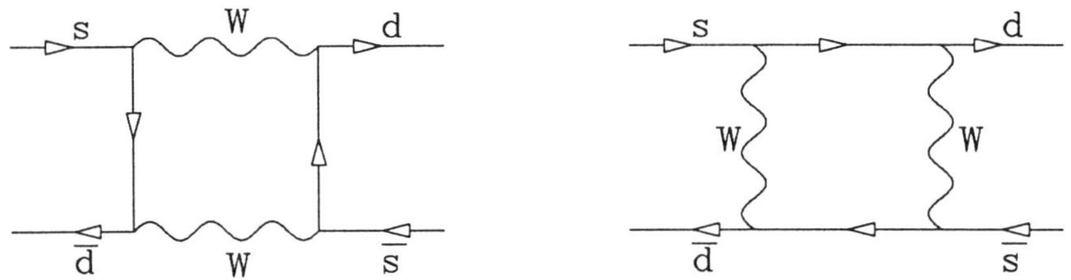

FIGURE 4. Box diagrams for mixing of K^0 and $\bar{K}^0$

The parameters taken for the present analysis are those chosen in Ref. *[1]*, and include the choices $m_t = 160 \pm 30$ GeV, $f_B = 180 \pm 30$ MeV, $|V_{ub}/V_{cb}| = 0.08 \pm 0.03$, and $A = 0.785 \pm 0.093$. The allowed region at 90% c.l. has $-0.4 \leq \rho \leq 0.5$ for $\eta \simeq 0.3$, while for $\rho \simeq 0$ one has $0.2 \leq \eta \leq 0.6$. For a broad range of parameters, CKM phases can describe CP violation in the kaon system. The question is whether this explanation of the observed CP violation is the correct one. A partial answer may be obtained by acquiring improved information about the top quark mass or about CKM elements. (See also Refs. *[21, 22]*.)

The choice of m_t mentioned above was based on an analysis of electroweak data parallel to that presented in Ref. *[23]* and reaching the same conclusions. We shall give more details in Section 5. The results of fixing the top quark mass at 160 or 190 GeV are shown in Table 1. The allowed region is shrunk only slightly, and there is not much difference between the two cases. The favored value of ρ increases by 0.03 for each 10 GeV increase in m_t.

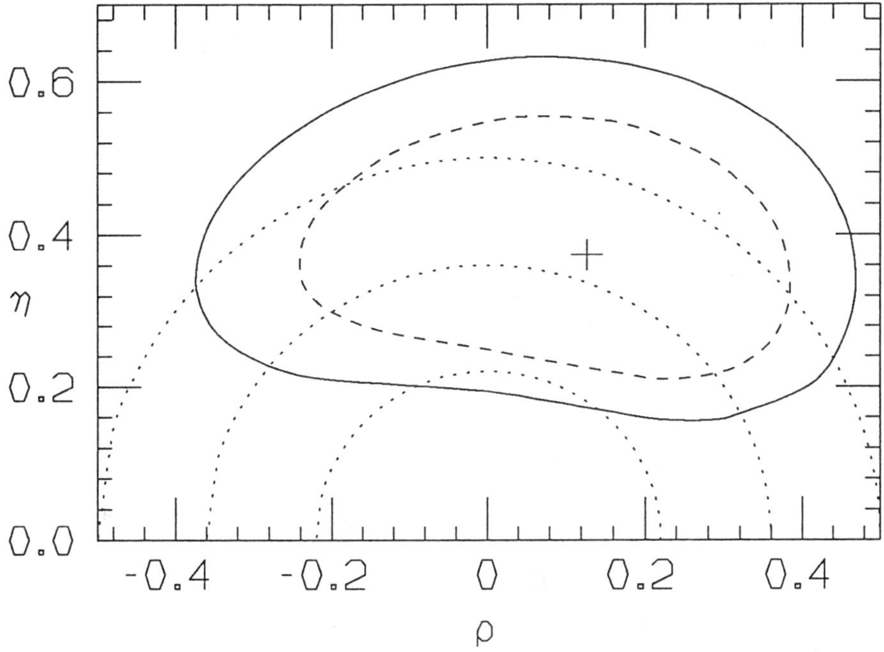

FIGURE 5. Contours of 68% (inner curve) and 90% (outer curve) confidence levels for regions in the (ρ, η) plane. Dotted semicircles denote central value and $\pm 1\sigma$ limits implied by $|V_{ub}/V_{cb}| = 0.08 \pm 0.03$. Plotted point corresponds to minimum $\chi^2 = 0.17$

TABLE 1.   Effects of changing parameters in fits to CKM matrix elements from

"nominal" values described in text

Parameters	ρ (χ^2_{min})	η (χ^2_{min})	χ^2_{min}	ρ range (90% c.l.)
"Nominal"	0.13	0.37	0.17	−0.37 to 0.47
$m_t = 160$ GeV	0.13	0.37	0.17	−0.31 to 0.45
$m_t = 190$ GeV	0.22	0.34	0.25	−0.19 to 0.50
$B_K = 0.80 \pm 0.02^{a)}$	0.13	0.38	0.23	−0.28 to 0.45
$V_{cb} = 0.038 \pm 0.002$	0.11	0.38	0.20	−0.37 to 0.45
$f_B = 180 \pm 10$ MeV	0.15	0.37	0.22	−0.28 to 0.43

[a)] m_t fixed at 160 GeV

A parameter known as B_K describes the degree to which the diagrams of Fig. 4 actually dominate the CP-violating $K^0 - \bar{K}^0$ mixing. In the fits described so far we took the nominal value of $B_K = 0.8 \pm 0.2$. If we take $m_t = 160\,\mathrm{GeV}/c^2$ and reduce the error on B_K to 0.02, we obtain the result shown in Table 1. The reduced errors on B_K are clearly not of much help. The major errors remaining are those of f_B, V_{cb}, and $|V_{ub}/V_{cb}|$.

We next tried reducing the error on V_{cb} to 0.002, keeping other parameters as in the original fit. Again, there is little shrinkage of the allowed parameter space. Reduction of the error on f_B from 30 to 10 MeV helps a little. The results of these exercises are shown in Table 1. The shapes of the allowed regions change very little. The conclusion is that one needs simultaneous reduction in the errors of several observables to significantly narrow down the range of CKM parameters. We explore these possibilities in Section 6. First, however, we concentrate on the top quark.

4   The top quark

4.1   Indirect evidence

Indirect evidence for the top quark has been around for a long time. The neutral-current couplings of the b quark (both flavor-conserving and the absence of flavor-violating ones) have persuaded us that the left-handed b quark is a member of a doublet $(t, b)_L$ of weak SU(2), while the right-handed b is a singlet of weak SU(2).

The expectation that the top quark is relatively heavy is more recent. A value of m_t of at least 70 GeV was needed in order to understand the unexpectedly large magnitude of $B^0 - \bar{B}^0$ mixing [19]. Even a higher lower bound is required to understand the size of CP-violating $K^0 - \bar{K}^0$ mixing [24]. The branching ratio of the W to (lepton) + (neutrino) of about 1/9 is compatible with there being no contribution from $W \to t + \bar{b}$, indicating that $m_t > M_W - m_b$.

An upper limit on the top quark mass is provided by its effects on W and Z self-energies. In the lowest-order electroweak theory, a measurement of the Z mass implies a specific value of M_W. The W and Z self-energies are affected by top quark and Higgs masses, so that now $M_W/M_Z = f(m_t/m_{Higgs})$. This function is quadratic in m_t but only logarithmic in M_{Higgs}. When the Higgs boson mass is allowed to range up to 1 TeV (above which the theory should generate a mass of that order dynamically in any case), the observed values of M_W and of many other electroweak observables allow one to conclude that $m_t \leq 200\,\mathrm{GeV}/c^2$.

4.2   Direct searches

The signature for top quark pair production in $\bar{p}p$ collisions is the simultaneous decay $t \to W^+ + b$, $\bar{t} \to W^- + \bar{b}$. One channel with little background involves the decay of one W to $e\nu$ and the other to $\mu\nu$. As of this workshop, the CDF Collaboration had identified two $e\mu$ candidates and the D0 Collaboration had observed one. On this basis, all that were quoted were lower limits on the top quark mass. Using a sample in which hadronic decays of one of the two W's were also searched for, D0 quoted a lower limit [25] of 131 GeV/c^2.

4.3 Postscript: evidence

Since this workshop, the CDF Collaboration has presented evidence for the production of a top quark *[26]* with $m_t = 174 \pm 10 \; ^{+13}_{-12}$ GeV/c^2. The cases we chose of $m_t = 160$ and 190 GeV/c^2 are compatible with this value. The main impact of this measurement is felt less on the determination of CKM parameters than on the interpretation of electroweak results, which we discuss next.

5 Impact of electroweak tests

In Fig. 6 we show the electroweak prediction for M_W as a function of m_t for various values of Higgs boson mass M_H. Also shown is the latest 1σ range of M_W, corresponding to the average over many experiments. (The latest results have been presented by the CDF and D0 collaborations *[27]*.) Even without a direct observation, one sees the upper bound of about 200 GeV/c^2 quite clearly. The recent (post-workshop) observation corresponds to a data point lying squarely in the allowed range. Greater precision on both M_W and m_t will be needed to distinguish among possibilities for Higgs boson masses.

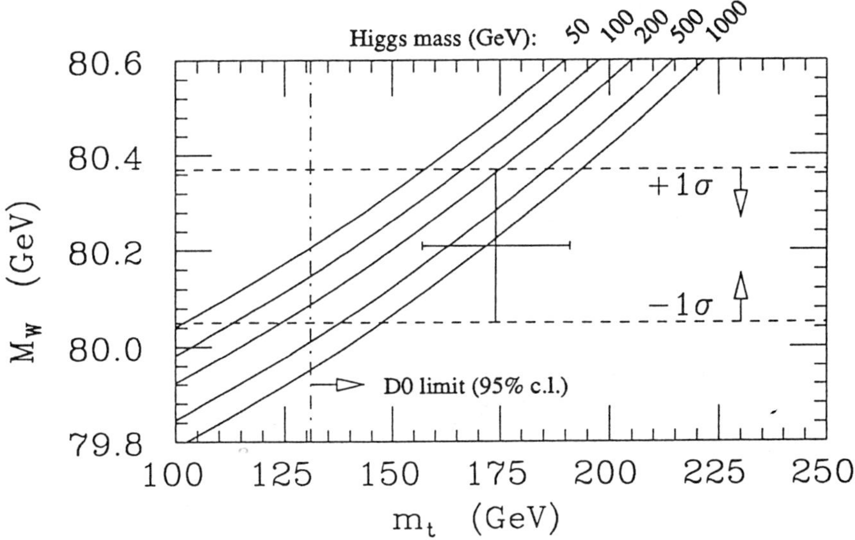

FIGURE 6. Dependence of W boson mass M_W on top quark mass m_t for various values of Higgs boson mass (labels on curves). [Postscript: The plotted point denotes the world average of direct W mass measurements, $M_W = 80.21 \pm 0.16$ GeV/c^2, and the recent CDF top mass value *[26]* of $m_t = 174 \pm 17$ GeV/c^2.]

A fit to the electroweak observables cited in Table 2 has been performed. In each case a prediction is made for the "nominal" values of $m_t = 140$ GeV and $M_H = 100$ GeV/c^2. The Higgs boson mass is held fixed, while the top quark mass is allowed to vary in such a way as to minimize the χ^2 of the fit. The results are shown in Fig. 7 and Table 3.

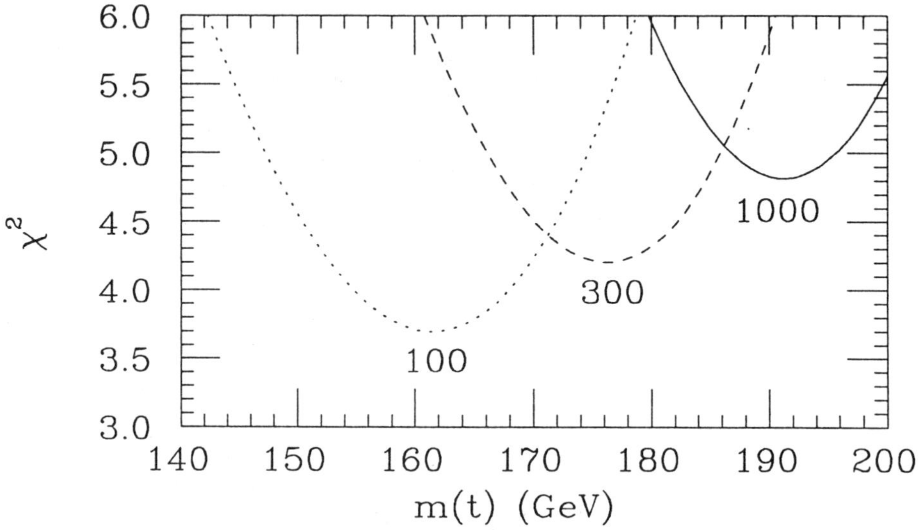

FIGURE 7. Behavior of χ^2 as function of top quark mass for Higgs boson masses of 100, 300, and 1000 GeV (labels on curves)

TABLE 2.   Electroweak observables described in fit

Quantity	Experimental value	Nominal value	Experiment/ Nominal
Q_W (Cs)	-71.0 ± 1.8 [a]	-73.2 [b]	0.970 ± 0.025
M_W (GeV/c^2)	80.22 ± 0.14 [c]	80.174 [d]	1.001 ± 0.002
$\Gamma_{\ell\ell}(Z)$ (MeV)	83.82 ± 0.27 [e]	83.6 [f]	1.003 ± 0.003
$\Gamma_{tot}(Z)$ (MeV)	2489 ± 7 [e]	2488 ± 6 [f]	1.000 ± 0.004
$\sin^2 \hat{\theta}_W^{eff}$	0.2318 ± 0.0008 [g]	0.2322 [f]	0.998 ± 0.003
$\sin^2 \hat{\theta}_W^{eff}$	0.232 ± 0.009 [h]	0.2322	0.999 ± 0.039
$\sin^2 \hat{\theta}_W^{eff}$	0.2287 ± 0.0010 [i]	0.2322	0.985 ± 0.004

[a] Weak charge in cesium. From Ref. *[28]*

[b] From Ref. *[29]*, incorporating corrections of Ref. *[30]*

[c] Average of direct measurements from Ref. *[27]* and indirect information from neutral/charged current ratio in deep inelastic neutrino scattering *[31]*

[d] As calculated in Ref. *[32]*

[e] LEP average as of August, 1993 *[23]*

[f] As calculated in Ref. *[29]*

[g)] From asymmetries at LEP, containing corrections of Ref. *[33]*

[h)] From $\nu_\mu e$ and $\bar{\nu}_\mu e$ scattering *[34]*

[i)] From left-right asymmetry in annihilations at SLC *[35]*, containing corrections of Ref. *[33]*

TABLE 3. Values of m_t for χ^2 minima in fits to electroweak observables

M_H (GeV/c^2)	m_t (GeV/c^2)	χ^2
100	161 ± 12	3.7
300	176 ± 11	4.2
1000	191 ± 10	4.8

A slight preference is shown for a light Higgs boson. This is driven in part by the low value of $\sin^2 \hat{\theta}_W^{eff}$ obtained at SLC/SLD.

The range of top quark masses we chose to discuss at the workshop is compatible both with the results of Table 3 and with the recently announced observation. Present errors on m_t do not allow a conclusion to be drawn yet about the Higgs boson mass.

6 Improved CKM Information

6.1 Meson decay constants

As we mentioned earlier, uncertainty in f_B is a major source of indeterminacy in extracting $|V_{td}|$ from $B^0 - \bar{B}^0$ mixing. Early compilations of predictions are contained in Ref. [FBR]. More recent information on meson decay constants has been provided by lattice gauge theory *[37]*, QCD sum rules *[38]*, and direct quark-model calculations *[39]* which make use of spin-dependent electromagnetic mass splittings in charmed and B mesons to estimate the wave function of a light-heavy system at zero interquark separation. One can expect the reliability of the lattice and QCD sum rule calculations to improve as they are tested on a wide range of properties of charmed and b-flavored hadrons, while the quark model estimate would be helped by a precision measurement of isospin splittings in B and B^* mesons. Modest improvements of recent measurements of the decay constant f_{D_s} *[40]* will allow one to check these schemes.

6.2 Rare kaon decays

Several rare kaon decays can provide information on CKM parameters. Here we discuss the decays of kaons to a pion and a lepton pair.

The rate for the process $K^+ \to \pi^+ \nu\bar{\nu}$ (summed over neutrino species) is sensitive to $|V_{td}|^2$. Very roughly, for a nominal top quark mass of 140 GeV/c^2, a branching ratio of less than 10^{-10} favors $\rho > 0$ while a branching ratio of greater than 10^{-10} favors $\rho < 0$. The branching ratio is an increasing function of top quark mass. Details have been given in Refs. *[1, 21]*.

The present experimental limit *[41]*, $B(K^+ \to \pi\nu\bar{\nu}) < 5 \times 10^{-9}$ (90% c.l.) is a factor of 50 above the expected level, but further improvements in data collection are foreseen.

Postscript: Here information on m_t is very welcome; the predictions were quoted at the workshop for $m_t = 100$, 140, and 180 GeV/c^2. An updated set of predictions may be found in Ref. *[22]*. It now seems unlikely that the branching ratio will be less than 10^{-10}.

The decays $K_L \to \pi^0 \ell^+ \ell^-$ are expected to proceed mainly through CP violation *[42]*, while key CP-conserving backgrounds to this process (see, e.g., Ref. *[43]*) are absent in $K_L \to \pi^0 \nu \bar{\nu}$. Present 90% c.l. upper limits on the branching ratios for these processes are shown in Table 4. The expected branching ratios are about 10^{-11}.

$$\textsc{Table 4.} \quad \text{Upper limits on branching ratios}$$
$$\text{for decays of neutral kaons to neutral pions and a lepton pair}$$

Process	90% c.l. upper limit	Reference
$K_L \to \pi^0 e^+ e^-$	1.8×10^{-9}	*[44, 45, 46]*
$K_L \to \pi^0 \mu^+ \mu^-$	5.1×10^{-9}	*[47]*
$K_L \to \pi^0 \nu \bar{\nu}$	2.2×10^{-4}	*[48]*

6.3 *CP violation in decays of neutral kaons*

One can search for a difference between the CP-violation parameters $\eta_{+-} = \epsilon + \epsilon'$ and $\eta_{00} = \epsilon - 2\epsilon'$ in the decays of neutral kaons to pairs of charged and neutral pions. A non-zero value of ϵ'/ϵ would confirm predictions of the CKM origin of CP violation in the kaon system, and has long been viewed as one of the most promising ways to disprove a "superweak" theory of this effect *[50, 51]*.

The latest estimates by A. Buras and collaborators *[49]* are equivalent to $[\epsilon'/\epsilon]|_{kaons} = (1/2 \ \ to \ \ 3) \times 10^{-3}\eta$, with smaller values for higher top quark masses. The Fermilab E731 Collaboration measures $\epsilon'/\epsilon = (7.4 \pm 6) \times 10^{-4}$, leading to no restrictions on η in comparison with the range (0.2 to 0.6) we have already specified. The CERN NA31 Collaboration *[52]* finds $\epsilon'/\epsilon = (23 \pm 7) \times 10^{-4}$, consistent only with $\eta \gtrsim 1/2$ and a light top quark. [Postscript: this scenario now appears less likely in view of the result of Ref. *[26]*.] Both groups are preparing new experiments, for which results should be available around 1996.

6.4 *Rare B decays*

The rate for the purely leptonic process $B \to \ell \bar{\nu}_\ell$ provides information on the combination $f_B |V_{ub}|$. One expects a branching ratio of about 10^{-4} for $\tau \bar{\nu}_\tau$ and $(1/2) \times 10^{-6}$ for $\mu \bar{\nu}_\mu$. A suggestion was made *[54]* for eliminating f_B by comparing the $B \to \ell \bar{\nu}_\ell$ rate with the $B^0 - \bar{B}^0$ mixing amplitude, and thereby measuring the ratio $|V_{ub}/V_{cb}|$ directly. While such a measurement is unlikely to tell whether the untarity triangle has nonzero area (and thus whether the CKM phase is the origin of CP violation in the kaon system), it *can* help resolve ambiguity regarding the value of ρ.

Another interesting ratio *[55]* is the quantity $\Gamma(B \to \rho\gamma)/\Gamma(B \to K^*\gamma)$, which, aside from small phase space corrections, should just be $|V_{td}/V_{ts}|^2 \simeq 1/20$.

6.5 $B_s - \bar{B}_s$ mixing

The mixing of B_s and $\bar{B}_s$ via diagrams similar to those in Fig. 2 involves the combination $f_{B_s}^2 |V_{ts}|^2$ instead of $f_B^2 |V_{td}|^2$. Since we expect $|V_{ts}| \approx |V_{cb}| \approx 0.04$, the main uncertainties in $x_s \equiv (\Delta m / \Gamma)|_{B_s}$ are associated with f_{B_s} and m_t. A range of 10 to 50 is possible for this quantity. Alternatively, one can estimate the ratio f_{B_s}/f_B using models for $SU(3)$ symmetry breaking, and one finds [1] $x_s = (19 \pm 4)/[(1 - \rho)^2 + \eta^2]$. With the values of ρ and η suggested by present fits to data, the most likely value of x_s seems to be around 20. This corresponds to many oscillations between B_s and $\bar{B}_s$ over the course of a B_s lifetime (about 1.5 ps), and represents a strong experimental challenge.

6.6   *CP violation in B systems*

Asymmetries in the rates for certain decays of B mesons can provide direct information about the angles in the unitarity triangle of Fig. 1. These decays involve final states which are eigenstates of CP, so that they can be reached both from an initial B^0 and from an initial $\bar{B}^0$.

We may define a time-integrated rate asymmetry $A(f)$ as

$$A(f) \equiv \frac{\Gamma(B^0_{t=0} \to f) - \Gamma(\bar{B}^0_{t=0} \to f)}{\Gamma(B^0_{t=0} \to f) + \Gamma(\bar{B}^0_{t=0} \to f)} \quad . \tag{7}$$

The angles β and α in Fig. 1 are related to the asymmetries in decays to $J/\psi K_S$ and $\pi^+\pi^-$ final states:

$$A(J/\psi K_S) = -\frac{x_d}{1 + x_d^2} \sin 2\beta \quad , \tag{8}$$

$$A(\pi^+\pi^-) = -\frac{x_d}{1 + x_d^2} \sin 2\alpha \quad , \tag{9}$$

where $x_d \equiv (\Delta m / \Gamma)|_{B^0}$, and we have neglected lifetime differences between eigenstates. Contours of the expected values of these asymmetries have been quoted in Refs. [1, 21] and will not be reproduced here. The *ratios* of these asymmetries can be useful in cancelling certain common (and sometimes hard-to-estimate) "dilution factors" associated with identification of the flavor of the decaying neutral B meson. Contours [54] of the ratio $\mathcal{R} \equiv \sin 2\alpha / \sin 2\beta$ are shown in Fig. 8. As long as $\rho^2 + \eta^2 < 1$, which certainly is true, a value of $\mathcal{R} > 1$ signifies $\rho < 0$, while $\mathcal{R} < 1$ signifies $\rho > 0$.

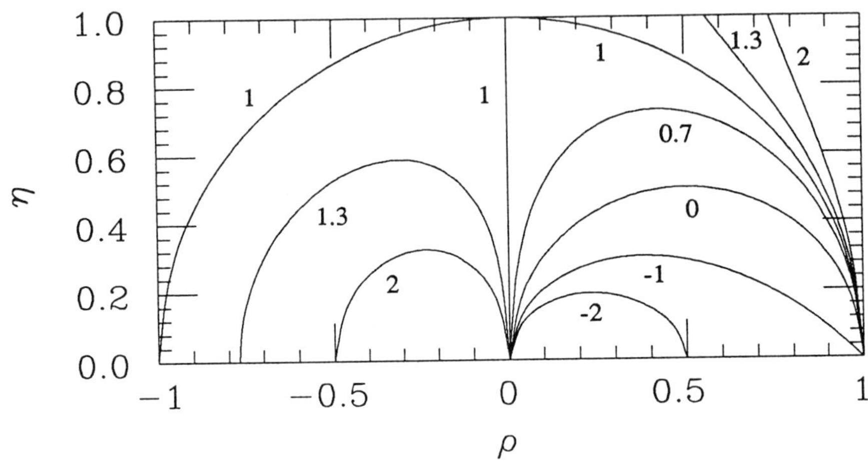

FIGURE 8. Contours of ratios of asymmetries $A(\pi^+\pi^-)/A(J/\psi K_S)$ $= \sin(2\alpha)/\sin(2\beta)$ (labels on curves) in the (ρ, η) plane

7   Recent results on tagging neutral B mesons

7.1   Why neutral B mesons?

The observation of a CP-violating asymmetry between the rate for a process and its charge-conjugate requires some sort of interference. Two examples serve to illustrate the major possibilities [56].

Self-tagging modes

The rates for such processes as $B^+ \to K^+\pi^0$ and $B^- \to K^-\pi^0$ can differ from one another. Under charge-conjugation, weak phases change sign but strong phases do not. One can see a CP-violating rate difference, but only if strong phases differ in the $I = 1/2$ and $I = 3/2$ channels. Interpretation of an effect requires knowledge of this final-state phase difference. [Postscript: with the help of SU(3) symmetry and some simplifying assumptions, it is possible to extract CKM phases from rates of self-tagging modes alone [57].]

Decays to a CP eigenstate

Final-state phase information is not needed if one compares the rates for a state which is produced as a B^0 and a state which is produced as a $\bar{B}^0$ to decay to an eigenstate f of CP. The relevant interference leading to a rate asymmetry occurs between amplitudes for decay and $B^0 - \bar{B}^0$ mixing. As mentioned above, decay rate asymmetries can directly probe the angles of the unitarity triangle, as long as a single amplitude contributes to each transition $B^0 \to f$ and $\bar{B}^0 \to f$. To make use of this method, one must identify the flavor of the decaying particle at the time of production: was it a B^0 or a $\bar{B}^0$?

7.2   Identifying neutral B's

At the $\Upsilon(4S)$

A peak in the cross section for $e^+e^- \to B^0\bar{B}^0$ occurs just above threshold at the $\Upsilon(4S)$ resonance. If one "tags" the flavor of the decaying state by observing the semileptonic decay of the "other" B, the existence of $B^0 - \bar{B}^0$ mixing and the correlation of the B^0 and $\bar{B}^0$ in a state of negative charge-conjugation lead to an asymmetry proportional to $\sin(t_1 - t_2)$, where t_1 is the proper time of the decay to the CP eigenstate and t_2 is the proper time of the tagging decay. This asymmetry vanishes when integrated over all decay times, so one needs information on $t_1 - t_2$ such as might be provided by an asymmetric B factory.

Away from the $\Upsilon(4S)$

In any reaction in which a $b\bar{b}$ pair is produced at high energy, such as a hadronic collision or the decay of the Z^0, the flavor of a neutral B meson decaying to a CP eigenstate can be tagged by looking at the flavor of the b-flavored particle produced in association with it. Such a particle might be another neutral nonstrange or strange B (in which case mixing would cause a dilution of tagging efficiency), or it could be a charged B or a b-flavored baryon. Here one has to find the tagging particle among the debris of the collision, and estimates of tagging efficiency are likely to be model-dependent.

Tagging using associated hadrons

A neutral B meson produced in a high-energy collision is likely to be accompanied by other hadrons as a result of the fragmentation of the initial quark or as a result of cascades from higher resonances. This feature could be useful for tagging the flavor of a produced B meson *[58, 59, 60]*.

   The correlation is easily visualized with the help of the quark diagrams shown in Fig. 9. By convention (the same as for kaons), a neutral B meson containing an initially produced $\bar{b}$ is a B^0. It also contains a d quark. The next charged pion down the fragmentation chain must contain a $\bar{d}$, and hence must be a π^+. Similarly, a $\bar{B}^0$ will be correlated with a π^-.

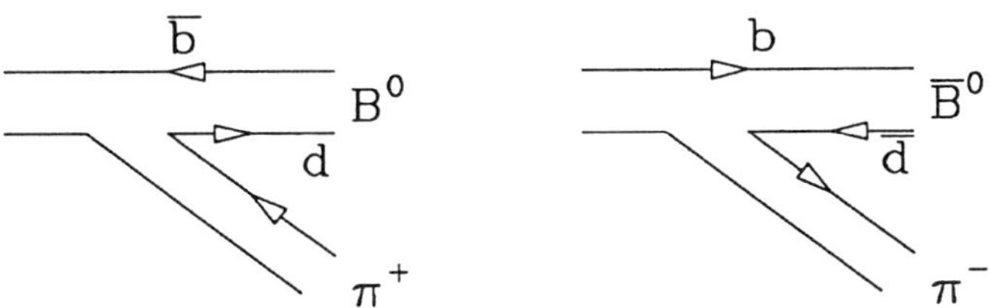

FIGURE 9. Quark graphs describing correlation between flavor of neutral B meson and charge of leading pion in fragmentation

The same conclusion can be drawn by noting that a B^0 can resonate with a positive pion to form an excited B^+, which we shall call B^{**+} (to distinguish it from the B^* lying less than 50 MeV/c^2 above the B). Similarly, a $\bar{B}^0$ can resonate with a negative pion to form a B^{**-}. The combinations $B^0\pi^-$ and $\bar{B}^0\pi^+$ are *exotic*, i.e., they cannot be formed as quark-antiquark states. No evidence for exotic resonances exists.

7.3 Results from simulation

We have asked the authors of a fragmentation Monte Carlo program to see if the correlation between pions and neutral B mesons is evident in their work. The result, based on 10^6 events generated using ARIADNE and JETSET [61],

shows a slight excess of the "right-sign" combinations $B^0\pi^+$ over the "wrong-sign" combinations $B^0\pi^-$. The ratio (right - wrong)/(right + wrong) varies from 0.17 for $M(B\pi) = 5.5$ GeV/c^2 to 0.27 for $M(B\pi) = 5.8 - 6.2$ GeV/c^2 (where there are fewer events). No explicit resonances were put into the simulation; their inclusion would strengthen the correlation.

7.4 B^{**} resonances

The existence of a soft pion in $D^* \to D\pi$ decays [62] has been a key feature in tagging the presence of D mesons since the earliest days of charmed particles. The mass of a D^* is just large enough that the decays $D^* \to D\pi$ can occur (except in the case of $D^0 \to \pi^- D^+$). In contrast, the B^* is only 46 MeV above the B, so it cannot decay via pion emission. The lightest states which can decay to $B\pi$ and/or $B^*\pi$ are P-wave resonances of a b quark and an $\bar{u}$ or $\bar{d}$. The expectations for masses of these states [60, 63], based on extrapolation from the known D^{**} resonances, are summarized in Table 5.

TABLE 5.   P-wave resonances of a b quark and a light ($\bar{u}$ or $\bar{d}$) antiquark

J^P	Mass (GeV/c^2)	Allowed final state(s)
2^+	~ 5.77	$B\pi$, $B^*\pi$
1^+	~ 5.77	$B^*\pi$
1^+	< 5.77	$B^*\pi$
0^+	< 5.77	$B\pi$

The known D^{**} resonances are a 2^+ state around 2460 MeV/c^2, decaying to $D\pi$ and $D^*\pi$, and a 1^+ state around 2420 MeV/c^2, decaying to $D^*\pi$. These states are relatively narrow, probably because they decay via a D-wave. In addition, there are expected to be much broader (and probably lower) D^{**} resonances: a 1^+ state decaying to $D^*\pi$ and a 0^+ state decaying to $D\pi$, both via S-waves.

The expected spectrum of nonstrange charmed meson resonances is shown in Fig. 10 [63, 64], as calculated in the potential of Ref. [65]. For strange states, one should add about 0.1 GeV, while for B's one should add about 3.32 GeV. The predicted narrow B^{**} resonances lie at 5767 MeV (2^+) and 5755

MeV (1^+). These are the states in which the light quark spin $s = 1/2$ and the orbital angular momentum $L = 1$ combine to form a total light-quark angular momentum $j = 3/2$. One also expects broad 1^+ and 0^+ B^{**} states with $j = 1/2$, decaying via S-waves.

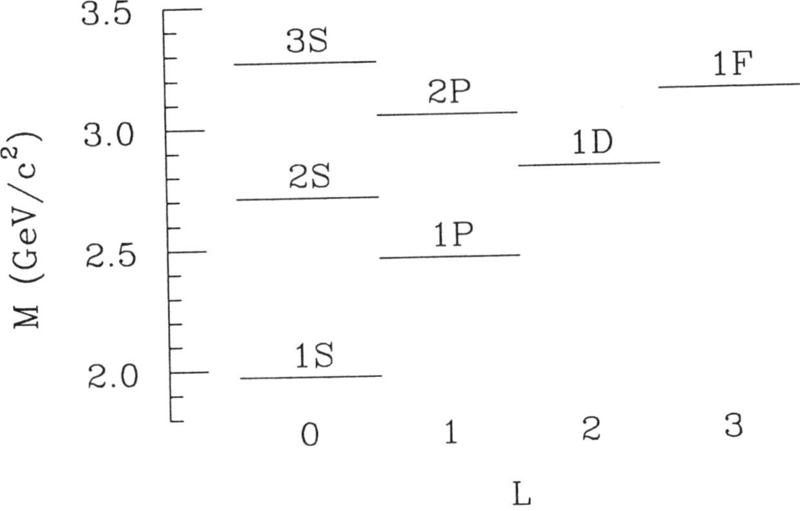

FIGURE 10. Predicted spectrum *[63,64]* of nonstrange charmed meson resonances. Observed states are labeled by a check mark

7.5 *The question of coherence*

As mentioned, one expects the $B^0 \bar{B}^0$ pair produced at the $\Upsilon(4S)$ resonance to be in a state of charge-conjugation eigenvalue $C = -1$. The particle which decays at a time t_1 to a CP eigenstate is then a coherent superposition of B^0 and $\bar{B}^0$ when the particle produced in association decays at a time t_2 to a tagging final state (e.g., to $D^{*+}e^-\bar{\nu}_e$). The fact that the time-dependent rate asymmetry is an odd function of $t_1 - t_2$ is why one has to stretch out the decay region using asymmetric kinematics.

On the other hand, in the high energy associated production of pairs of b-flavored hadrons, one usually assumes no coherence between B^0 and $\bar{B}^0$ on one side of the reaction when tagging on the other. Thus, the decaying state is assumed to be an incoherent *mixture* of B^0 and $\bar{B}^0$ with specific probabilities of each.

M. Gronau and I *[59]* have proposed a way to test for coherence using a density-matrix formalism. One can measure the elements of the density matrix using time-dependences of appearance of specific final states.

We work in a two-component basis labeled either by B^0 and $\bar{B}^0$ or, more conveniently, by mass eigenstates. In this last basis, in which components of the density matrix are labeled by primed quantities, we can denote an arbitrary coherent or incoherent state by the density matrix

$$\rho = \frac{1}{2}\left(1 + Q' \cdot \sigma\right) \tag{10}$$

where σ_i are the Pauli matrices. The intensities for decays to states of identified flavor can be written

$$I\left(\frac{B^0}{\bar{B}^0}\right) = \frac{1}{2}|A|^2 e^{-\Gamma t}\left[1 \pm Q'_\perp \cos(\Delta m t + \delta)\right] \quad . \tag{11}$$

where

$$Q'_1 = Q'_\perp \cos\delta \quad , \quad Q'_2 = Q'_\perp \sin\delta \quad . \tag{12}$$

In order to measure Q'_3 one needs also to see decays to CP eigenstates, such as $J/\psi K_S$. One then learns not only the components of the density matrix, but also one of the angles of the unitarity triangle (such as β). Then, one can look at other final states to learn other angles. For example, if penguin diagrams are not important in the decays $B \to \pi\pi$, the final state $\pi^+\pi^-$ provides information on the angle α.

8   Conclusions

The present knowledge about magnitudes and phases of CKM matrix elements allows lots of "wiggle room" for inventive schemes. Choosing among these schemes will require progress on many fronts. Among these, we have discussed improved knowledge of meson decay constants, decays of neutral kaons, rare B decays, mixing of strange neutral B mesons with their antiparticles, and the observation of CP violation in decays of neutral B mesons. The identification of the initial flavor of a neutral B meson may profit from the study of hadrons produced in association with it, and we have described ways in which our knowledge of such correlations may be improved.

9   Acknowledgments

I would like to thank several people for enjoyable collaborations on aspects of the work presented here: Geoff Harris on the work which permitted several of the figures to be drawn, and Alex Nippe and Michael Gronau on the topics mentioned in Section 7. This work was supported in part by the United States Department of Energy under Grant No. DE FG02 90ER40560.

References

[1]   J. L. Rosner, Enrico Fermi Institute report 93-62, to be published in *B Decays*, edited by S. Stone (Singapore: World Scientific, 1994)

[2]   N. Cabibbo, *Phys. Rev. Lett.* **10** (1963) 531

[3]   S. L. Glashow, J. Iliopoulos, and L. Maiani, *Phys. Rev.* D **2** (1970) 1285

[4]   M. Kobayashi and T. Maskawa, *Prog. Theor. Phys.* **49** (1973) 652

[5]   S. Stone, in *B Decays*, edited by S. Stone (Singapore: World Scientific, 1994)

[6]   S. Stone, this workshop

[7] J. D. Bjorken and I. Dunietz, *Phys. Rev.* D **36** (1987) 2109

[8] L. Wolfenstein, *Phys. Rev. Lett.* **51** (1983) 1945

[9] ARGUS Collaboration, H. Albrecht *et al.*, *Phys. Lett.* B **234** (1990) 409; *ibid.* **255** (1991) 297

[10] CLEO Collaboration, as reported by F. Muheim in *The Fermilab Meeting - DPF 92* (Division of Particles and Fields Meeting, American Physical Society, Fermilab, 10 – 14 November, 1992), ed. by C. H. Albright *et al.* (World Scientific, Singapore, 1993), p. 614, and by D. Besson in *Proceedings of the XVI International Symposium on Lepton and Photon Interactions*, Cornell University, August 10–15, 1993

[11] G. Altarelli *et al.*, *Nucl. Phys.* **B208** (1982) 365

[12] M. Wirbel, B. Stech, and M. Bauer, *Zeit. Phys.* C **29** (1985) 637; J. G. Körner and G. A. Schuler, *ibid.* **38** (1988) 511; N. Isgur, D. Scora, B. Grinstein, and M. B. Wise, *Phys. Rev.* D **39** (1989) 799.

[13] L.-L. Chau and W.-Y. Keung, *Phys. Rev. Lett.* **53** (1984) 1802; M. Gronau and J. Schechter, *ibid.* **54** (1985) 385; M. Gronau, R. Johnson, and J. Schechter, *Phys. Rev.* D **32** (1985) 3062; C. Jarlskog, in *Physics at LEAR with Low Energy Antiprotons,* proceedings of the workshop, Villars-sur-Ollon, Switzerland, 1987, edited by C. Amsler *et al.* (Harwood, Chur, Switzerland, 1988), p. 571; J. D. Bjorken, private communication; Bjorken and Dunietz, Ref. *[7]*

[14] W. J. Marciano, *Ann. Rev. Nucl. Part. Sci.* **41** (1991) 469

[15] C. S. Kim and A. D. Martin, *Phys. Lett.* B **225** (1989) 186

[16] J. L. Rosner, in *Testing the Standard Model* (Proceedings of the 1990 Theoretical Advanced Study Institute in Elementary Particle Physics), edited by M. Cvetič and P. Langacker (World Scientific, Singapore, 1991), p. 91

[17] S. Nussinov and W. Wetzel, *Phys. Rev.* D **36** (1987) 130; M. Neubert, *Phys. Lett.* B **264** (1991) 455 and in *B Decays,* edited by S. Stone (Singapore: World Scientific, 1994); N. Isgur and M. Wise, in *B Decays,* edited by S. Stone (Singapore: World Scientific, 1994)

[18] C. Ramirez, J. F. Donoghue, and G. Burdman, *Phys. Rev.* D **41** (1990) 1496

[19] ARGUS Collaboration, H. Albrecht *et al.*, *Phys. Lett.* B **192** (1987) 245

[20] ARGUS Collaboration, H. Albrecht *et al.*, *Zeit. Phys.* C **55** (1992) 357; CLEO Collaboration, J. Bartelt *et al.*, *Phys. Rev. Lett.* **71** (1993) 1680. The average is based on tagged events as quoted in Ref. *[10]*

[21] G. Harris and J. L. Rosner, *Phys. Rev.* D **45** (1992) 946

[22] A. Buras, M. E. Lautenbacher, and G. Ostermaier, Max-Planck-Institut report MPI-PH-94-14, 1994

[23] M. Swartz, in *Proceedings of the XVI International Symposium on Lepton and Photon Interactions* (see Ref. *[10]*)

[24] A. Buras, *Phys. Lett.* B **317** (1993) 449

[25] D0 Collaboration, S. Abachi *et al.*, *Phys. Rev. Lett.* **72** (1994) 2138

[26] CDF Collaboration, F. Abe *et al.*, Fermilab report FERMILAB-PUB-94-097-E, April, 1994, submitted to Phys. Rev. D

[27] CDF Collaboration, F. Abe *et al.*, *Phys. Rev. Lett.* **65** (1990) 2243; *Phys. Rev.* D **43** (1991) 2070; UA2 Collaboration, J. Alitti *et al.*, *Phys. Lett.* B **276** (1992) 354; CDF Collaboration, F. Abe *et al.*, presented at Fermilab, October 15, 1993; D0 Collaboration, presented by B. Klima at Fermilab, October 11, 1993

[28] M. C. Noecker, B. P. Masterson, and C. E. Wieman, *Phys. Rev. Lett.* **61** (1988) 310

[29] W. J. Marciano and J. L. Rosner, *Phys. Rev. Lett.* **65** (1990) 2963; *ibid.* **68** (1992) 898(E)

[30] S. A. Blundell, J. Sapirstein, and W. R. Johnson, *Phys. Rev.* D **45** (1992) 1602

[31] CDHSW Collaboration, H. Abramowicz *et al.*, *Phys. Rev. Lett.* **57** (1986) 298; A. Blondel *et al.*, *Zeit. Phys.* C **45** (1990) 361; CHARM II Collaboration, J. V. Allaby *et al.*, *Phys. Lett.* B **177** (1986) 446; *Zeit. Phys.* C **36** (1987) 611; CCFRR Collaboration, P. G. Reutens *et al.*, *Phys. Lett.* **152B** (1985) 404; *Zeit. Phys.* C **45** (1990) 539; P. G. Reutens, thesis, University of Chicago, 1986 (unpublished); CCFR Collaboration, C. G. Arroyo *et al.*, presented at XVI International Symposium on Lepton and Photon Interactions (see Ref. *[10]*); FMM Collaboration, D. Bogert *et al.*, *Phys. Rev. Lett.* **55** (1985) 1969

[32] G. DeGrassi, B. A. Kniehl, and A. Sirlin, *Phys. Rev.* D **48** (1993) 3963

[33] P. Gambino and A. Sirlin, New York University report NYU-TH-93-09-07, Sept., 1993

[34] CHARM-II Collaboration, D. Geiregat *et al.*, presented at XVI International Symposium on Lepton and Photon Interactions (see Ref. *[10]*)

[35] SLD Collaboration, K. Abe *et al.*, SLAC report SLAC-PUB-6456, March, 1994 (submitted to Phys. Rev. Lett.

[36] J. L. Rosner, *Phys. Rev.* D **42** (1990) 3732; in *Research Directions for the Decade* (Proceedings of the 1990 DPF Snowmass Workshop), edited by E. L. Berger (World Scientific, Singapore, 1991), p. 255; M. Schmidtler and K. R. Schubert, *Zeit. Phys.* C **53** (1992) 347

[37] See, e.g., P. B. Mackenzie, in *Proceedings of the XVI International Symposium on Lepton and Photon Interactions* (Ref. *[10]*); C. W. Bernard, J. N. Labrenz, and A. Soni, *Phys. Rev.* D **49** (1994) 2536; 1993; UKQCD Collaboration, R. M. Baxter *et al.*, Univ. of Edinburgh report no. 93-526, Sept., 1993; A. S. Kronfeld, Fermilab report FERMILAB-CONF-93/277-T, presented at Workshop on *B* Physics at Hadron Accelerators, Snowmass, Colo., 21 June – 2 July, 1993; P. B. Mackenzie and A. S. Kronfeld, *Ann. Rev. Nucl. Part. Sci.* **43** (1993) 793

[38] M. Neubert, *Phys. Rev.* D **46** (1992) 1076; V. Rieckert, *Phys. Rev.* D **47** (1993) 3053; M. Paulini, DESY report DESY-F15-93-02, 1993 (doctoral thesis)

[39] J. F. Amundson *et al.*, *Phys. Rev.* D **47** (1993) 3059; J. L. Rosner, in *The Fermilab Meeting* (see *[10]*), p. 658; L. H. Chan, Louisiana State University report, 1993, submitted to 1993 Lepton-Photon Symposium (see *[10]*)

[40] WA75 Collaboration, S. Aoki *et al.*, *Prog. Theor. Phys.* **89** (1993) 131; CLEO Collaboration, D. Acosta *et al.*, Cornell University report CLNS 93/1238, August, 1993; F. Muheim and S. Stone, *Phys. Rev.* D **49** (1994) 3767

[41] BNL E787 Collaboration, M. S. Atiya *et al.*, *Phys. Rev. Lett.* **64** (1990) 21; *ibid.* **70** (1993) 2521

[42] M. K. Gaillard and B. W. Lee, *Phys. Rev.* D **10** (1974) 897

[43] H. B. Greenlee, *Phys. Rev.* D **42** (1990) 3724

[44] E845 Collaboration, K. E. Ohl *et al.*, *Phys. Rev. Lett.* **64** (1990) 2755

[45] Fermilab E799 Collaboration, D. Harris *et al.*, *Phys. Rev. Lett.* **71** (1993) 3918

[46] Fermilab E731 Collaboration, A. Barker *et al.*, *Phys. Rev.* D **41** (1990) 3546

[47] Fermilab E799 Collaboration, D. Harris *et al.*, *Phys. Rev. Lett.* **71** (1993) 3914

[48] Fermilab E799 Collaboration, G. Graham *et al.*, *Phys. Lett.* B **295** (1992) 169

[49] A. J. Buras, M. Jamin, and M. E. Lautenbacher, Max-Planck-Institut report MPI-Ph/93-11, March, 1993

[50] L. Wolfenstein, *Phys. Rev. Lett.* **13** (1964) 562

[51] B. Winstein and L. Wolfenstein, *Rev. Mod. Phys.* **65** (1993)

[52] CERN NA31 Collaboration, G. D. Barr *et al.*, *Phys. Lett.* B **317** (1993) 233

[53] Fermilab E731 Collaboration, L. K. Gibbons *et al.*, *Phys. Rev. Lett.* **70** (1993) 1203

[54] P. F. Harrison and J. L. Rosner, *J. Phys.* G **18** (1992) 1673

[55] A. Ali, *J. Phys.* G **18** (1992) 1065

[56] A. B. Carter and A. I. Sanda, *Phys. Rev. Lett.* **45** (1980) 952; *Phys. Rev.* D **23** (1981) 1567; I. I. Bigi and A. I. Sanda, *Nucl. Phys.* **B193** (1981) 85; *ibid.* **B281** (1987) 41; I. Dunietz and J. L. Rosner, *Phys. Rev.* D **34** (1986) 1404; I. Dunietz, *Ann. Phys. (N.Y.)* **184** (1988) 350

[57] M. Gronau, J. L. Rosner, and D. London, March, 1994, submitted to Phys. Rev. Lett; M. Gronau, O. F. Hernandez, D. London, and J. L. Rosner, April, 1994, submited to Phys. Rev. D; O. F. Hernandez, D. London, M. Gronau, and J. L. Rosner, April, 1994, submitted to Phys. Lett. B

[58] M. Gronau, A. Nippe, and J. L. Rosner, *Phys. Rev.* D **47** (1992) 1988

[59] M. Gronau and J. L. Rosner, *Phys. Rev. Lett.* **72** (1994) 195

[60] M. Gronau and J. L. Rosner, *Phys. Rev.* D **49** (1994) 254

[61] G. Gustafson and J. Hakkinen, private communication

[62] S. Nussinov, *Phys. Rev. Lett.* **35** (1975) 1672

[63] E. Eichten, C. T. Hill, and C. Quigg, *Phys. Rev. Lett.* **71** (1994) 4116

[64] E. Eichten, C. T. Hill, and C. Quigg, private communication

[65] W. Buchmüller and S.-H. H. Tye, *Phys. Rev.* D **24** (1981)

Yukawa Coupling Thresholds in the MSSM and the Minimal SUSY-SU(5) GUT [1]

BRIAN D. WRIGHT

PHYSICS DEPARTMENT, UNIVERSITY OF WISCONSIN
MADISON, WI 53706, USA

Abstract: We discuss the treatment of Yukawa coupling thresholds at the unification scale M_{GUT} and the effective supersymmetry scale M_{SUSY} and apply them to corrections to the tree-level prediction $y_b(M_{GUT}) = y_\tau(M_{GUT})$ in minimal supersymmetric SU(5). We discuss the dependencies of M_{GUT} and colored Higgs triplet mass, M_{H_3}, on $\alpha_s(M_Z)$, M_t and the sparticle spectrum. The effect of the Yukawa coupling thresholds on M_t are given for representative sparticle spectra. We describe the quantitative differences between these effects for low and high $\tan\beta$ and discuss the robustness of the M_t solution. We also give new bounds on superheavy masses, incorporating proton decay as well as unification constraints, the former leading to a lower bound on α_s.

1   Introduction

In light of the current renewal of interest in renormalization group (RG) constraints arising from supersymmetric grand unified theories (SUSY-GUTs), it has become increasingly important to quantify corrections to GUT scale predictions. Recently several two loop renormalization group analyses of SUSY-GUTs with soft supersymmetry-breaking induced via minimal N=1 supergravity have been performed [1]-[7]. In certain scenarios, these analyses have made predictions for the sparticle spectrum and low energy parameters arising from GUT scale constraints. Emphasis should be placed on determining the uncertainties in these predictions and their dependence on the unknown heavy mass spectrum as well as the details of supersymmetry breaking. The purpose of this talk is to present the Yukawa threshold corrections in the Standard Model (SM), the Minimal Supersymmetric Standard Model (MSSM) and the minimal supersymmetric SU(5) GUT and apply them in the context of gauge and third generation Yukawa unification in minimal SU(5) (see Ref. [8] for more details).

The importance of Yukawa coupling thresholds has come to the attention of several authors recently [9, 10, 11]. Loop corrections to low energy mass parameters have been treated in certain cases such as the well-known case of radiative corrections to the Higgs boson masses in the MSSM [12] although usually not explicitly in the form of threshold conditions on running mass parameters except in the case of the Standard Model [13] and for y_b and y_τ in ordinary GUTs [14]. However, in the context of renormalization group analyses of high scale predictions they have not been treated generally and usually

---

[1]This research was supported in part by the University of Wisconsin Research Committee with funds granted by the Wisconsin Alumni Research Foundation and in part by the U. S. Department of Energy under contract no. DE-AC02-76ER00881.

only the gauge coupling thresholds are considered. We find that the Yukawa coupling thresholds are typically at least as important as the gauge coupling threshold effects.

We would like to advocate a particular philosophy for the treatment of threshold corrections in minimal subtraction (MS) schemes. From the Appelquist-Carrazzone decoupling theorem [15] we expect the physics at energies below a given mass scale to be independent of the particles with masses higher than this threshold. However, MS schemes are not physical in the sense that they are scale dependent and mass independent so that the decoupling theorem is not manifest. As described in Refs. [16, 17, 18], one implements the decoupling in MS schemes by formulating a low energy effective theory obtained by integrating out the heavy fields to one loop. The effect of this procedure is to give relations between renormalized parameters just above and below the particle mass(es), the so-called matching functions, and to modify the various β function coefficients so that in the lower scale theory the contribution of the particle(s) to these coefficients is removed. The running parameters are typically discontinuous at the boundary at which the matching function is applied unless one tunes the boundary scale. In either case reliable values for the parameters of the theory are obtained asymptotically away from the boundary.

We advocate the use of the scheme of Ref. [17] in which one integrates out together all particles with similar masses at a single scale. If one uses one loop matching functions then this is justified as long as one loop β functions can be used between the different particle masses [17]. This is simpler both from the analytic and numerical standpoint when it is applied to the MSSM and its grand unified extensions. For example, in this analysis in which one has a light SM Higgs doublet below M_{SUSY}, the top quark as well as the W, Z, and SM Higgs are integrated out at M_Z, all sparticles including the heavier Higgs doublet can be integrated out at a fixed scale, M_{SUSY} (with complicated matching functions incorporating the details of the spectrum) and all superheavy particles at M_{GUT}.

2 Yukawa and Gauge Thresholds in Minimal SUSY-SU(5)

The treatment of Yukawa coupling and mass parameter thresholds in minimal subtraction (MS) schemes is analogous to the treatment of gauge thresholds discussed above [16, 17, 18]. We apply this formalism to the one loop GUT threshold corrections to the tree-level relation $y_b(M_{GUT}) = y_\tau(M_{GUT})$ in minimal supersymmetric SU(5). This is of particular interest since this condition strongly constrains the allowed range of the top quark mass in an experimentally accessible region [19, 20, 21, 22].

We start with a superpotential of the form

$$P = \sqrt{2}\,\Psi_a Y^{(d)} \chi^{ab} H_b^{(1)} - \tfrac{1}{4}\epsilon_{abcde}\chi^{ab} Y^{(u)} \chi^{cd} H^{(2)e}$$
$$+ M_2 H^{(2)a} H_a^{(1)} + \lambda_2 H^{(2)a}\Sigma_a^b H_b^{(1)} + \frac{\lambda_1}{3}Tr\Sigma^3 + \frac{M_1}{2}Tr\Sigma^2 , \qquad (1)$$

where Ψ, $H^{(1)}$, $H^{(2)}$, χ and Σ are SU(5) superfields transforming as the 5, 5, $\bar{5}$, 10 and 24 dimensional representations, respectively. In the usual way we associate these superfields with their $SU(3) \times SU(2) \times U(1)$ decompositions.

In addition to the component field interactions obtained from (1), one also can add the most general set of SU(5) invariant soft supersymmetry breaking terms. After SU(5) breaking we obtain the following mass spectrum in the minimal model: two degenerate Higgs triplet superfields of mass M_{H_3}, degenerate X and Y leptoquark gauge superfields of mass M_V, color octet and SU(2) triplet superfields of mass M_Σ along with a singlet of mass $0.2 M_\Sigma$. The electroweak doublets remain light as long as the fine-tuning constraint $\lambda_1 M_2 - 3\lambda_2 M_1 \approx M_{EW}$ is satisfied, where M_{EW} is a typical electroweak mass scale[2].

The GUT scale threshold corrections are obtained by integrating out the superheavy X and Y gauge supermultiplets, the Higgs triplet, and the Higgs adjoint superfields to obtain an effective MSSM. Since the soft supersymmetry breaking mass parameters are much smaller than typical GUT scale masses we can work in an approximately supersymmetric formalism in which both light fields along with their superpartners are treated as massless in loops involving superheavy fields. We therefore use supergraph methods [24] to simplify the calculations. Due to nonrenormalization theorems [24, 25], the only modifications to the parameters of the superpotential in the effective action arise through superfield wavefunction renormalizations. Thus, for example, the threshold corrections to the top Yukawa coupling, y_t, involve evaluating supergraph two point functions with external top and light Higgs superfields and superheavy gauge and Higgs superfields in the loop. In general one then obtains matching conditions of the form $y_\alpha(\mu) = \bar{y}_\alpha(\mu)(1 + \Delta_{y_\alpha}^{GUT})$ $(\alpha = t, b, \tau)$. The y_b/y_τ threshold matching function is obtained by evaluating analogous supergraphs, with the result

$$\frac{y_b}{y_\tau}(\mu) = 1 + \frac{1}{16\pi^2}\left(2\bar{g}^2(\mu)(2\ln\frac{M_V}{\mu} - 1) - \tfrac{1}{2}\bar{y}_t^2(\mu)(2\ln\frac{M_{H_3}}{\mu} - 1)\right) , \qquad (2)$$

where the (un)barred couplings are (low) GUT scale parameters $(\bar{y}_b = \bar{y}_\tau)$. This condition is applied at a scale $\mu \simeq M_{GUT}$.

We also quote the gauge matching conditions for minimal SUSY-SU(5) in the $\overline{DR}$ scheme [26, 27]:

$$\frac{1}{\alpha_i(\mu)} = \frac{1}{\alpha_G(\mu)} - \Delta_i^{GUT}(\mu) , \qquad (3)$$

where $\alpha_G = \bar{g}^2/4\pi$ and

$$\Delta_1^{GUT}(\mu) = -\frac{5}{\pi}\ln\frac{M_V}{\mu} + \frac{1}{5\pi}\ln\frac{M_{H_3}}{\mu} ,$$

$$\Delta_2^{GUT}(\mu) = -\frac{3}{\pi}\ln\frac{M_V}{\mu} + \frac{1}{\pi}\ln\frac{M_\Sigma}{\mu} , \qquad (4)$$

$$\Delta_3^{GUT}(\mu) = -\frac{2}{\pi}\ln\frac{M_V}{\mu} + \frac{3}{2\pi}\ln\frac{M_\Sigma}{\mu} + \frac{1}{2\pi}\ln\frac{M_{H_3}}{\mu} .$$

As noted in Ref. [28], differences of the Δ_i depend on M_V and M_Σ in the combination $M_{GUT} = (M_V^2 M_\Sigma)^{\frac{1}{3}}$ which we will take as GUT scale at which the matching functions are applied.

---

[2]This can of course be relaxed in the missing doublet model for example [23], but we limit our discussion to the minimal model for simplicity.

3   Electroweak and Supersymmetric Thresholds

In addition to the GUT scale thresholds there are also thresholds at the electroweak scale and the effective supersymmetry scale M_{SUSY}. At the electroweak scale, M_Z, we integrate out the top quark, the weak gauge bosons and the SM Higgs. These thresholds corrections are reviewed in Ref. [29]. The definition of the inverse electromagnetic coupling $\alpha^{-1}(M_Z) = 127.9 \pm 0.1$ and the Weinberg angle $s_0^2(M_Z) = 0.2324 \pm 0.0003$ includes one loop corrections from electroweak gauge bosons and the top quark [30] for a pole mass $M_{t0} = 143$ GeV. This must be corrected for different top masses above M_Z using $s^2(M_Z) = s_0^2(M_Z) + \Delta_{s2}^{top}$, where $\Delta_{s2}^{top} \approx -0.92 \times 10^{-7} GeV^{-2}(M_t^2 - M_{t0}^2)$ [31]. We also incorporate the mass thresholds of Ref. [13] to determine the relation between the running fermion masses defined in the effective $SU(3) \times U(1)$ low energy theory and those above M_Z. The largest effect occurs in the case of the bottom mass, although, for pole masses M_t up to 200 GeV, this effect corresponds to a shift of $m_b(M_Z)$ downwards by at most 0.6% from its value below M_Z.

The supersymmetric threshold is a potentially more important correction to the RG evolution of the couplings. It is sensitive to the details of the sparticle spectrum and, in the case of the Yukawa couplings, has quite different effects in different regions of parameter space.

We define our superpotential in the MSSM as

$$P = \epsilon_{ij}(Y_d \hat{H}_1^i \hat{Q}^j \hat{D} + Y_e \hat{H}_1^i \hat{L}^j \hat{E} + Y_u \hat{H}_2^j \hat{Q}^i \hat{U} + \mu_H \hat{H}_1^i \hat{H}_2^j) \,, \tag{5}$$

where the hats denote superfields. A general set of soft supersymmetry breaking parameters is introduced without explicit regard to their origin. These include soft trilinear scalar couplings mimicking those of the superpotential as well as explicit mass terms for the gauginos, squarks, sleptons and Higgs fields.

$$V_{soft} = -\epsilon_{ij}(Y_d A_d H_1^i \tilde{Q}^j \tilde{D} + Y_e A_e H_1^i \tilde{L}^j \tilde{E} + Y_u A_u H_2^j \tilde{Q}^i \tilde{U} + \mu_H B H_1^i H_2^j) \,,$$

$$\mathcal{L}_{soft}^{mass} = \tfrac{1}{2}\mathcal{M}_1 \overline{\lambda}_B \lambda_B + \tfrac{1}{2}\mathcal{M}_2 \overline{\lambda}_W^i \lambda_W^i + \tfrac{1}{2}\mathcal{M}_3 \overline{\lambda}_g^A \lambda_g^A - m_Q^2 \tilde{Q}^{i*}\tilde{Q}^i - m_U^2 \tilde{U}^*\tilde{U} \tag{6}$$

$$- m_D^2 \tilde{D}^*\tilde{D} - m_L^2 \tilde{L}^{i*}\tilde{L}^i - m_E^2 \tilde{E}^*\tilde{E} - m_{H_1}^2 \tilde{H}_1^{i*}\tilde{H}_1^i - m_{H_2}^2 \tilde{H}_2^{i*}\tilde{H}_2^i \,,$$

where we denote superpartners of ordinary particles with a tilde, and λ is used to denote the Majorana gaugino fields. The threshold corrections will in general involve squark and slepton mixing matrices for the third generation. For example, denoting the stop squark mass eigentates as $\tilde{t}_{1,2}$, we can relate them to weak eigenstates $\tilde{t}_{L,R}$ by an orthogonal matrix O^t such that $O^t M_{\tilde{t}}^2 O^{tT} = \bar{M}_{\tilde{t}}^2$ is diagonal. We define O^b and O^τ similarly.

The general result for the gauge threshold matching conditions is

$$\frac{1}{\alpha_i^-(\mu)} = \frac{1}{\alpha_i^+(\mu)} - \Delta_i^{SUSY}(\mu) - \Delta_i^{DR} \,, \tag{7}$$

where $\alpha_i^{\pm}$ denotes the gauge couplings just above/below M_{SUSY}. We convert from $\overline{MS}$ to $\overline{DR}$ couplings above M_{SUSY} with the conversion factor given by $\Delta_i^{DR} = -C_2(G_i)/12\pi$, where the quadratic Casimir $C_2(G)$ is N for SU(N) and 0 for U(1) groups. The cumulative effect of these thresholds can be parametrized in terms of mass scales M_i [32] via

$$\Delta_i^{SUSY}(\mu) = \frac{1}{2\pi}(b_i - b_i^{SM}) \ln \frac{M_i}{\mu} , \tag{8}$$

where $b_i - b_i^{SM} = (\frac{5}{2}, \frac{25}{6}, 4)$ are the differences between the gauge β function coefficients in the MSSM and the SM. To get some idea of the dependence of the M_i on the sparticle spectrum we compute them for the case of separate degeneracies among squarks, sleptons, gauginos, higgsinos, and heavy Higgs particles:

$$
\begin{aligned}
M_1 &= m_{\tilde{\ell}}^{\frac{9}{25}} m_{\tilde{q}}^{\frac{11}{25}} m_{\tilde{H}}^{\frac{4}{25}} m_H^{\frac{1}{25}} , \\
M_2 &= m_{\tilde{\ell}}^{\frac{3}{25}} m_{\tilde{q}}^{\frac{9}{25}} m_{\tilde{H}}^{\frac{4}{25}} m_{\tilde{W}}^{\frac{8}{25}} m_H^{\frac{1}{25}} , \\
M_3 &= m_{\tilde{q}}^{\frac{1}{2}} m_{\tilde{g}}^{\frac{1}{2}} .
\end{aligned}
\tag{9}
$$

The complete Yukawa thresholds at M_{SUSY} are quite complicated, however we can give some indication of their form and estimate their effects. The most tractable form useful for making estimates occurs in the limit of small gaugino-Higgsino mixing and $m_A \simeq m_{H^0}$ is given by

$$
\begin{aligned}
y_t^{SM}(\mu) &= y_t(\mu) \sin\beta(1 + \Delta_{y_t}^{SUSY}) , \\
y_b^{SM}(\mu) &= y_b(\mu) \cos\beta(1 + \Delta_{y_b}^{SUSY}) , \\
y_\tau^{SM}(\mu) &= y_\tau(\mu) \cos\beta(1 + \Delta_{y_\tau}^{SUSY}) ,
\end{aligned}
\tag{10}
$$

where

$$
\begin{aligned}
16\pi^2 \Delta_{y_t}^{SUSY} \simeq \frac{8}{3}g_3^2 \Bigg(&-2m_{\tilde{g}} \sum_i m_t G_2(m_{\tilde{g}}^2, m_{\tilde{t}_i}^2, m_{\tilde{t}_i}^2) O_{i1}^t O_{i2}^t \\
&+ \sum_{i,j} \Big((\mu_H \cot\beta + A_t) G_2(m_{\tilde{g}}^2, m_{\tilde{t}_i}^2, m_{\tilde{t}_j}^2)(P_{ij}(t))^2 \Big) \\
&+ \frac{1}{4} \sum_i F_2(m_{\tilde{t}_i}^2, m_{\tilde{g}}^2) \Bigg) \\
&+ \frac{1}{4}y_t^2 \Bigg(\sum_i F_2(m_{\tilde{b}_i}^2, m_{\tilde{H}^\pm}^2)(O_{i1}^b)^2 \\
&+ \sum_i F_2(m_{\tilde{t}_i}^2, m_{\tilde{H}^0}^2) \\
&+ \cos^2\beta \Big(F_2(m_{H^\pm}^2, m_b^2) + F_2(m_{H^0}^2, m_t^2) \\
&+ F_2(m_A^2, m_t^2) + 6\ln(\frac{m_t^2}{\mu^2}) \Big) \Bigg) \\
&+ \frac{1}{4}y_b^2 \Bigg(\sum_i F_2(m_{\tilde{b}_i}^2, m_{\tilde{H}^\pm}^2)(O_{i2}^b)^2 + \sin^2\beta\, F_2(m_{H^\pm}^2, m_b^2)
\end{aligned}
$$

$$
\begin{aligned}
&+\cos^2\beta\left(4F_1(m_b^2,m_{H\pm}^2)-6\ln(\tfrac{m_b^2}{\mu^2})\right)\bigg) \\
&-\frac{1}{2}y_\tau^2\cos^2\beta\,\ln(\tfrac{m_\tau^2}{\mu^2}) \\
&+\frac{1}{4}g_2^2\bigg(\sum_i\Big(F_2(m_{\tilde{b}_i}^2,m_{\tilde{W}\pm}^2)(O_{i1}^b)^2+\frac{1}{2}\big(F_2(m_{\tilde{t}_i}^2,m_{\tilde{W}^3}^2)(O_{i1}^t)^2 \\
&\qquad+\frac{t^2}{9}F_2(m_{\tilde{t}_i}^2,m_{\tilde{B}}^2)((O_{i1}^t)^2+16(O_{i2}^t)^2))\big)\bigg) \\
&+2F_3(m_{\tilde{W}\pm}^2,m_{\tilde{H}\pm}^2)+F_3(m_{\tilde{W}^3}^2,m_{\tilde{H}^0}^2)+t^2F_3(m_{\tilde{B}}^2,m_{\tilde{H}^0}^2) \\
&-2\sum_i\Big(G_1(m_{\tilde{t}_i}^2,m_{\tilde{W}^3}^2,m_{\tilde{H}^0}^2)(O_{i1}^t)^2 \\
&\qquad+\frac{t^2}{3}G_1(m_{\tilde{t}_i}^2,m_{\tilde{B}}^2,m_{\tilde{H}^0}^2)(-(O_{i1}^t)^2+4(O_{i2}^t)^2) \\
&\qquad+2G_1(m_{\tilde{b}_i}^2,m_{\tilde{W}\pm}^2,m_{\tilde{H}\pm}^2)(O_{i1}^b)^2\Big)\bigg)\ ,
\end{aligned}
\tag{11}
$$

$$
\begin{aligned}
16\pi^2\Delta_{y_b}^{SUSY}\simeq&\ \frac{8}{3}g_3^2\bigg(-2m_{\tilde{g}}(\mu_H\tan\beta+A_b)\sum_{i,j}G_2(m_{\tilde{g}}^2,m_{\tilde{b}_i}^2,m_{\tilde{b}_j}^2)(P_{ij}(b))^2 \\
&+\frac{1}{4}\sum_i F_2(m_{\tilde{b}_i}^2,m_{\tilde{g}}^2)\bigg) \\
&+\frac{1}{4}y_t^2\bigg(-8\mu_H\tan\beta\Big(m_t\sum_i G_2(m_{\tilde{H}\pm}^2,m_{\tilde{t}_i}^2,m_{\tilde{t}_i}^2)O_{i1}^t O_{i2}^t \\
&\qquad+(\mu_H\cot\beta+A_t)\sum_{i,j}G_2(m_{\tilde{H}\pm}^2,m_{\tilde{t}_i}^2,m_{\tilde{t}_j}^2)(P_{ij}(t))^2\Big) \\
&+\sum_i F_2(m_{\tilde{t}_i}^2,m_{\tilde{H}\pm}^2)(O_{i2}^t)^2+\cos^2\beta\,F_2(m_{H\pm}^2,m_t^2) \\
&+\sin^2\beta\left(4F_1(m_t^2,m_{H\pm}^2)-6\ln(\tfrac{m_t^2}{\mu^2})\right)\bigg) \\
&+\frac{1}{4}y_b^2\bigg(\sum_i F_2(m_{\tilde{t}_i}^2,m_{\tilde{H}\pm}^2)(O_{i1}^t)^2+\sum_i F_2(m_{\tilde{b}_i}^2,m_{\tilde{H}^0}^2) \\
&\qquad+\sin^2\beta\left(F_2(m_{H\pm}^2,m_t^2)+F_2(m_{H^0}^2,m_b^2)+F_2(m_A^2,m_b^2)\right) \\
&\qquad+6\ln(\tfrac{m_b^2}{\mu^2})\bigg) \\
&+\frac{1}{2}y_\tau^2\sin^2\beta\,\ln(\tfrac{m_\tau^2}{\mu^2}) \\
&+\frac{1}{4}g_2^2\bigg(2\mu_H\mathcal{M}_2\tan\beta\sum_i\Big(G_2(m_{\tilde{b}_i}^2,m_{\tilde{H}^0}^2,m_{\tilde{W}^3}^2)(O_{i1}^b)^2 \\
&\qquad+G_2(m_{\tilde{t}_i}^2,m_{\tilde{H}\pm}^2,m_{\tilde{W}}^2)(O_{i1}^t)^2\Big)
\end{aligned}
\tag{12}
$$

$$+\sum_i\Big(F_2(m_{\tilde{t}_i}^2,m_{\tilde{W}\pm}^2)(O_{i1}^t)^2+\frac{1}{2}\big(F_2(m_{\tilde{b}_i}^2,m_{\tilde{W}3}^2)(O_{i1}^b)^2$$

$$+\frac{t^2}{9}F_2(m_{\tilde{b}_i}^2,m_{\tilde{B}}^2)((O_{i1}^b)^2+4(O_{i2}^b)^2))\Big)$$

$$+2F_3(m_{\tilde{W}\pm}^2,m_{\tilde{H}\pm}^2)+F_3(m_{\tilde{W}3}^2,m_{\tilde{H}^0}^2)+t^2F_3(m_{\tilde{B}}^2,m_{\tilde{H}^0}^2)$$

$$-2\sum_i\Big(G_1(m_{\tilde{b}_i}^2,m_{\tilde{W}3}^2,m_{\tilde{H}^0}^2)(O_{i1}^b)^2$$

$$+\frac{t^2}{3}G_1(m_{\tilde{b}_i}^2,m_{\tilde{B}}^2,m_{\tilde{H}^0}^2)((O_{i1}^b)^2+2(O_{i2}^b)^2)$$

$$+2G_1(m_{\tilde{t}_i}^2,m_{\tilde{W}\pm}^2,m_{\tilde{H}\pm}^2)(O_{i1}^t)^2\Big)\Big)\,,$$

$$16\pi^2\Delta_{y_\tau}^{SUSY}\simeq\frac{3}{2}\sin^2\beta\,(-y_t^2\ln(\frac{m_t^2}{\mu^2})+y_b^2\ln(\frac{m_b^2}{\mu^2}))$$

$$+\frac{1}{4}y_\tau^2\Big(F_2(m_{\tilde{\nu}_\tau}^2,m_{\tilde{H}\pm}^2)+\sum_i F_2(m_{\tilde{\tau}_i}^2,m_{\tilde{H}^0}^2)$$

$$+\sin^2\beta\,(F_2(m_{H\pm}^2,0)+F_2(m_{H^0}^2,m_\tau^2)+F_2(m_A^2,m_\tau^2)$$

$$+2\ln(\frac{m_\tau^2}{\mu^2})\Big)$$

$$+\frac{1}{4}g_2^2\Big(2\mu_H\mathcal{M}_2\tan\beta\sum_i(G_2(m_{\tilde{\tau}_i}^2,m_{\tilde{H}^0}^2,m_{\tilde{W}3}^2)(O_{i1}^\tau)^2$$

$$+G_2(m_{\tilde{\nu}_\tau}^2,m_{\tilde{H}\pm}^2,m_{\tilde{W}}^2))$$

$$-4t^2\mathcal{M}_1(\mu_H\tan\beta+A_\tau)\sum_{i,j}G_2(m_{\tilde{B}}^2,m_{\tilde{\tau}_i}^2,m_{\tilde{\tau}_j}^2)(P_{ij}(\tau))^2\,(13)$$

$$+F_2(m_{\tilde{\nu}_\tau}^2,m_{\tilde{W}\pm}^2)+\frac{1}{2}\sum_i\Big(F_2(m_{\tilde{\tau}_i}^2,m_{\tilde{W}3}^2)(O_{i1}^\tau)^2$$

$$+t^2F_2(m_{\tilde{\tau}_i}^2,m_{\tilde{B}}^2)((O_{i1}^\tau)^2+4(O_{i2}^\tau)^2)\Big)$$

$$+2F_3(m_{\tilde{W}\pm}^2,m_{\tilde{H}\pm}^2)+F_3(m_{\tilde{W}3}^2,m_{\tilde{H}^0}^2)+t^2F_3(m_{\tilde{B}}^2,m_{\tilde{H}^0}^2)$$

$$-2\sum_i\Big(G_1(m_{\tilde{\tau}_i}^2,m_{\tilde{W}3}^2,m_{\tilde{H}^0}^2)(O_{i1}^\tau)^2$$

$$+t^2G_1(m_{\tilde{\tau}_i}^2,m_{\tilde{B}}^2,m_{\tilde{H}^0}^2)(-(O_{i1}^\tau)^2+2(O_{i2}^\tau)^2)$$

$$+2G_1(m_{\tilde{\nu}_\tau}^2,m_{\tilde{W}\pm}^2,m_{\tilde{H}\pm}^2)\Big)\Big)\,,$$

where $t=\tan\theta_W$, μ_H is the supersymmetric Higgs mass parameter defined in Eq. (5) and $A_{t,b,\tau}$ and $\mathcal{M}_{1,2}$ are the soft supersymmetry breaking parameters defined in Eqs. (6). These expressions include the dominant finite corrections as well as the full leading $\ln\mu$ dependence so that the evolution of the Yukawas does not depend on a particular choice of M_{SUSY}. The explicit logarithmic terms arise from h^0, H^0 mixing. All parameters as well as the function F_2

have an implicit dependence on the renormalization scale μ ($\sim M_{SUSY}$). The functions F_2 and G_2 are defined by

$$F_2(M_A^2, M_B^2) = \frac{1}{(M_A^2 - M_B^2)^2}\left((M_A^4 - 2M_A^2 M_B^2)\ln\frac{M_A^2}{\mu^2} + M_B^4 \ln\frac{M_B^2}{\mu^2}\right)$$
$$+ \frac{M_B^2}{M_A^2 - M_B^2} - \tfrac{1}{2}, \tag{14}$$
$$G_2(M_A^2, M_B^2, M_C^2) = \frac{1}{M_A^2 - M_B^2}\left(\frac{M_A^2}{M_A^2 - M_C^2}\ln\frac{M_A^2}{M_C^2} - \frac{M_B^2}{M_B^2 - M_C^2}\ln\frac{M_B^2}{M_C^2}\right).$$

where G_2 is positive. The P matrix involves products of the squark and slepton mixing matrices $O^{t,b,\tau}$ and is defined by

$$P_{ij}(f) = \tfrac{1}{2}(O_{i1}^f O_{j2}^f + O_{i2}^f O_{j1}^f), \tag{15}$$

where $f = (t, b, \tau)$. Note that all terms proportional to the products $O_{i1}^f O_{i2}^f$ drop out in the limit of no squark or slepton mixing.

For small $\tan\beta$ one may further neglect squark and slepton mixing except for the stop mixing effects. In this case the gluino contributions to the quark wavefunction renormalization and to the vertex diagrams of Fig. 1a give the dominant effect. Note that for $\mu_H < 0$ one typically finds larger stop squark mixing, so that the threshold effect in y_t should be largest in that case.

At large $\tan\beta$, where the corrections to y_b and y_τ tend to be more significant, the sbottom and stau mixing must be included. In this case enhancements proportional to $\tan\beta$ occur from vertex graphs containing gluinos, charginos and neutralinos. The gluino contribution from Fig. 1a dominates the correction to y_b for large $\tan\beta$. The chargino and neutralino contributions of Fig. 1b are also large in this regime. Depending on the relative signs of μ_H and A_t, the part of the finite diagram of Fig. 1b proportional to $\mu_H A_t y_t^2 \tan\beta$ can either enhance or decrease the gluino contribution. The divergent wino exchange graph tends to depress the gluino contribution slightly. The enhanced y_τ vertex corrections come from the order $g_2^2 \tan\beta$ contributions of Fig. 1c. In this case the bino exchange graph can be important due to the maximal hypercharge of the leptons. It is opposite in sign to wino and Higgsino vertex contributions. In analyzing the possible large $\tan\beta$ enhancements, it is important to note that the sign of most of the relevant vertex contributions is controlled by the sign of μ_H (or μ_H and A_t) and the various large contributions can appear with opposite sign. In addition, if we take $M_{SUSY} \simeq M_Z$ then[3] the wavefunction renormalization contributions tend to be positive. These facts justify the need for a complete analysis.

---

[3]The choice of M_{SUSY} is not important when the full threshold corrections are included since they incorporate the effects of one loop running between different effective supersymmetry scales.

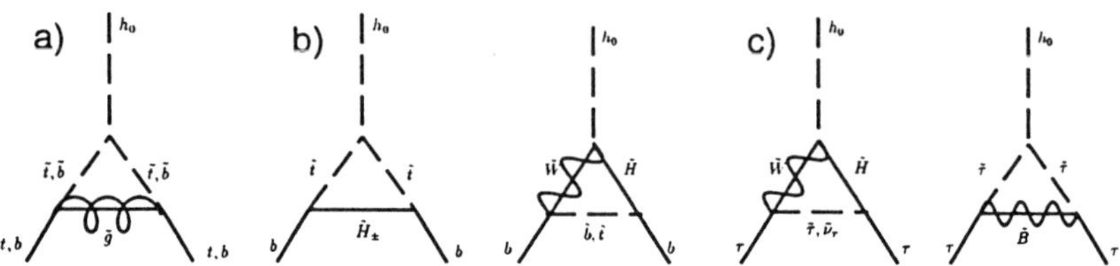

FIGURE 1.   Vertex diagrams involving a) dominant gluino contributions to y_t and y_b threshold corrections and dominant contributions to b) y_b and c) y_τ threshold corrections which are enhanced in the large $\tan\beta$ region.

4   Robustness of the M_t Solution

One may use Yukawa unification to predict either M_t or $m_b(M_Z)$. We point out that in certain regions of parameter space, the former is more robust in the presence of threshold corrections. The one loop β functions for the Yukawa couplings are not solvable analytically, however we shall give some semi-analytic solutions. The one loop Yukawa β functions for the third generation in the MSSM are

$$\beta_{y_\alpha} = \frac{dy_\alpha}{dt} = \frac{y_\alpha}{16\pi^2}(-c_{\alpha i}g_i^2 + b_{\alpha\beta}y_\beta^2) \,, \tag{16}$$

where $\alpha = (t, b, \tau)$ and $b_{\alpha\beta}$ and $c_{\alpha i}$ in the MSSM and the SM are given in Ref. [33]. respectively. At one loop, in the naive step approximation, the SM Yukawa couplings are matched to those in the MSSM at M_{SUSY} via $y_t = y_t^{SM}/\sin\beta$, $y_b = y_b^{SM}/\cos\beta$ and $y_\tau = y_\tau^{SM}/\cos\beta$, where $\tan\beta = \frac{v_2}{v_1}$ is the ratio of the VEVs of the two Higgs doublets H_1 and H_2 in the MSSM. Using Eqs. (16), the one loop solution for m_b/m_τ at M_Z can be written as

$$\frac{m_b^0(M_Z)}{m_\tau^0(M_Z)} = A_{b/\tau}^{0\,SM} A_{b/\tau}^0 B_t^{0\,SM\,-\frac{3}{2}} B_b^{0\,SM\,\frac{3}{2}} B_\tau^{0\,SM\,-\frac{7}{2}} B_t^0 \left(\frac{B_b^0}{B_\tau^0}\right)^3 \,, \tag{17}$$

where

$$A_\alpha^0 = \exp\left(\frac{c_{\alpha i}}{4\pi}\int_{t_s}^{t_G^0} dt\, \alpha_i^0\right) \,, \tag{18}$$

and

$$B_\alpha^0(M_{SUSY}, M_{GUT}^0) = exp\left(-\frac{1}{16\pi^2}\int_{t_s}^{t_G^0} dt\,(y_\alpha^0)^2\right) \,, \tag{19}$$

with similar expressions for the SM quantities. Here the superscript zeros indicate the naive one loop quantities with step function thresholds. The parameters $t_s = \ln\frac{M_{SUSY}}{M_Z}$ and $t_G^0 = \ln\frac{M_{GUT}^0}{M_{SUSY}}$. On the other hand the full two loop solution including the Yukawa thresholds is

$$\frac{m_b(M_Z)}{m_\tau(M_Z)} = A_{b/\tau}^{SM} A_{b/\tau} B_t^{SM\,-\frac{3}{2}} B_b^{SM\,\frac{3}{2}} B_\tau^{SM\,-\frac{7}{2}} B_t \left(\frac{B_b}{B_\tau}\right)^3$$

$$\times \Theta_{b/\tau}(1 + \Delta_{b/\tau}^{SM} + \Delta_{b/\tau}^{SUSY} + \Delta_{b/\tau}^{GUT}) \,, \tag{20}$$

$$\approx \frac{m_b^0(M_Z)}{m_\tau^0(M_Z)} \Theta_{b/\tau}(1 + \Delta_{b/\tau}^{SUSY} + \Delta_{b/\tau}^{GUT} + \Delta_{b/\tau}^{\alpha}) \,,$$

where the A's and B's without superscript zeroes are those obtained by running the gauge and Yukawa couplings between threshold boundaries. The two loop corrections are contained in $\Theta_{b/\tau}$ which we will take to be unity here. The indirect correction term $\Delta_{b/\tau}^{\alpha}$ arises when $A_{b/\tau}$ is rewritten in terms of the naive one loop results and primarily involves the gauge coupling thresholds. We define $A_{b/\tau} = A_{b/\tau}^0(1 + \Delta_{b/\tau}^{\alpha})$. Equation (20) may be used to determine $m_b(M_Z)$ using the Yukawa unification condition and the value for $m_\tau(M_Z)$[4]. However it is important to note that such a solution will be sensitive to potentially large threshold corrections both at M_{SUSY} and M_{GUT}.

If we instead use Eq. (20) to solve for $m_t(M_Z)$, the sensitivity to such corrections is diminished. Here we use a good approximate analytic solution for y_t for small to moderate $\tan\beta$ (see also Ref. [35]),

$$B_t(M_{SUSY}^+, M_{GUT}^-) \simeq (1 - y_t^2(M_{SUSY}^+)K_t(M_{SUSY}^+, M_{GUT}^-))^{\frac{1}{12}} \,, \tag{21}$$

where

$$K_t(M_{SUSY}^+, M_{GUT}^-) = \frac{3}{4\pi^2} \int_{t_s^+}^{t_G^-} dt \, A_t^{-2}(t_s, t) \,, \tag{22}$$

and the $\pm$ superscripts indicate parameters just above/below the relevant threshold boundary. We then solve (20) for $y_t(M_Z)$ using $B_t = k(\frac{B_\tau}{B_b})^3$ where $k = k^0(1 - \delta k)$,

$$k^0 = \frac{m_b(M_Z)}{m_\tau(M_Z)}(A_{b/\tau}^{0\,SM} A_{b/\tau}^0)^{-1} B_t^{0\,SM\,\frac{3}{2}} B_b^{0\,SM\,-\frac{3}{2}} B_\tau^{0\,SM\,\frac{7}{2}} \,, \tag{23}$$

and $\delta k \simeq \Delta_{b/\tau}^{SUSY} + \Delta_{b/\tau}^{GUT} + \Delta_{b/\tau}^{\alpha}$. The solution for $m_t(M_Z)$ is

$$m_t(M_Z) \approx m_t^0(M_Z)\left(1 + \Delta_{y_t}^{SUSY} - \frac{1}{2}\Delta_{K_t} + C\delta k\right) \,. \tag{24}$$

The factor C is given by

$$C = \frac{6(k^0)^{12}\left(\frac{B_\tau^0}{B_b^0}\right)^{36}}{1 - (k^0)^{12}\left(\frac{B_\tau^0}{B_b^0}\right)^{36}} \,, \tag{25}$$

We have also introduced a threshold correction factor Δ_{K_t}, defined by $K_t = K_t^0(1 + \Delta_{K_t})$, which must be determined numerically.

---

[4]The pole mass for the τ lepton is given experimentally by $M_\tau = 1.7771 \pm 0.0005$ GeV [34], which, using the QED relation between the pole and running masses gives $m_\tau(M_Z) = 1.7476 \pm 0.0006$ just below the electroweak threshold. With the loop effects of W and Z bosons included, one has $m_\tau^+(M_Z) = 1.7494 \pm 0.0006$ just above the threshold.

For $\alpha_s(M_Z) = 0.120$, $M_b = 4.9$ GeV (pole mass), $\tan\beta = 1.5$ and $M_{SUSY} = M_Z$ we find $\Delta^{\alpha}_{b/\tau} = -0.06$ and $\Delta_{K_t} = -0.03$ with $K_t^0 = 0.77$. The gauge coupling sparticle spectrum parameters are taken to be $M_1 = 368$ GeV, $M_2 = 256$ GeV and $M_3 = 375$ GeV. These results are insensitive to M_t and $\tan\beta$. For low $\tan\beta$ we also find $m_t^0(M_Z) = (195.7\,GeV)\sin\beta$, consistent with the infrared quasi-fixed point solution [36, 37] for m_t. This behaviour is modified for larger $\tan\beta$ since $B_\tau^0/B_b^0 \propto \exp(d\sec^2\beta)$ $(d \ll 1)$. For low $\tan\beta$ the fixed point solution implies that this factor is very close to 1. For the case discussed above we find $k^0 = 0.72$ and $C = 0.12$. Hence we see explicitly how the fixed point solution inhibits the influence of potentially large GUT or MSSM threshold corrections to y_b/y_τ on the M_t solution. In this sense the solution for $m_t(M_Z)$ tends to be more robust than the alternative $m_b(M_Z)$ solution entertained by some authors. Of course, away from the fixed point B_τ^0/B_b^0 tends to be larger than 1, and the factor C need not be small. Still, for moderate values of $\tan\beta$, one finds $C \lesssim 0.3$, and to the extent that the approximation of Eq. (21) is still valid, the robustness of the M_t solution is maintained except for extremely large $\tan\beta$.

5 Renormalization Group Results and Conclusions

A complete two loop numerical analysis is necessary to properly determine the influence of threshold effects on the solutions for M_t, M_{H_3} and M_{GUT}. We will consider the effect of turning on and off various thresholds as well as the effect of varying various parameters appearing in the threshold formulae.

We first integrate the relevant two loop β functions (see Refs. [13, 38, 33] and references therein) to find the solutions for M_{H_3} and M_{GUT} consistent with the mismatch of the gauge couplings at M_{GUT}. They depend on M_t only through two loop effects in the gauge running and through threshold effects, primarily at the weak scale. By utilizing the threshold corrections given in Eqs. (3), (4), (7) and (8), one obtains the approximate solutions

$$t_G = \ln\frac{M_{GUT}}{M_Z} = \frac{\pi}{6}\left(\frac{1-2s^2}{\alpha} - \frac{2}{3}\frac{1}{\alpha_s}\right) - \frac{2}{9}\ln\frac{U_{SUSY}}{M_Z} + +\delta^{SM}_{t_G} + \delta^{\overline{DR}}_{t_G} + \text{2-loop} ,$$

$$t_H = \ln\frac{M_{H_3}}{M_Z} = \frac{\pi}{2}\left(\frac{-1+6s^2}{\alpha} - \frac{10}{3\alpha_s}\right) + \frac{5}{6}\ln\frac{V_{SUSY}}{M_Z} + \delta^{SM}_{t_H} + \text{2-loop} , \qquad (26)$$

where the δ^{SM} incorporate the electroweak threshold effects and are dominated by the top dependent uncertainty in $\sin^2\theta_W$. In the above all low scale parameters are implicitly evaluated at M_Z. The effective scales replacing the naive scale M_{SUSY} are

$$U_{SUSY} = M_1^{-\frac{25}{16}} M_2^{\frac{25}{16}} M_3 \simeq \left(\frac{m_{\tilde{q}}}{m_{\tilde{\ell}}}\right)^{\frac{3}{8}} m_{\tilde{W}}^{\frac{1}{2}} m_{\tilde{g}}^{\frac{1}{2}} ,$$

$$V_{SUSY} = M_1^{-\frac{5}{4}} M_2^{\frac{25}{4}} M_3^{-4} \simeq \left(\frac{m_{\tilde{\ell}}}{m_{\tilde{q}}}\right)^{\frac{3}{10}} \left(\frac{m_{\tilde{W}}}{m_{\tilde{g}}}\right)^2 m_H^{\frac{1}{5}} m_{\tilde{H}}^{\frac{4}{5}} , \qquad (27)$$

We can approximate a range for these effective scales in the context of the canonical soft breaking of supersymmetry induced from a hidden sector of N= 1 supergravity. In this case there is a common SU(5) invariant gaugino mass $M_{\frac{1}{2}}$, a common soft supersymmetry breaking scalar mass M_0 and a common trilinear scalar coupling A at M_{GUT}. Then there is a simple one loop relation between the gaugino masses:

$$\frac{m_{\tilde{W}}}{m_{\tilde{g}}} = \frac{\alpha_2(M_{SUSY})}{\alpha_3(M_{SUSY})} \approx 0.28 \ . \tag{28}$$

Assuming $m_{\tilde{\ell}} \simeq m_{\tilde{q}}$ we obtain the approximate relations

$$U_{SUSY} \approx 0.53 \, m_{\tilde{g}} \ , \qquad V_{SUSY} \approx 0.08 \, m_H^{\frac{1}{5}} m_{\tilde{H}}^{\frac{4}{5}} \ . \tag{29}$$

Alternatively, in the no-scale supergravity case ($M_0 = A = 0$) one has $m_{\tilde{q}} \simeq m_{\tilde{g}} \simeq 3 m_{\tilde{W}} \simeq 3 m_{\tilde{\ell}}$ with slightly different coefficients. As pointed out in Ref. [28], the supersymmetric threshold corrections to M_{GUT} depend mainly on the wino and gluino masses while M_{H_3} is most sensitive to the Higgsino mass. If we allow a range of $100 - 1000$ GeV for m_H and $m_{\tilde{H}}$ while the gluino mass is taken to range between $120 - 1000$ GeV[5] then for degenerate squarks and sleptons $64 \lesssim U_{SUSY} \lesssim 530$ GeV and $8 \lesssim V_{SUSY} \lesssim 80$ GeV, and bear little relation to the naive $M_Z \lesssim M_{SUSY} \lesssim 1$ TeV. Note that the approximate upper limit on the sparticle masses of 1 TeV is the generic order of magnitude expectation if supersymmetry is to avoid introducing a new low scale hierarchy problem.

The influences of corrections from thresholds, two loop terms and the experimental error bars in are summarized in Table 1.

Table 1: Corrections to M_{GUT} and M_{H_3}.

Correction	Parameter Variations		% Deviations	
	δ_{t_G}	δ_{t_H}	M_{GUT}	M_{H_3}
SM (top) threshold	$(+0.008, +0.35)$	$(+.06, -3.0)$	$(0.8, 41)$	$(6, -95)$
MSSM threshold	$(+0.08, -0.39)$	$(-2.0, -.11)$	$(8, -32)$	$(-86, -10)$
	$(-0.01, -0.48)$	$(-2.3, -.35)$	$(-1, -38)$	$(-90, -30)$
s_0^2 error bar	± 0.04	± 0.36	± 4	$(-30, 43)$
α error bar	± 0.03	± 0.06	± 3	± 6
α_s error bar	$(-0.19, +0.17)$	$(-2.8, +2.5)$	$(-17, 18.5)$	$(-94, +1120)$
two loop ($y_t = 0$)	0.18	-3.8	20	-98
two loop (y_t only)	0	0.4	0	49

The top dependent thresholds are summed together with $\overline{DR}$ conversion factors in the SM threshold entries for $130 \leq M_t \leq 200$. The MSSM threshold entries include the supersymmmetric threshold as well as the effect of running SM parameters from M_Z to M_{SUSY} for the previously mentioned ranges taken for U_{SUSY} and V_{SUSY}. The upper range corresponds to the degenerate squark-slepton case while the lower entry is for the no-scale case. The α_s error bar

---

[5]The lower limit on the gluino mass is near the experimental lower bound [39] if the light gluino window is closed.

entries are given first for the lower, then the upper limit for $\alpha_s = 0.120 \pm 0.007$. We see that M_{H_3} decreases with M_t and α and increases with s^2, α_s and V_{SUSY} $(m_{\tilde{H}})$ while M_{GUT} increases with M_t and α_s and decreases with s^2, α and U_{SUSY} $(m_{\tilde{W}}, m_{\tilde{g}})$. For extreme ranges of these parameters and $0.5 \lesssim \tan\beta \lesssim 60$ gauge unification gives

$$1.0(0.96) \times 10^{16} < M_{GUT} < 3.9(3.7) \times 10^{16} \ GeV \ ,$$
$$1.7(1.2) \times 10^{13} < M_{H_3} < 1.6(0.9) \times 10^{17} \ GeV \ , \tag{30}$$

for the case of approximately degenerate squarks and sleptons, with the no-scale limits given in parentheses. These results are comparable with those of Ref. [28].

The magnitudes of the effects on M_{GUT} due to variations of M_t, α_s and $m_{\tilde{g}}$ are comparable, ranging up to 40%. Note that the $\Delta_{s^2}^{top}$ correction dominates the SM threshold effects and, as expected, the "correction" terms for M_{H_3} can be large. The α_s error bar correction is the largest, changing M_{H_3} over three orders of magnitude. The SM and MSSM thresholds can be comparable in magnitude, with the SM effects changing M_{H_3} by up to two orders of magnitude, corresponding to a change in the proton decay lifetime of four orders of magnitude. Clearly, when we also solve for M_t from the condition of Yukawa unification, the solution for M_{H_3} will be substantially correlated with the M_t solution. The tendency to predict large values of M_t means that the effect of the SM threshold will generally be to depress M_{H_3}. The opposite is true for the lower bound on M_{H_3} from proton decay. The combination of these results may be useful in strengthening proton decay constraints on the parameter space [40]. Indeed, the application of the most conservative proton decay bound, $M_{H_3} > 5.3 \times 10^{15}$ GeV [28], gives a lower bound on the strong coupling constant. By tuning the remaining parameters so that M_{H_3} is as large as possible we find $\alpha_s(M_Z) \gtrsim 0.118 \, (0.119)$, where the limit in parenthesis corresponds to the no-scale case.

To get some idea of the size of the GUT scale Yukawa coupling threshold corrections, we combine the renormalization group constraints (30) with perturbativity constraints of on the superpotential parameters of Eq. (1) We then find

$$M_V > 7.0(6.1) \times 10^{15} \ GeV \qquad M_\Sigma < 0.9(1.2) \times 10^{17} \ GeV \ , \tag{31}$$

where the two values correspond to perturbativity of λ_1 and λ_2 up to $M_{Planck}/\sqrt{8\pi} = 2.4 \times 10^{18}$ GeV and 10^{17} GeV, respectively. If we further assume that no superheavy masses lie above the relevant Planck mass scale, we find in addition

$$\frac{M_V}{M_\Sigma} \lesssim 1.2 \times 10^7 \ . \tag{32}$$

Although this may appear to be an extreme choice there is nothing to rule it out phenomenologically. From Eqs. (2), (23) and (24) we see that such a large splitting tends to increase the top mass prediction. In many cases no perturbative solution can be found, however, away from the infrared quasi-fixed point of y_t, sizable enhancements of the predictions can occur. In Fig. 2a we plot the M_t solution *vs.* $\tan\beta$ for various values of the ratio M_V/M_Σ for $M_b = 5.2$ GeV, $\alpha_s(M_Z) = 0.120$ and $M_{SUSY} = M_Z$. The largest deviation from the naive case is a 15% increase in M_t and corresponds to $\Delta_{b/\tau}^{GUT} \simeq 0.1$. The sparticle spectrum parameters are fixed at $m_{\tilde{g}} = M_3 = 1$ TeV and $V_{SUSY} = 80$ GeV. The typical GUT masses are then $M_{GUT} = 10^{16.1\pm0.1}$ GeV and $M_{H_3} = 10^{15.7\pm0.4}$ GeV for these solutions. We also plot the bound from the nonperturbative limit on y_t, indicating that for this particular example, one is not in the domain of attraction of the fixed point.

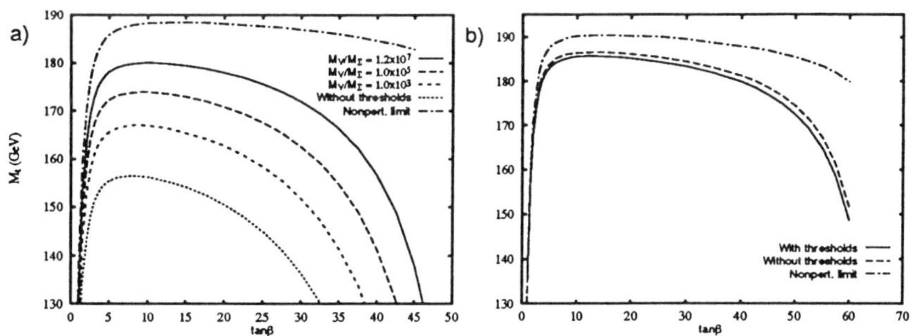

FIGURE 2.   The dependence of the M_t solution on $\tan\beta$ showing the effect of a) a large splitting of M_V and M_Σ for $M_b = 5.2$ GeV, $\alpha_s(M_Z) = 0.120$ and b) a splitting of M_{H_3} above M_{GUT} for $M_b = 4.9$ GeV, $\alpha_s(M_Z) = 0.127$. We take $M_{SUSY} = M_Z$ and the sparticle spectrum parameters are fixed at $m_{\tilde{g}} = M_3 = 1$ TeV and $V_{SUSY} = 80$ GeV. We also indicate the bound from the nonperturbative limit on y_t.

The GUT scale matching function for y_b/y_τ of Eq. (2) can also be negative for large y_t and M_{H_3}. We set $M_V/M_\Sigma = 0.3$, near the minimum required by the perturbativity argument. In Fig. 2b we give an example for this case for $M_b = 4.9$ GeV, $\alpha_s(M_Z) = 0.127$ and $M_{SUSY} = M_Z$. Using the same set of sparticle spectrum parameters as in the previous figure, we obtain GUT masses in the ranges $M_{GUT} = 10^{16.2\pm0.1}$ GeV and $M_{H_3} = 10^{16.8\pm0.3}$ GeV. The deviations here are smaller: $\Delta_{b/\tau}^{GUT} \simeq -0.02$ since M_{H_3} cannot be larger than M_{GUT} by more than an order of magnitude. The important thing to note is that the lack of knowledge of the GUT scale spectrum does not preclude extracting reliable predictions for M_t over much of the parameter space and the uncertainties in these predictions can be reasonably estimated. In regions where predictions are not robust, *i.e.* away from the attraction of the infrared fixed point for y_t, one must worry about 10 - 20% effects in the minimal model.

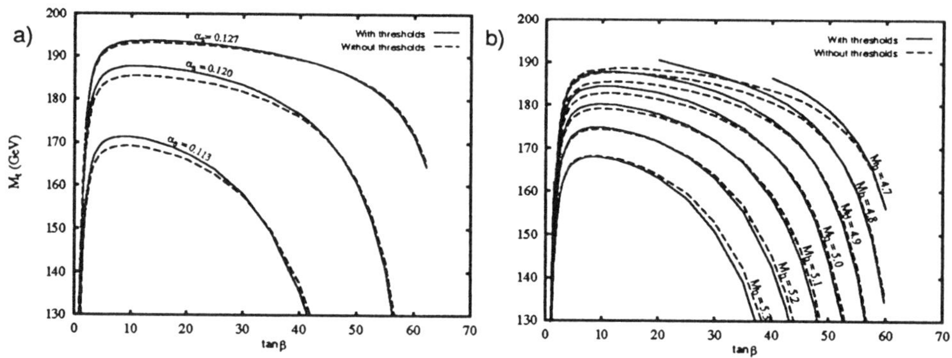

FIGURE 3. The dependence of the M_t solution from Yukawa unification on $\tan\beta$ for a) the allowed range of $\alpha_s(M_Z)$ for $M_b = 4.9$ GeV and b) bottom pole masses in the range 4.7 GeV $\leq M_b \leq 5.3$ GeV for $\alpha_s(M_Z) = 0.120$. We take $M_{SUSY} = M_Z$ and generic sparticle spectrum parameters: $m_{\tilde{g}} = 400$ GeV, $M_3 = 375$ GeV and $V_{SUSY} = 35$ GeV. We give the solutions with and without GUT scale Yukawa thresholds for the generic case $M_V = M_\Sigma$.

We also depict the most significant uncertainty in the M_t solution: the error bars on α_s and M_b. Fig. 3a shows the M_t vs. $\tan\beta$ contours for different α_s for the generic case $M_V = M_\Sigma$, $M_b = 4.9$ GeV and sparticle spectrum parameters $m_{\tilde{g}} = 400$ GeV, $V_{SUSY} = 35$ GeV and $M_3 = 375$ GeV. In Fig. 3b we show these contours for different M_b ranging from 4.7 to 5.3 GeV[6] for $\alpha_s(M_Z) = 0.120$ and the same remaining inputs as in Fig. 3a. Note that for small M_b the curves are cut off due to the lack of a perturbative solution for M_t for intermediate values of $\tan\beta$. Clearly, in the absence of additional theoretical cuts such as proton decay or radiative electroweak symmetry breaking there remains a large parameter space. Even if one imposes the restricted M_t limits from fits to electroweak data, one can find solutions for all regions of $\tan\beta$, at least for larger M_b. A better determination of M_b could eliminate the intermediate $\tan\beta$ solutions, although this subject is frought with theoretical uncertainties.

Next we turn to a discussion of the Yukawa threshold corrections in the MSSM. Using the approximate matching functions of Eqs. (11) – (13) we incorporate M_t into the numerical routine together with M_{GUT} and M_{H_3} and compute the threshold effects on the top mass solution for representative sparticle spectra consistent with universal soft supersymmetry breaking parameters at the GUT scale as well as radiative electroweak symmetry breaking [6]. We consider the low, intermediate and high $\tan\beta$ cases separately. Table 2 gives

---

[6]Pole masses M_b = 4.7, 4.9, 5.1, 5.3 GeV correspond to $m_b(M_Z)$ = 2.81, 2.96, 3.11, 3.26 ± 0.16 GeV and $m_b(M_b) = 4.06$, 4.24, 4.43, 4.61 ± 0.08 GeV, respectively, where we have included the uncertainty for the strong coupling $\alpha_s(M_Z) = 0.120 \pm 0.007$.

the effect of the threshold corrections for low $\tan\beta = 1.47$ and some sample soft breaking parameters including the generic no-scale case. We also consider both signs of μ_H and take $A = 0$ throughout. All masses in the tables are in GeV.

Table 2: Low $\tan\beta$ Threshold Corrections to M_t in the MSSM ($\tan\beta = 1.47$, $\alpha_s(M_Z) = 0.120$, $M_b = 4.8$ GeV, $A = 0$).

M_0	$M_{\frac{1}{2}}$	μ_H	$\Delta_{y_t}^{SUSY}$	$\Delta_{y_b}^{SUSY}$	$\Delta_{y_\tau}^{SUSY}$	M_t (naive)	M_t
100	140	+375	0.02	0.006	−0.006	164.1	167.6
100	140	−375	0.03	0.03	−0.01	164.1	170.5
0	150	+353	0.02	0.007	−0.007	164.2	167.8
0	200	−466	0.07	0.04	−0.01	163.0	175.4

Note that the increased stop squark splitting in the case of $\mu_H < 0$ increases the y_t corrections. The SUSY thresholds increase the top mass prediction relative to the naive result which includes only the gauge coupling thresholds. The dominant effect arises from the gluino-induced vertex corrections of Fig. 1a (for $\mu_H < 0$) and the gluino contribution to the top quark wavefunction renormalization. These are also the dominant contributions to the y_b threshold, although for $\mu_H > 0$ the gluino-induced vertex and wavefunction renormalization contributions almost cancel, while they add constructively for $\mu_H < 0$. Since y_t is near the fixed point in this region, the order 5% corrections to y_b/y_τ translate into a negligible effect on the M_t solution as discussed in Sec. 4.

For intermediate and high $\tan\beta$ the bottom and τ Yukawa corrections become more important. For intermediate $\tan\beta = 15$ and $M_{\frac{1}{2}} = 100$ GeV, Table 3 summarizes the typical threshold effects.

Table 3: Intermediate $\tan\beta$ Threshold Corrections to M_t in the MSSM ($\tan\beta = 15$, $\alpha_s(M_Z) = 0.118$, $M_b = 4.9$ GeV, $M_{\frac{1}{2}} = 100$ GeV, $A = 0$).

M_0	μ_H	$\Delta_{y_t}^{SUSY}$	$\Delta_{y_b}^{SUSY}$	$\Delta_{y_\tau}^{SUSY}$	M_t (naive)	M_t
100	164	0.01	−0.1	0.025	194.5	178.3
200	316	0.03	−0.03	0.025	190.3	187.5
300	462	0.035	−0.01	0.02	187.3	188.7

It is interesting to see the source of the enhanced y_b and y_τ corrections. For $M_0 = 100$ GeV the gluino induced vertex of Fig. 1a contributes −0.13, while the chargino induced vertices of Fig. 1b with a stop squark internal lines contribute 0.0175. Accounting for other small corrections leaves a 10% effect. On the other hand for $M_0 = 300$ GeV and correspondingly larger squark and slepton masses, many contributions come into play. Now the chargino contributions add to 0.03, the gluino vertex correction is -0.09 and the total wavefunction renormalization contribution is 0.03. Accounting for heavy Higgs effects and the neutralino contributions leaves a net 1% effect in y_b. Clearly a complete analysis is important even for intermediate values of $\tan\beta$. The dominant corrections to y_τ come from the neutral and charged wino induced graphs of Fig. 1c.

For larger $\tan\beta = 40$ and $M_{\frac{1}{2}} = 100$ GeV, we give typical threshold effects in Table 4. Here the heavier spectrum chosen enhances all the wavefunction

Table 4: Large $\tan\beta$ Threshold Corrections to M_t in the MSSM ($\tan\beta = 40$, $\alpha_s(M_Z) = 0.118$, $M_b = 4.9$ GeV, $M_0 = 400$ GeV, $A = 0$).

$M_{\frac{1}{2}}$	μ_H	$\Delta^{SUSY}_{y_t}$	$\Delta^{SUSY}_{y_b}$	$\Delta^{SUSY}_{y_\tau}$	M_t (naive)	M_t
200	205	0.04	-0.035	0.04	177.6	154.5
400	450	0.05	-0.045	0.06	169.0	78.2
600	660	0.055	-0.03	0.065	162.3	45.2

renormalization contributions and the gluino vertex contributions to y_b approach 20%. For the positive μ_H chosen however, this is partially cancelled as described in the intermediate $\tan\beta$ case. Even so, the effect on M_t is dramatic. Here one is far from the attraction of the infrared fixed point. The typically large M_t predictions in SUSY-GUTs normally arise because the QCD running of y_b typically places it below y_τ at M_{GUT}. A large top Yukawa coupling counters this behaviour. However, here the large, order 10%, threshold correction increases y_b relative to y_τ at M_{GUT} and smaller values of M_t are required. In fact, for some spectra the situation becomes unstable in the sense that no positive y_t can be found such that the $y_b(M_{GUT}) = y_\tau(M_{GUT})$ unification occurs.

We conclude that the Yukawa thresholds corrections are important and are sensitive to $\tan\beta$ and the soft supersymmetry breaking parameters. These effects should be included when making arguments concerning the allowed range of $\tan\beta$ consistent with a definite measurement of the top mass. In particular, one may need to reconsider the possibility of intermediate $\tan\beta$ solutions. We have also seen that a consistent treatment of the threshold corrections puts stronger constraints on the parameter space when proton decay is considered. The same techniques presented here can be applied to other mass and mixing angle relations coming from more complicated GUTs. Such relations may be classified in terms of their robustness in the presence of high scale threshold corrections in the same way that we argued for the robustness of the M_t solution. Finally, the determination of the threshold sensitivities of weak scale parameters together with improvements in precision electroweak scale measurements gives us a window onto unknown particle spectra and may yield clues to new physics close to and beyond the scope of present day colliders.

References

[1]   R. Arnowitt and P. Nath, Phys. Rev. Lett. **69**, 725 (1992); Phys. Lett. **289B**, 368 (1992).

[2]   G. G. Ross and R. G. Roberts, Nucl. Phys. **B377**, 571 (1992).

[3]   S. Kelley, J. L. Lopez, D. V. Nanopoulos, H. Pois, and K. Yuan, Nucl. Phys. **B398**, 3 (1993).

[4]   M. Olechowski and S. Pokorski, Nucl. Phys. **B404**, 590 (1993).

[5]   P. Ramond, in *Proceedings of the International Workshop on Recent Advances in the Superworld, Woodlands, Texas, 1993*, edited by J. L. Lopez (World Scientific, River Edge, N.J., 1994); D. J. Castaño, E. J. Piard, and P. Ramond, Institute for Fundamental Theory Report No. UFIFT-HEP-93-18, Aug. 1993 (unpublished).

[6]   V. Barger, M. S. Berger and P. Ohmann, University of Wisconsin Report No. MAD/PH/801, Nov. 1993 (unpublished).

[7]   G. L. Kane, C. Kolda, L. Roszkowski, and J. D. Wells, University of Michigan Report No. UM-TH-93-24, 1993 (unpublished).

[8]   B. D. Wright, University of Wisconsin Report No. MAD/PH/812, March 1994 (submitted to Physical Review D).

[9]   L. Hall, R. Rattazzi, and U. Sarid, Lawrence Berkeley Lab Report No. LBL-333997, June 1993 (unpublished).

[10]   R. Hempfling, DESY Report No. DESY-93-092, July 1993 (unpublished).

[11]   W. A. Bardeen, M. Carena, S. Pokorski, and C. E. M. Wagner, Max Planck Institute Report No. MPI-PH-93-103, Feb. 1994 (unpublished).

[12]   H. E. Haber and R. Hempfling, Phys. Rev. D **48**, 4280 (1993); R. Hempfling, in *Proceedings of the 23rd Workshop of the INFN Eloisatron Project: The Decay Properties of SUSY Particles, Erice, Italy, 1992*, edited by L. Cifarelli and V. A. Khoze (World Scientific, River Edge, N.J., 1993).

[13]   H. Arason, D. Castaño, B. Keszthelyi, S. Mikaelian, E. Piard, P. Ramond, and B. D. Wright, Phys. Rev. D **46**, 3945 (1992).

[14]   J. Oliensis and M. Fishler, Phys. Rev. D **28**, 194 (1983).

[15]   T. W. Appelquist and J. Carrazzone, Phys. Rev. D **11**, 2856 (1975).

[16]   S. Weinberg, Phys. Lett. **91B**, 51 (1980).

[17]   L. Hall, Nucl. Phys. **B178**, 75 (1981).

[18]   B. A.Ovrut and H. J. Schnitzer, Phys. Rev. D **21**, 3369 (1980); *ibid.* **22**, 2518 (1980); Nucl. Phys. **B179**, 381 (1981); Phys. Lett. **100B**, 403 (1981); Nucl. Phys. **B184**, 109 (1981).

[19]   H. Arason, D. Castaño, B. Keszthelyi, S. Mikaelian, E. Piard, P. Ramond, and B. D. Wright, Phys. Rev. Lett. **67**, 2933 (1991).

[20]   S. Kelly, J. L. Lopez, and D. V. Nanopoulos, Phys. Lett. **274B**, 387 (1992).

[21]   B. Ananthanarayan, G. Lazarides, and Q. Shafi, Phys. Rev. D **44**, 1613 (1991).

[22]   A. Giveon, L. Hall, and U. Sarid, Phys. Lett. **271B**, 138 (1991).

[23]   A. Masiero, D. V. Nanopoulos, K. Tamvakis, and T. Yanagida, Phys. Lett. **115B**, 380 (1982).

[24] M. T. Grisaru, W. Siegel, and M. Roček, Nucl. Phys. **B159**, 429 (1979).

[25] J. Wess and B. Zumino, Phys. Lett. **49B**, 52 (1974); J. Iliopoulos and B. Zumino, Nucl. Phys. **B76**, 310 (1974); S. Ferrara, J. Ilioupoulos, and B. Zumino, *ibid.* **B77**, 413 (1974); B. Zumino, *ibid.* **B89**, 535 (1975); S. Ferrara and O. Piguet, *ibid.* **B93**, 261 (1975).

[26] M. B. Einhorn and D. R. T. Jones, Nucl. Phys. **B196**, 475 (1982).

[27] I. Antoniadis, C. Kounnas, and K. Tamvakis, Phys. Lett. **119B**, 377 (1982).

[28] J. Hisano, H. Murayama, and T. Yanagida, Nucl. Phys. **B402**, 46 (1993); Phys. Rev. Lett. **69**, 1014 (1992).

[29] P. Langacker and N. Polonsky, Phys. Rev. D **47**, 4028 (1993).

[30] G. Degrassi, S. Fanchiotti, and A. Sirlin, Nucl. Phys. **B351**, 49 (1991).

[31] P. Langacker and N. Polonsky, University of Pennsylvania Report No. UPR-0594T, Feb. 1994 (unpublished).

[32] P. Langacker and N. Polonsky, Phys. Rev. D **49**, 1454 (1994).

[33] V. Barger, M. S. Berger, and P. Ohmann, Phys. Rev. D **47**, 1093 (1993).

[34] H. Marsiske, in *Proceedings of the Second International Workshop on Tau Lepton Physics, Ohio State University, 1992*, edited by K. K. Gan (World Scientific, River Edge, N.J., 1993).

[35] S. G. Naculich, Phys. Rev. D **48**, 5293 (1993).

[36] V. Barger, M. S. Berger, and P. Ohmann, Phys. Rev. D **47**, 2038 (1993); V. Barger, M. S. Berger, P. Ohmann, and R. J. Phillips, Phys. Lett. **314B**, 351 (1993).

[37] M. Carena, M. Olechowski, S. Pokorski, and C. E. M. Wagner, Phys. Lett. **320B**, 110 (1994).

[38] M. E. Machacek and M. T. Vaughn, Nucl. Phys. **B222**, 83 (1983); *ibid.* **B236**, 221 (1984); *ibid.* **B249**, 70 (1985).

[39] F. M. Borzumati, DESY Report No. 93-090, 1993 (unpublished).

[40] V. Barger, M. S. Berger, P. Ohmann, and B. Wright, in preparation.